T0192348

Flexible AC Transmission Systems (FACTS)

Newton Power-Flow Modeling of Voltage-Sourced Converter-Based Controllers

Flexible AC Transmission Systems (FACTS)

Newton Power-Flow Modeling of Voltage-Sourced Converter-Based Controllers

Suman Bhowmick

CRC Press
Taylor & Francis Group
Boca Raton London New York

CRC Press is an imprint of the
Taylor & Francis Group, an **informa** business

MATLAB® is a trademark of The MathWorks, Inc. and is used with permission. The MathWorks does not warrant the accuracy of the text or exercises in this book. This book's use or discussion of MATLAB® software or related products does not constitute endorsement or sponsorship by The MathWorks of a particular pedagogical approach or particular use of the MATLAB® software.

CRC Press
Taylor & Francis Group
6000 Broken Sound Parkway NW, Suite 300
Boca Raton, FL 33487-2742

First issued in paperback 2023

© 2016 by Taylor & Francis Group, LLC
CRC Press is an imprint of Taylor & Francis Group, an Informa business

No claim to original U.S. Government works

Version Date: 20160325

ISBN-13: 978-1-4987-5619-8 (hbk)
ISBN-13: 978-1-138-32267-7 (pbk)
ISBN-13: 978-1-315-22243-1 (ebk)

DOI: 10.1201/9781315222431

This book contains information obtained from authentic and highly regarded sources. Reasonable efforts have been made to publish reliable data and information, but the author and publisher cannot assume responsibility for the validity of all materials or the consequences of their use. The authors and publishers have attempted to trace the copyright holders of all material reproduced in this publication and apologize to copyright holders if permission to publish in this form has not been obtained. If any copyright material has not been acknowledged please write and let us know so we may rectify in any future reprint.

Except as permitted under U.S. Copyright Law, no part of this book may be reprinted, reproduced, transmitted, or utilized in any form by any electronic, mechanical, or other means, now known or hereafter invented, including photocopying, microfilming, and recording, or in any information storage or retrieval system, without written permission from the publishers.

For permission to photocopy or use material electronically from this work, please access www.copyright.com (http://www.copyright.com/) or contact the Copyright Clearance Center, Inc. (CCC), 222 Rosewood Drive, Danvers, MA 01923, 978-750-8400. CCC is a not-for-profit organization that provides licenses and registration for a variety of users. For organizations that have been granted a photocopy license by the CCC, a separate system of payment has been arranged.

Trademark Notice: Product or corporate names may be trademarks or registered trademarks, and are used only for identification and explanation without intent to infringe.

Publisher's Note
The publisher has gone to great lengths to ensure the quality of this reprint but points out that some imperfections in the original copies may be apparent.

Library of Congress Cataloging-in-Publication Data

Names: Bhowmick, Suman author.
Title: Flexible AC Transmission Systems (FACTS) : Newton power-flow modeling of voltage-sourced converter based controllers / Suman Bhowmick.
Description: Boca Raton : Taylor & Francis, 2016. | Includes bibliographical references and index.
Identifiers: LCCN 2015046458 | ISBN 9781498756198 (hardcover : alk. paper)
Subjects: LCSH: Flexible AC transmission systems. | Electric power distribution--Mathematical models. | Newton-Raphson method.
Classification: LCC TK3148 .B46 2016 | DDC 621.319/13--dc23
LC record available at http://lccn.loc.gov/2015046458

Visit the Taylor & Francis Web site at
http://www.taylorandfrancis.com

and the CRC Press Web site at
http://www.crcpress.com

Dedicated to my parents

Contents

Preface

SINCE THE PAST THREE decades, new economic, social, and legislative developments have demanded a review of traditional power transmission theory and practice and the creation of new concepts to allow full utilization of existing power generation and transmission facilities. In this respect, the vision of *Flexible AC Transmission System* (*FACTS*) was formulated by the Electric Power Research Institute, Palo Alto, California, in the late 1980s. FACTS emerged as a technology in which various power electronics-based static controllers enhance controllability and increase power transfer capability over existing transmission corridors.

FACTS controllers can be broadly classified into two categories. The first one comprises thyristors. Examples include the static VAR compensator, the thyristor-controlled series capacitor, and the phase shifter. The second category comprises self-commutated static converters as controlled voltage sources. Examples include the static compensator (STATCOM), the static synchronous series compensator (SSSC), the unified power flow controller (UPFC), the interline power flow controller (IPFC), and the generalized unified power flow controller (GUPFC). Compared to thyristor-based FACTS controllers, voltage-sourced converter (VSC)-based FACTS controllers generally possess superior performance characteristics. As a consequence, VSC-based FACTS controllers have gained in popularity over the years.

Now, for proper planning, design, and operation of power system networks incorporating VSC-based FACTS controllers, a power flow solution of the network(s) incorporating them is required. Therefore, development of suitable power flow models of VSC-based FACTS controllers is of fundamental importance.

Although a number of books on modeling and applications of FACTS controllers exist, very few books dwell on their power-flow modeling. In addition, I have experienced that slow learners face difficulty in comprehending Newton's method, particularly when proceeding from a single variable to multiple variables. The same holds true for the concept of exchange of real and reactive powers by FACTS controllers, especially the independency of reactive power exchange by individual VSCs, unlike the exchange of real power. This provided the motivation for writing this book.

This book begins by introducing different VSC-based FACTS controllers and their working principle. The way in which these FACTS controllers exchange real and reactive power with the system is explained in detail. The concepts are supplemented with a few solved problems on some typical FACTS controllers. Subsequently, the book presents an introduction to Newton's method and its application in solving nonlinear algebraic equation(s), proceeding in a lucid, step-by-step way from a single variable toward multiple variables. It also introduces the reader to the concept of the power flow problem and the application of Newton's method to the solution of the power flow problem. This is followed by a systematic and generalized approach for the Newton power flow modeling of VSC-based FACTS controllers, which is developed from the first principles. Because of the unique modeling strategy, existing Newton power flow codes can be reused even after inclusion of FACTS controllers. Practical device constraint limits of these FACTS controllers are also accommodated in the power flow models. A large number of case studies have also been included for the validation of the power flow model of each of the FACTS controllers.

This book is intended for senior undergraduate and graduate students in electrical engineering in general and electrical power systems in particular. The reader is expected to have an undergraduate-level background in engineering mathematics, network analysis, electrical machines, electrical power systems, and power electronics.

The book is organized into eight chapters. Chapter 1 provides a brief introduction to FACTS technology and various VSC-based FACTS controllers.

Chapter 2 introduces the reader to the Newton–Raphson method and the power flow problem. It also presents the application of the Newton–Raphson method for solving the power flow problem.

Chapters 3 through 7 detail the Newton power flow modeling of SSSC, UPFC, IPFC, GUPFC, and STATCOM, respectively.

Chapter 8 presents a Newton power flow model of multiterminal VSC-HVDC systems with pulse width modulation (PWM) control schemes.

A unified power flow model of a hybrid AC–DC system is developed. The model is suitable for an AC power system incorporating multiterminal DC network(s) with arbitrary topologies.

The Appendix at the end of the book details the derivations of all the difficult formulae used in the chapters.

In writing this book, I have been greatly influenced by Professor Biswarup Das, Department of Electrical Engineering, Indian Institute of Technology, Roorkee, Uttarakhand, India, who has been my friend, philosopher, and guide. I particularly thank Professors Pragati Kumar and Madhusudan Singh of my department who constantly encouraged me. I thank my PhD student, Ms. Shagufta Khan, for helping me with the manuscript and the figures in this book. I also thank the publisher and my family for their efforts in pursuing me to take up the project of writing this book.

Suman Bhowmick
Delhi Technological University

MATLAB® is a registered trademark of The MathWorks, Inc. For product information, please contact:

The MathWorks, Inc.
3 Apple Hill Drive
Natick, MA 01760-2098, USA
Tel: +1 508 647 7000
Fax: +1 508 647 7001
E-mail: info@mathworks.com
Web: www.mathworks.com

Author

Suman Bhowmick, PhD, is an associate professor of electrical engineering at the Department of Electrical Engineering, Delhi Technological University (formerly Delhi College of Engineering), Delhi, India. He has more than 23 years of experience in both the industry and academia. He is a member of the IEEE.

His research interests include the area of flexible AC transmission system, voltage-sourced converter-HVDC systems, and their control.

List of Abbreviations

BTB	Back to back
CO	Control objectives
CT	Computational time in seconds taken by the algorithm to converge to a specified tolerance
FACTS	Flexible ac transmission systems
GTO	Gate-turn-off thyristor
GUPFC	Generalized unified power flow controller
HVDC	High voltage dc transmission
IGBT	Insulated gate bipolar transistor
IPFC	Interline power flow controller
MTDC	Multiterminal dc
MVA	Mega voltampere
MVAR	Mega voltampere reactive
MW	Mega watt
NI	Number of iterations taken by the algorithm to converge to a specified tolerance
NR	Newton–Raphson
OPF	Optimal power flow
PSASP	Power system analysis software package
PTP	Point-to-point
PWM	Pulse-width modulation
RE	Receiving end
SE	Sending end
SSSC	Static synchronous series compensator
STATCOM	Static compensator
SVC	Static VAR compensator
TCSC	Thyristor-controlled series compensator
UPFC	Unified power flow controller
VSC	Voltage-sourced converter

List of Symbols

CAPITALS

$\mathbf{B}'$	Matrix constituted from the negative of the susceptances of the bus admittance matrix after omitting the effects of all line series resistances, shunt capacitors, and reactors, setting all off-nominal transformer tap settings to unity and accounting for the slack bus
$\mathbf{B}''$	Matrix constituted from the negative of the susceptances of the bus admittance matrix after omitting the effects of all phase shifters and accounting for the voltage controlled buses
B_{ij}	Susceptance of the element in the ith row and jth column of the bus admittance matrix
G	Conductance
G_C^{eq}	Effective conductance because of switching loss in a SSSC
$\mathbf{I}$	Phasor current
$\mathrm{I}_{\mathrm{DC}h}$	Net dc current injection at the hth dc terminal
$\mathbf{I}_{g0}$	Line charging current of the transmission line in series with the gth series converter of the FACTS device
$\mathbf{I}_{\mathrm{i}}$	Net phasor current injection at bus "**i**"
$I_{\mathrm{se}k}$	Magnitude of the line current through the kth series converter (or the kth series FACTS device)
$I_{\mathrm{se}m}^{\mathrm{Lim}}$	Limit constraint on the magnitude of the line current through the mth series converter (or the mth series FACTS device)
I_{sh}	Magnitude of the shunt converter current of the FACTS device
$I_{\mathrm{sh}m}^{\mathrm{Lim}}$	Limit constraint on the magnitude of the current through the shunt converter of the mth FACTS device
$\mathbf{J}^{\mathrm{A}}, \mathbf{J}^{\mathrm{B}}$	Intermediate Jacobian submatrices
$\mathbf{J}_k$	kth Jacobian submatrix related to FACTS device(s)
$\mathbf{J}^{\mathrm{new}}$	Intermediate Jacobian matrix (amenable for computation using existing Jacobian codes) in the proposed model

$\mathbf{J}_{\mathbf{old}}$	Conventional load flow Jacobian matrix
$\mathbf{J}_{\mathbf{se}\,k}$	kth Jacobian submatrix related to SSSC(s)
JX	Jacobian submatrices
P	Bus active power injection vector
$P_{\mathrm{DC}}^{\mathrm{Lim}}$	Limit constraint on the real power transfer through the dc link(s) of the FACTS device
$\mathrm{P}_{\mathrm{DC}k}$	Real power transfer through the dc link of the kth FACTS device (or the kth dc link of the FACTS device)
$\mathbf{P}_{\mathrm{G}}$	Vector comprising real powers delivered by GUPFCs
P_i	Net injected active power at bus "i"
P_{ij}	Active power flow in the line between buses "i" and "j"
P_{ij}^{SP}	Specified active power flow in the line between buses "i" and "j"
$\mathbf{P}_{\mathbf{IP}}$	Vector comprising real powers delivered by IPFCs
$P_{\mathrm{IPFC}\,k}$	Real power delivered by the kth IPFC
$\mathbf{P}_{\mathrm{L}}$	Line active power flow vector
$P_{\mathrm{LINE}\,k}$	Active power flow in the transmission line connected to the kth series converter (or FACTS device)
$\mathbf{P}^{\mathbf{new}}$	Vector comprising net active power injections at the load flow buses in the proposed model
$\hat{P}^{\mathbf{new}}$	Modified vector comprising ratios of net bus active power injections to the voltage magnitudes for all load flow buses in the proposed model
$\mathbf{P}_{\mathbf{se}}$	Vector comprising real powers delivered by SSSCs
$P_{\mathrm{se}\,k}$	Real (active) power delivered by the kth SSSC
$\mathbf{P}_{\mathbf{STAT}}$	Vector comprising real powers delivered by STATCOMs
$\mathrm{P}_{\mathrm{STAT}\,g}$	Real power delivered by gth STATCOM
$\mathbf{P}_{\mathrm{U}}$	Vector comprising real powers delivered by UPFCs
$P_{\mathrm{UPFC}\,k}$	Real power delivered by the kth UPFC
Q	Bus reactive power injection vector
$\hat{\mathbf{Q}}$	Modified vector comprising ratios of net bus reactive power injections to the voltage magnitudes for all load buses
Q_i	Net injected reactive power at bus "i"
Q_{ij}	Reactive power flow in the line between buses "i" and "j"
Q_{ij}^{SP}	Specified reactive power flow in the line between buses "i" and "j"
$\mathbf{Q}_{\mathrm{L}}$	Line reactive power flow vector
$Q_{\mathrm{LINE}\,k}$	Reactive power flow in the transmission line connected to the kth series converter (or FACTS device)
Q_{STAT}	Reactive power delivered by the STATCOM
R	Resistance; control objective specification for the SSSC

R Vector comprising control objective specifications and/or device limit constraint specifications

R_C Effective resistance because of switching loss in a SSSC

R_{DChj} Resistance of dc link between dc buses "*h*" and "*j*"

V Vector comprising magnitudes of bus voltages

$\mathbf{V_B}$ Vector comprising bus voltage magnitudes (sending end) to which shunt converters of FACTS devices are connected

$\mathbf{V_{BUS}}$ Vector comprising magnitudes of voltages at the buses to which STATCOMs are connected

V_{BUS} Magnitude of voltage at the bus to which a STATCOM is connected

V_{BUSk} Bus voltage magnitude (sending end) for the *k*th FACTS device; magnitude of voltage at the bus to which *k*th STATCOM is connected

V_{DC} dc side voltage (across capacitor) in a FACTS device or a VSC

V_{DCg} dc side voltage of the *g*th VSC

$\mathbf{V_i}$ Voltage (phasor) at bus "**i**"

V_i Magnitude of voltage at bus "*i*"

V_i^{SP} Specified magnitude of voltage at bus "*i*"

V_{mk}^{Lim} Limit constraint on the magnitude of voltage on the line side of the *k*th series converter of a GUPFC

$\mathbf{V^{new}}$ Vector comprising magnitudes of voltages at the load flow buses in the proposed model

$\mathbf{V_{se}}$ Injected voltage because of the series converter of a FACTS device; vector comprising magnitudes of series injected voltages of FACTS device(s)

V_{se}^0 Initial value of the magnitude of the injected voltage of the series converter of the FACTS device

V_{sek} Magnitude of the injected voltage because of the *k*th series converter (or the *k*th series FACTS device)

V_{sem}^{Lim} Limit constraint on the voltage magnitude of the *m*th series converter (or the *m*th series FACTS device)

$\mathbf{V_{sh}}$ Injected voltage because of the shunt converter of a FACTS device; vector comprising magnitudes (rms) of shunt converter voltages of FACTS devices

V_{sh} Magnitude (rms) of the injected voltage of the shunt converter (of a FACTS device); magnitude (rms) of the VSC ac side voltage

V_{sh}^0 Initial value of the magnitude of the injected voltage of the shunt converter of the FACTS device

V_{sh}^{Lim} Limit constraint on the voltage magnitude of the shunt converter of a STATCOM

X	Reactance of transmission line
X_{ij}^{SP}	Specified (effective) reactance of the line between buses "i" and "j"
X_{se}	Effective reactance offered by the SSSC
$\mathbf{Y}_{ii}^{\text{old}}$	Self-admittance of bus "i" for the existing "n" bus power system network without any FACTS device connected
$\mathbf{Y}_{ij}$	Element in ith row and jth column of the bus admittance matrix
Y_{ij}	Magnitude of the element "$\mathbf{Y}_{ij}$" in the bus admittance matrix
$\mathbf{Z}$	Complex impedance
$\mathbf{Z}_{ij}$	Complex series impedance of the transmission line connected between buses "i" and "j"
$\mathbf{Z}_{se}$	Complex series impedance of the coupling transformer for a UPFC series converter
$\mathbf{Z}_{sh}$	Complex series impedance of the coupling transformer for a UPFC shunt converter; complex series impedance of the coupling transformer for STATCOM
$\mathbf{Z}_{T}$	Complex series impedance of the coupling transformer for SSSC

LOWERCASE

c	Constant (relating the ac and dc side voltages in a VSC)
c_k	Derived function of voltage magnitude and phase angle state variables
f	function
$\mathbf{f}$	Vector comprising functions related to power balance equations of VSCs in a VSC–HVDC system
f_g	function corresponding to gth VSC
g	function; variable indicating number of converters of FACTS device(s) or number of FACTS devices or number of VSC
h, j	Variables indicating the dc bus number in a VSC–HVDC system
k	Variable indicating ac bus number
k_{se}	Constant (accounting for the type of converter in a FACTS device)
m	Number of generators; number of the FACTS device considered; node at which a series converter of a GUPFC and a transmission line are interconnected (GUPFC series converters are connected at the sending end of the line)
$\mathbf{m}$	Vector comprising modulation indices of VSCs
m_g	Modulation index of the PWM control scheme for the gth VSC
n	Total number of system buses (without any FACTS device)
p	Number of SSSCs; number of UPFCs; number of converters (series) in a single IPFC; number of series converters in a single GUPFC; number of VSCs in a VSC–HVDC system

p.u.	Per unit
q	VSC that acts as the master converter
r	Number of generators in a hybrid ac–dc power system
w	Total number of transmission lines containing IPFC converters (series) with specified line reactive power flows; total number of converters (both series and shunt) in all the GUPFCs combined
x	Number of IPFCs ; number of GUPFCs
$\mathbf{y}$	Complex admittance
$\mathbf{y}_{g0}$	Half-line charging shunt admittance of the transmission line in series with the *g*th series converter of the FACTS device
$\mathbf{y}_{\mathbf{ij0}}$	Half-line charging shunt admittance of the line between buses "*i*" and "*j*"
$\mathbf{y}_{i0}^{p}$	Effective half-line charging shunt admittance because of all transmission lines connected to bus "*i*," except the line between buses "*i*" and "*j*"
$\mathbf{y}_{\mathbf{se}k}$	Admittance of the coupling transformer of the *k*th series converter (or the series coupling transformer of the *k*th FACTS device)
$y_{\mathrm{se}k}$	Magnitude of the admittance $\mathbf{y}_{\mathbf{sek}}$
$\mathbf{y}_{\mathbf{sh}k}$	Admittance of the coupling transformer (of the shunt converter of the *k*th FACTS device or the *k*th VSC)
$y_{\mathrm{sh}k}$	Magnitude of the admittance $\mathbf{y}_{\mathbf{shk}}$
z	Total number of converters (series) in "*x*" IPFCs ; total number of series converters in "*x*" GUPFCs

UPPERCASE GREEK

Σ	Summation sysmbol
Δ	Mismatch in the electrical quantity of interest
$\boldsymbol{\Delta}$	Vector comprising mismatches in electrical quantities of interest

LOWERCASE GREEK

$\boldsymbol{\alpha}_{\mathbf{k}}, \boldsymbol{\beta}_{\mathbf{k}}, \boldsymbol{\gamma}_{\mathbf{k}}$	Effective admittances because of the transmission line—series coupling transformer combination (of *k*th series converter or *k*th FACTS device)
$\alpha_k, \beta_k, \gamma_k$	Magnitudes of admittances $\boldsymbol{\alpha}_{\mathbf{k}}, \boldsymbol{\beta}_{\mathbf{k}}, \boldsymbol{\gamma}_{\mathbf{k}}$
$\partial\mathbf{A} / \partial\mathbf{B}$	Matrix comprising partial derivatives of elements in vector **A** with respect to elements in vector **B**

$\partial A_p/\partial B_q$	Partial derivative of the *p*th element of vector **A** with respect to *q*th element of vector **B**; element in the *p*th row and *q*th column of the matrix $\partial \mathbf{A}/\partial \mathbf{B}$
θ_i	Phase angle of the rms phasor voltage at bus "*i*"
$\boldsymbol{\theta}$	Vector comprising phase angles of rms bus voltage phasors
$\boldsymbol{\theta}_{\mathbf{se}}$	Vector comprising phase angles of series voltage sources of FACTS device(s)
$\theta_{se\,k}$	Phase angle of the *k*th series voltage source (of FACTS device)
$\boldsymbol{\theta}_{\mathbf{sh}}$	Vector comprising phase angles of shunt voltage sources of FACTS devices or VSCs
θ_{sh}	Phase angle of the shunt voltage source of a FACTS device or VSC
θ_{sh}^0	Initial value of the phase angle of the series voltage source of the FACTS device
θ_{sh}^0	Initial value of the phase angle of the shunt voltage source of the FACTS device
$\boldsymbol{\theta}^{\mathbf{new}}$	Vector comprising phase angles of rms voltage phasors of the load flow buses in the proposed model
ϕ_{yij}	Phase angle of the element "$\mathbf{Y_{ij}}$" in the bus admittance matrix
$\phi_{y\alpha k}, \phi_{y\beta k}$	Phase angles of admittances $\boldsymbol{\alpha}_{\mathbf{k}}$, $\boldsymbol{\beta}_{\mathbf{k}}$
$\phi_{y\gamma k}$	Phase angle of admittance $\boldsymbol{\gamma}_{\mathbf{k}}$
ϕ_{ysek}	Phase angle of the admittance $\mathbf{y_{sek}}$
ϕ_{yshk}	Phase angle of the admittance $\mathbf{y_{shk}}$

SUBSCRIPTS

B	Bus quantities (vector)
BUS	Bus quantities (vector)
BUS*m*	Quantity corresponding to the bus at which the *m*th FACTS device is connected
c	*c*th FACTS device quantity (for admittances $\boldsymbol{\alpha}_{\mathbf{c}}$, $\boldsymbol{\beta}_{\mathbf{c}}$, etc.)
C	dc side capacitor quantity
dc	dc side quantity
dc*m*	dc side quantity of the *m*th FACTS device or the *m*th VSC
g	quantity corresponding to the *g*th VSC
G	GUPFC quantities (vector)
GUPFC	GUPFC quantity
GUPFC*m*	Quantity corresponding to the *m*th GUPFC
i	Bus "*i*" quantity

ij	Transmission line series quantity (of the line connected between bus "*i*" and bus "*j*"); quantity (for magnitude) of the admittance element $\mathbf{Y}_{ij}$
*ij*0	Line shunt quantity (of line connected between bus "*i*" and bus "*j*")
IP	IPFC quantities (vector)
IPFC	IPFC quantity
IPFC*m*	Quantity corresponding to the *m*th IPFC
j	Bus "*j*" quantity
L	Transmission line quantities (vector)
LINE, line	Transmission line quantity
LINE*m*	Quantity corresponding to the mth transmission line
m	Quantity at the node at which a series converter of a GUPFC and a transmission line are interconnected
old	Quantity (usually for Jacobian) related to the existing network (without any FACTS device)
se	SSSC quantity; FACTS device series converter quantity
se*c*	*c*th SSSC quantity ; *c*th series converter quantity
sh	FACTS device shunt converter quantity; VSC quantity
sh*g*	Quantity of the shunt converter of the *g*th FACTS device or the *g*th VSC; quantity (magnitude) of admittance element $\mathbf{y}_{shc}$
STAT	STATCOM quantities (vector)
STAT*m*	Quantity corresponding to the *m*th STATCOM
SW	Switching loss quantity in a SSSC
T	FACTS device coupling transformer quantity
U	UPFC quantities (vector)
UPFC	UPFC quantity
UPFC*m*	Quantity corresponding to the *m*th UPFC
yab	Quantity (for phase angle] of admittance element $\mathbf{Y}_{ab}$
yshc	Quantity (for phase angle] of admittance element $\mathbf{y}_{shc}$
yαc, yβc	Quantities (for phase angles] of admittance elements $\boldsymbol{\alpha}_c$, $\boldsymbol{\beta}_c$

SUPERSCRIPTS

$(\ )^{eq}$	Effective or equivalent quantity (to account for loss etc.)
$(\ )^{Lim}$	Limit constraint of quantity in a FACTS device
$(\ )^{new}$	Quantities in the proposed model (vector); matrix (usually for Jacobian) in the proposed model

$(\ )^{old}$	Quantity in the existing network without any FACTS device (usually for bus admittance matrix element)
$(\ )^{SP}$	Specified or known quantity
$(\ )^{T}$	Transpose of a matrix
$(\ )^{*}$	Conjugate of a complex quantity
$(\ )^{0}$	Initial value of quantity (for voltage magnitudes or angles)
$\hat{(\ )}$	Modified quantity (vector)
$(\ )'$	Modified quantity (usually the Jacobian matrix) for fast decoupled power flow (active power–phase angle relation)
$(\ )''$	Modified quantity (usually the Jacobian matrix) for fast decoupled power flow (reactive power–voltage magnitude relation)

CHAPTER 1

FACTS and FACTS Controllers

1.1 INTRODUCTION

Since the past three decades, the construction of both generation facilities and, in particular, new transmission lines has been delayed due to energy cost, environmental concerns, right-of-way restrictions, and other legislative and cost problems. Recently, the philosophy of open access transmission has facilitated the development of competitive electric energy markets. New economic, social, and legislative developments have demanded a review of traditional power transmission theory and practice and the creation of new concepts to allow full utilization of existing power generation and transmission facilities, without compromising system availability and security. In the late 1980s, the Electric Power Research Institute (EPRI), Palo Alto, California, USA formulated the vision of flexible AC transmission system (FACTS) in which various power electronics-based static controllers enhance controllability and increase power transfer capability while maintaining sufficient steady-state and transient margins [1–6]. The main objectives of FACTS technology are to increase the usable transmission capacity of lines and control power flow over designated transmission corridors.

The FACTS controllers achieve these objectives by controlling the interrelated parameters that govern the operation of transmission systems including series impedance, shunt impedance, current, voltage, phase angle and the damping of power system oscillations. The normal power transfer over the transmission lines can be estimated to increase significantly by using FACTS controllers (about 50%, according to some studies conducted) [4].

The development of FACTS controllers has followed two distinctly different technical approaches, both resulting in a comprehensive group of controllers able to address targeted transmission problems. The first group employs reactive impedances or a tap changing transformer with thyristor switches (i.e., having no intrinsic turn-off ability) as controlled elements. Examples include the static VAr compensator (SVC), the thyristor-controlled series capacitor, and the phase shifter. The second group uses self-commutated static converters as controlled voltage sources. Examples include the static compensator (STATCOM), the static synchronous series compensator (SSSC), unified power flow controller (UPFC), the interline power flow controller (IPFC), and the generalized unified power flow controller (GUPFC) [1–6].

Now, compared to the thyristor-based FACTS devices, the voltage-sourced converter (VSC)-based FACTS controllers generally provide superior performance characteristics and uniform applicability for transmission voltage, effective line impedance, and angle control. They also offer the unique potential to exchange real power directly with the AC system, in addition to providing the independently controllable reactive power compensation [1,3]. Consequently, these controllers are evoking a lot of interest from both the industry and the academia worldwide.

Among the VSC-based FACTS controllers, the STATCOM is the earliest device to be conceived. It operates as a shunt-connected static VAr compensator whose capacitive or inductive output current can be controlled independent of the AC system voltage [7–10]. The first STATCOM, with a rating of ±100 MVAR, was commissioned in late 1995 at the Sullivan substation of the Tennessee Valley Authority (TVA) in the United States, jointly sponsored by the EPRI and the TVA and manufactured by the Westinghouse Electric Corporation [11,12]. The STATCOM is described in Section 1.2.

1.2 THE STATCOM

As already discussed in Section 1.1, the STATCOM is a VSC-based shunt FACTS controller. It is primarily used for voltage control of buses. The schematic diagram of a STATCOM is shown in Figure 1.1.

From the figure, it can be observed that the STATCOM is connected to any bus j of a power system through a step-up transformer, also known as the coupling transformer. For power system analysis, we will focus on fundamental frequency, positive sequence voltages and currents of

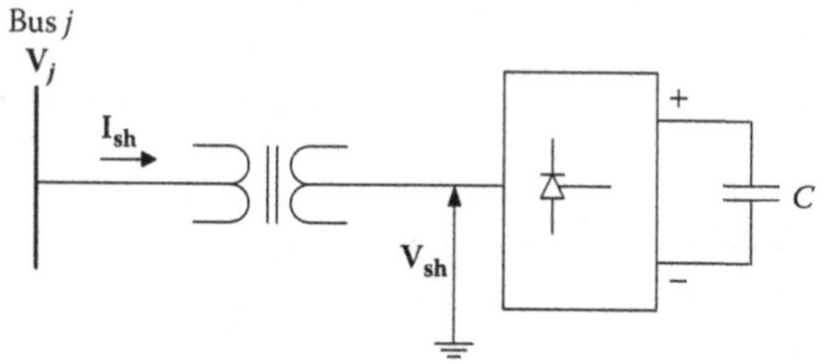

FIGURE 1.1 A STATCOM connected to any bus j of an n-bus power system.

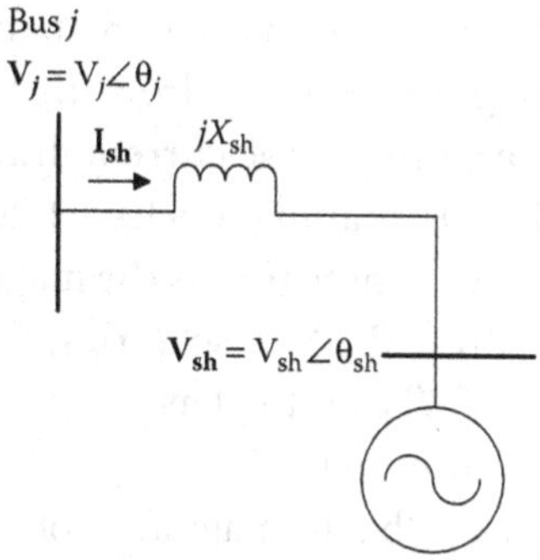

FIGURE 1.2 Equivalent circuit of Figure 1.1.

the STATCOM. The (fundamental frequency, positive sequence) output voltage of the STATCOM is shown as $\mathbf{V}_{sh}$. The STATCOM draws a current $\mathbf{I}_{sh}$ from the system. For ease of analysis, we will neglect the losses in the coupling transformer and the STATCOM. Also, the STATCOM rating limits are ignored. The equivalent circuit of Figure 1.1 is shown in Figure 1.2.

From the figure, using Kirchhoff's voltage law (KVL), we can write $\mathbf{V}_j = \mathbf{V}_{sh} + j\mathbf{I}_{sh}X_{sh}$, where X_{sh} represents the leakage reactance of the coupling transformer. Now, the STATCOM output voltage $\mathbf{V}_{sh}$ is controllable in both its magnitude and its phase. Since we have assumed a lossless STATCOM (and also due to the fact that the STATCOM cannot generate active power by itself), the phase angle of $\mathbf{V}_{sh}$ is kept equal to that of $\mathbf{V}_j$. This can be explained as follows. From Figure 1.2, for a lossless STATCOM, the active power transfer from bus j to the STATCOM will be

$$P = \frac{\mathbf{V}_j\mathbf{V}_{sh}}{X_{sh}}\sin(\theta_j - \theta_{sh}) = 0$$

This implies $\theta_{sh} = \theta_j$. Thus, in the phasor diagram, $\mathbf{V}_{sh}$ and $\mathbf{V}_j$ will be collinear phasors, having only two possibilities, depending upon whether the magnitude of $\mathbf{V}_{sh}$ is made more or less than $\mathbf{V}_j$. These two cases correspond

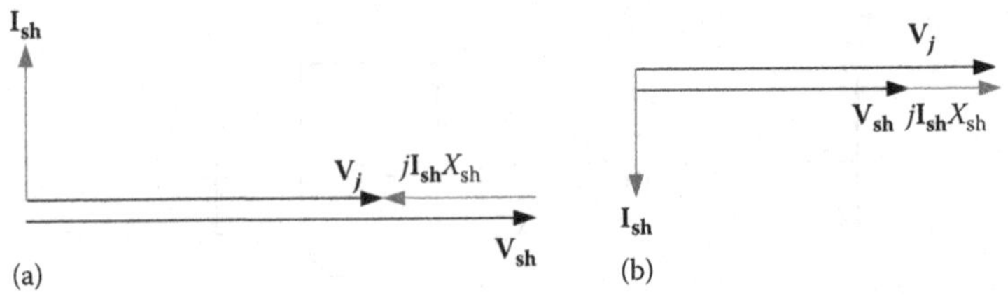

FIGURE 1.3 (a) Capacitive and (b) inductive modes of operation of STATCOM.

to the two phasor diagrams shown in Figure 1.3a and b. In the first case (Figure 1.3a), the magnitude of $\mathbf{V}_{sh}$ is made more than $\mathbf{V}_j$. By virtue of KVL, in the phasor diagram (Figure 1.3a), the drop $j\mathbf{I}_{sh}X_{sh}$ is in phase opposition to both $\mathbf{V}_{sh}$ and $\mathbf{V}_j$, resulting in the current drawn by the STATCOM leading the bus voltage by 90°, making it behave like a capacitor from the bus end. Also, it is important to note that as the magnitude of $\mathbf{V}_{sh}$ can vary (assuming $\mathbf{V}_j$ fixed), the drop $j\mathbf{I}_{sh}X_{sh}$ also varies, that is, the STATCOM current varies. Thus, the STATCOM behaves as a variable capacitor and delivers variable reactive power to the bus.

In the second case (Figure 1.3b), the magnitude of $\mathbf{V}_{sh}$ is made less than $\mathbf{V}_j$. Again, by virtue of KVL, in the phasor diagram (Figure 1.3b), the drop $j\mathbf{I}_{sh}X_{sh}$ is in phase with both $\mathbf{V}_{sh}$ and $\mathbf{V}_j$, resulting in the current drawn by the STATCOM lagging the bus voltage by 90°, making it behave like a inductor from the bus end. As the magnitude of $\mathbf{V}_{sh}$ can vary (assuming $\mathbf{V}_j$ fixed), the drop $j\mathbf{I}_{sh}X_{sh}$ also varies, that is, the STATCOM current varies. Thus, the STATCOM behaves as a variable inductor and absorbs variable reactive power from the bus.

1.3 THE SSSC

As described in Section 1.2, unlike the STATCOM, the SSSC is a VSC-based series FACTS controller. It acts as a series compensator whose output voltage is in quadrature with and controllable, independently of the line current for the purpose of altering the overall line reactive drop and thereby controlling the transmitted electric power [13,14]. It is primarily used for power flow control. The schematic diagram of an SSSC is shown in Figure 1.4. Because the SSSC is in series with the transmission line, it can be installed at the sending or receiving end substation.

The equivalent circuit of the diagram shown in Figure 1.4 is shown in Figure 1.5. For ease of analysis, the resistances of both the transmission line and the SSSC coupling transformer are neglected. In addition, the SSSC losses are also neglected. The shunt admittances of the transmission line are

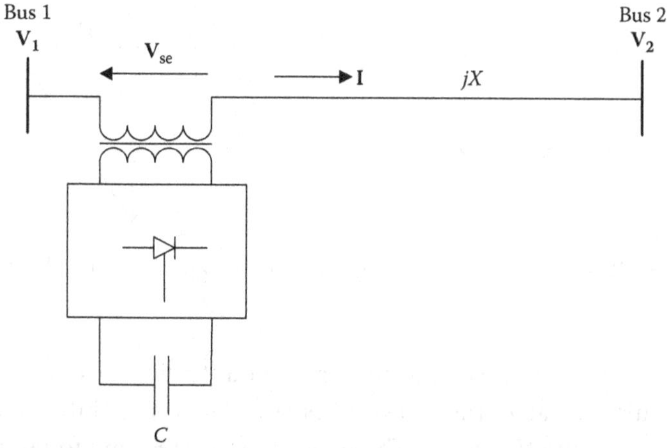

FIGURE 1.4 SSSC connected at the sending end of a transmission line.

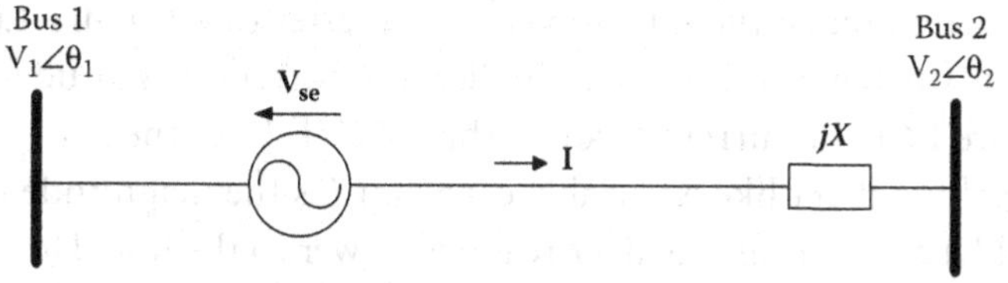

FIGURE 1.5 Equivalent circuit of the diagram shown in Figure 1.4.

also neglected. X denotes the combined reactance of the line along with the leakage reactance of the coupling transformer. The (fundamental frequency, positive sequence) output voltage of the SSSC is shown as $\mathbf{V}_{se}$. The current flowing through the transmission line and the SSSC is represented by **I**.

From the figure, the relationship among the sending end, the receiving end and the SSSC output voltages along with the line current can be expressed using KVL as $\mathbf{V}_1 = \mathbf{V}_2 + \mathbf{V}_{se} + j\mathbf{I}X$. The phasor diagrams corresponding to this are shown in Figure 1.6a and b. It is important to note certain important aspects of the phasor diagram. For ease of analysis, it is assumed that the sending and receiving end bus voltage magnitudes are same. It is also assumed that some nonzero active power is flowing through the transmission line from the sending end to the receiving end, making the difference in phase angles between the sending and receiving end bus voltages $\theta_1 - \theta_2 > 0$, that is, the bus voltage phasors $\mathbf{V}_1$ and $\mathbf{V}_2$ are not collinear and subtend a nonzero angle between each other. Due to the assumption that the SSSC is lossless, it does not absorb any active power. Nor it can supply any active power, as there is no active power source.

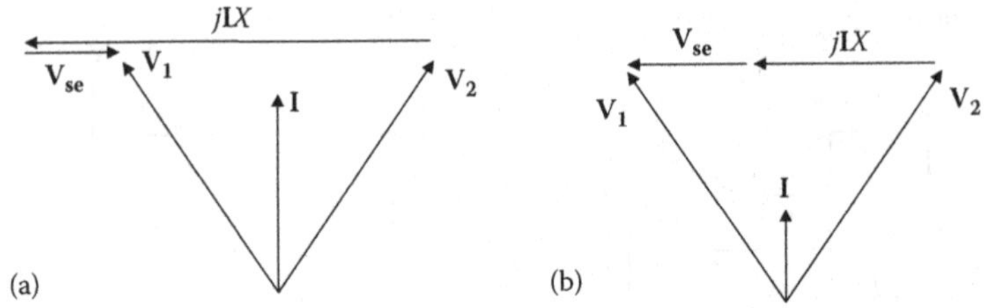

FIGURE 1.6 Phasor diagram of voltages in (a) the capacitive and (b) inductive modes.

Hence, the SSSC output voltage phasor $\mathbf{V}_{se}$ and the line current **I** must be perpendicular to each other. This gives rise to two possibilities, that is, **I** leading or lagging $\mathbf{V}_{se}$ by 90°. These two cases correspond to Figure 1.6a and b, respectively. In the former case (Figure 1.6a), the disposition of $\mathbf{V}_{se}$ makes the line voltage drop phasor *j***I**X larger resulting in an increase of the line current magnitude. However, in the latter case (Figure 1.6b), the line drop is smaller, resulting in a smaller line current magnitude.

In Figure 1.6a, the current through the SSSC (**I**) leads the voltage across it ($\mathbf{V}_{se}$), making it act like a variable capacitor (as the magnitude of $\mathbf{V}_{se}$ is controllable) and letting it deliver reactive power to the line. However, in Figure 1.6b, the SSSC acts like a variable inductor and absorbs the reactive power from the line.

The direction of the reactive powers corresponding to the phasor diagrams in Figure 1.6a and b is shown in Figure 1.7a and b, respectively, in thick white (unshaded) arrows (*Q*).

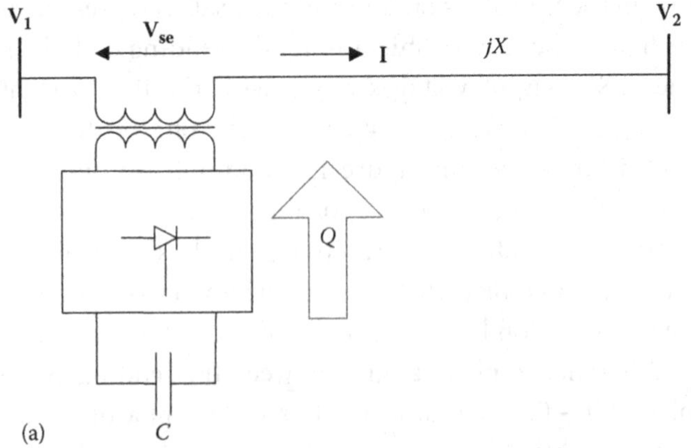

FIGURE 1.7 (a) Direction of reactive power flow of SSSC in the capacitive mode.

(Continued)

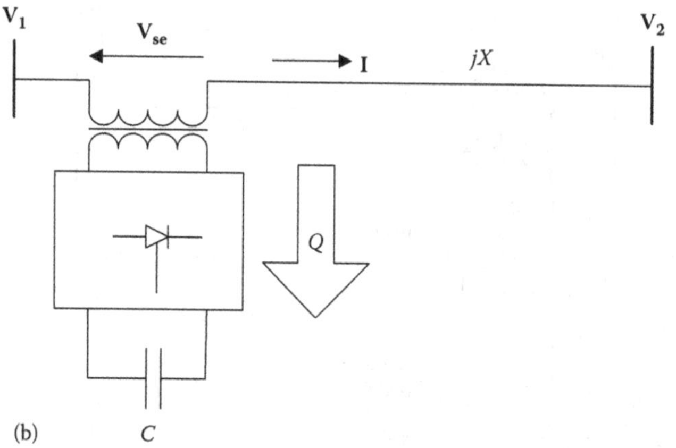

FIGURE 1.7 (Continued) (b) Direction of reactive power flow of SSSC in the inductive mode.

Both STATCOM and SSSC are single-converter-based devices. For better control of transmission networks, multiple VSC-based FACTS devices have been conceptualized, manufactured, and deployed in power systems. Among these multiple VSC-based devices, the UPFC is one of the most comprehensive and versatile controllers. It is discussed in Section 1.4.

1.4 THE UPFC

A UPFC consists of two back-to-back linked self-commutating converters operated from a common DC link and connected to the AC system through series and shunt coupling transformers. Within its operating limits, a UPFC can independently control three power system parameters [15–17]. The first UPFC, with a total rating of ±320 MVA, was commissioned in mid-1998 at the Inez Station of the American Electric Power in Kentucky for voltage support and power flow control [18–21].

The schematic diagram of a UPFC connected at the sending end of a transmission line is shown in Figure 1.8. The equivalent circuit of the diagram is shown in Figure 1.9.

In Figure 1.9, $\mathbf{V}_{se}$ and $\mathbf{V}_{sh}$ represent the fundamental frequency, positive sequence output voltages of the series and shunt converters. X represents the total reactance of the line and the leakage reactance of the series coupling transformer. X_{sh} represents the leakage reactance of the coupling transformer of the shunt converter. $\mathbf{I}$ represents the line current and $\mathbf{I}_{sh}$ represents the current drawn by the shunt converter.

For ease of analysis, we assume that the sending and receiving end bus voltage magnitudes are the same and the converters are lossless. The rating

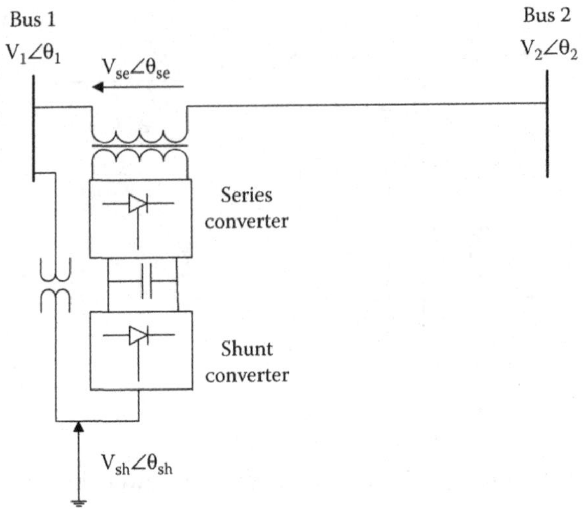

FIGURE 1.8 UPFC connected in a transmission line between two buses.

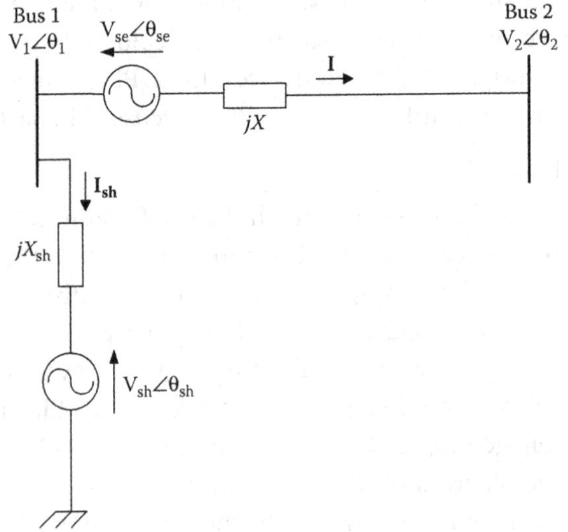

FIGURE 1.9 Equivalent circuit of the diagram shown in Figure 1.8.

limits of both the converters are ignored. We also assume that before the UPFC was installed, a nonzero active power ($P_{12} > 0$) was flowing from the sending end of the line to the receiving end, which makes $\theta_1 > \theta_2$ (by virtue of the active power transfer equation $P_{12} = (V_1V_2/X)\sin(\theta_1 - \theta_2)$). Under such a condition, with the UPFC, the phasor diagrams representing the line end bus voltages, the series converter output voltage and the line current can be drawn. We first assume the simple mode in which the series converter is not absorbing or delivering any active power. Thus, in this mode, for the series converter, only two phasor diagrams are possible: (1) $\mathbf{V}_{se}$ lagging the line current **I** by 90° or (2) $\mathbf{V}_{se}$ leading the line current **I** by 90°. The phasor diagrams will be identical to those shown in Figure 1.6a and b, respectively. Now, as it is assumed that the series converter is not exchanging any active power with the line, the shunt converter cannot do so either, with the sending end bus (since no converter can consume or deliver active power by itself). Thus, the phasor diagrams for the shunt converter will be similar to those already drawn for the STATCOM in Figure 1.3a and b except that the bus voltage $\mathbf{V}_j$ in Figure 1.3a and b will be replaced by $\mathbf{V}_1$. Absence of any active power exchange between bus 1 and the shunt converter can be observed from the fact that the phase angle difference between $\mathbf{V}_{sh}$ and $\mathbf{I}_{sh}$ is ±90° and that $\mathbf{V}_{sh}$ and $\mathbf{V}_1$ are collinear.

However, it is important to note that both the converters can independently supply reactive power. Hence, independent of the series converter absorbing or delivering reactive power, the shunt converter will be delivering reactive power to the bus if the magnitude of the shunt converter output voltage $\mathbf{V}_{sh}$ is more than that of $\mathbf{V}_1$ (Figure 1.3a) or vice versa (Figure 1.3b).

Next, we consider the more general mode in which the converters can exchange active power with the power system. Under such a condition, many possibilities exist. However, only four typical cases with their corresponding phasor diagrams are considered here, which are given below. It is important to note that these phasor diagrams are not to scale and are only figurative. All losses and rating limits of converters have been ignored. For active power, we follow the dictum $\mathrm{Re}\left[\mathbf{V}_{se}(-\mathbf{I}^*)+\mathbf{V}_{sh}(-\mathbf{I}_{sh}^*)\right]=0$, that is, the UPFC does not have the capability to deliver or consume active power by itself. Also, while drawing the phasor diagrams, it is followed that active power flows from a bus with a higher phase angle to bus with a lower phase angle and reactive power flows from a bus at a higher voltage to a bus at a lower voltage, in transmission systems.

Case 1: UPFC series converter delivers active power to the line and absorbs reactive power from the line. The phasor diagrams for the bus and converter voltages and currents are shown in Figure 1.10a through c. Figure 1.10a shows the phasors of the sending and receiving end bus voltages, the series converter output voltage, the line drop, and the line current.

First, we consider the active power. Figure 1.10b shows the phasors of the series converter voltage $\mathbf{V}_{se}$ and the line current **I** separately. It can be observed (not shown explicitly) that the line current has a component in phase opposition to the series converter voltage. This implies that the series

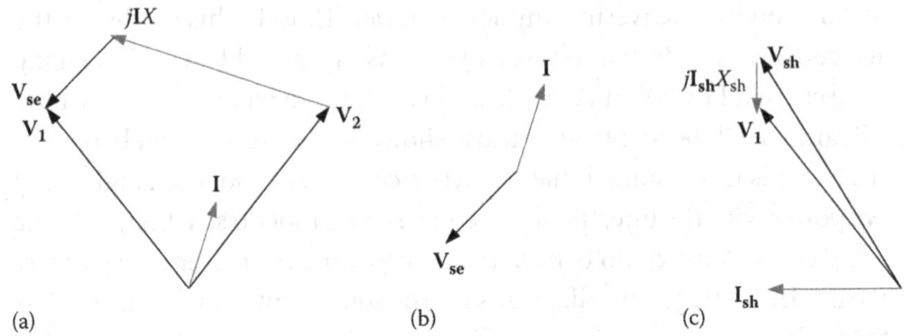

FIGURE 1.10 (a–c) Phasor diagrams of UPFC voltages and currents in general mode (case 1).

converter is delivering active power to the line. As the series converter cannot produce active power by itself, this active power must be coming from the shunt converter through the common DC link. Now, similar to the series converter, the shunt converter cannot generate this active power by itself. Hence, this active power must be coming from the sending end bus to the shunt converter through its coupling transformer. Because the shunt converter receives this active power from the sending end bus, the sending end bus voltage phasor $\mathbf{V}_1$ must be leading the shunt converter output voltage $\mathbf{V}_{sh}$ (active power flows from a bus having a higher phase angle to a lower one in transmission systems), as shown in Figure 1.10c. This is also reiterated by the shunt converter current phasor $\mathbf{I}_{sh}$ that has a component in phase with the shunt converter voltage $\mathbf{V}_{sh}$, which implies that the shunt converter is absorbing active power from the sending end bus. The direction of the active power is shown by thick black (shaded) arrows (P) in Figure 1.11.

Next, we consider the reactive powers. From Figure 1.10b, it can be observed (not shown explicitly) that the line current $\mathbf{I}$ has a quadrature component lagging the series converter voltage $\mathbf{V}_{se}$ by 90°. This means that the series converter is absorbing reactive power from the line. The direction of the reactive power flow of the series converter is shown by a thick thick white (unshaded) arrow (Q_1) in Figure 1.11. Also, from Figure 1.10c, it is observed that the magnitude of the shunt converter voltage $\mathbf{V}_{sh}$ is more than the sending end bus voltage $\mathbf{V}_1$ and the current drawn by the shunt converter $\mathbf{I}_{sh}$ has a quadrature component, which leads the shunt converter output voltage $\mathbf{V}_{sh}$ by 90°. Both of these imply that the shunt converter is supplying reactive power to the sending end bus. The direction of the reactive power flow of the shunt converter is shown by a thick white (unshaded) arrow (Q_2) in Figure 1.11.

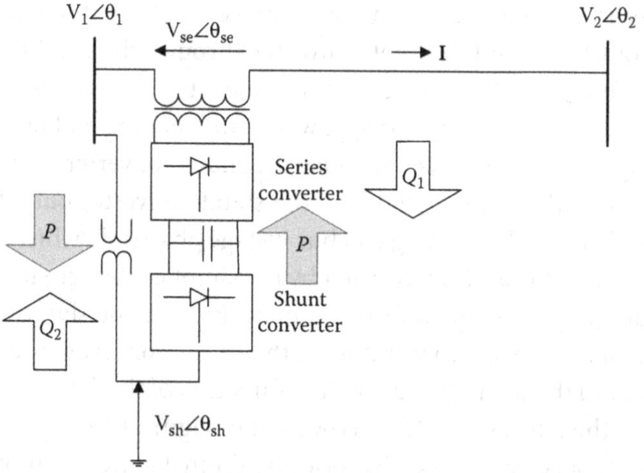

FIGURE 1.11 Directions of active and reactive powers of UPFC series and shunt converters.

Case 2: UPFC series converter absorbs both active and reactive powers from the line. The phasor diagrams for the bus and converter voltages and currents are shown in Figure 1.12a through c. Figure 1.12a shows the phasors of the sending and receiving end bus voltages, the series converter output voltage, the line drop, and the line current.

Figure 1.12b shows the phasors of the series converter voltage $\mathbf{V}_{se}$ and the line current **I** separately. First we consider the active power. From Figure 1.12b, it can be observed (not shown) that the line current **I** has a component in phase with the series converter voltage $\mathbf{V}_{se}$. This implies that the series converter is absorbing active power from the line. Because the

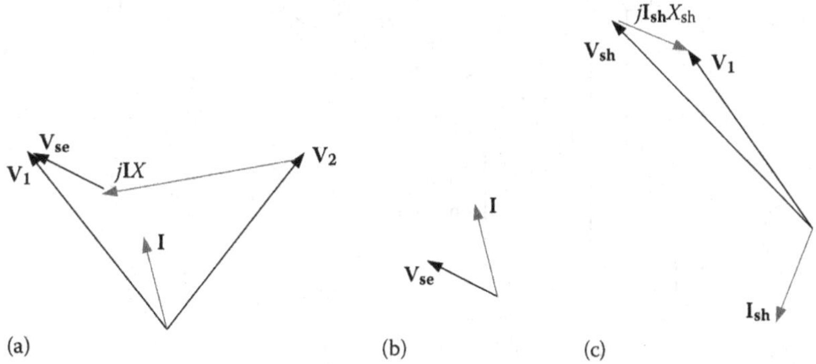

FIGURE 1.12 (a–c) Phasor diagrams of UPFC voltages and currents in general mode (case 2).

series converter does not consume active power by itself, this active power must be transferred to the shunt converter through the DC link. Again, the shunt converter does not consume this active power by itself and, therefore, must transfer this active power to the sending end bus through its coupling transformer. Also, because the shunt converter transfers this active power to the sending end bus, the shunt converter output voltage $\mathbf{V}_{sh}$ must be leading the sending end bus voltage phasor $\mathbf{V}_1$, which is shown in Figure 1.12c. This is also reiterated by the shunt converter current phasor $\mathbf{I}_{sh}$ that has a component in phase opposition to the shunt converter output voltage $\mathbf{V}_{sh}$, which implies that the shunt converter is delivering active power to the sending end bus. The direction of the active power flow is shown by thick black (shaded) arrows (*P*) in Figure 1.13.

Next, we consider the reactive powers. From Figure 1.12b, it can be observed that the line current **I** has a quadrature component lagging the series converter voltage $\mathbf{V}_{se}$. This implies that the series converter is absorbing reactive power from the line. The direction of the reactive power flow of the series converter is shown by a thick white (unshaded) arrow (Q_1) in Figure 1.13. However, from Figure 1.12c, it can be observed that the current drawn by the shunt converter $\mathbf{I}_{sh}$ has a quadrature component leading the shunt converter output voltage $\mathbf{V}_{sh}$. Also, the magnitude of the shunt converter output voltage is more than the sending end bus voltage. Both of these imply that the shunt converter is delivering reactive power to the

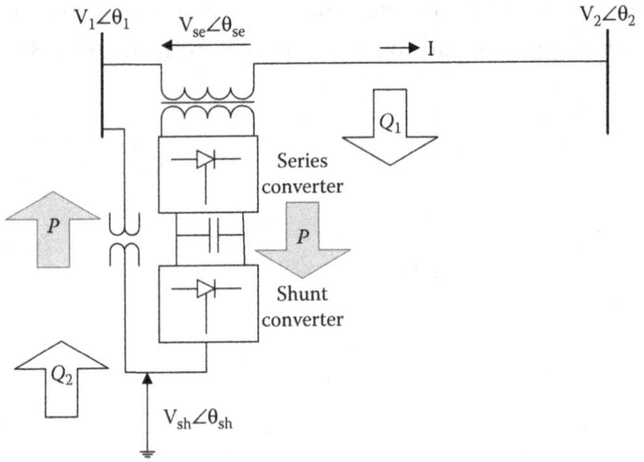

FIGURE 1.13 Directions of active and reactive powers of UPFC series and shunt converters.

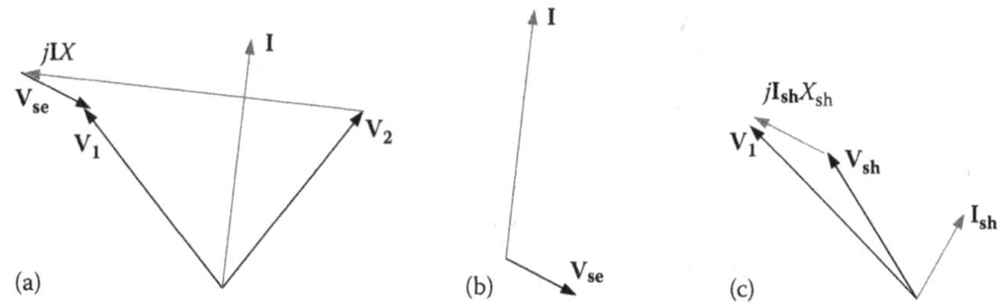

FIGURE 1.14 (a–c) Phasor diagrams of UPFC voltages and currents in general mode (case 3).

sending end bus. The direction of the reactive power flow of the shunt converter is shown by a thick white (unshaded) arrow (Q_2) in Figure 1.13.

Case 3: UPFC series converter delivers both active and reactive powers to the line. The phasor diagrams for the bus and converter voltages and currents are shown in Figure 1.14a through c.

First we consider the active power. From Figure 1.14b, it can be observed that the line current **I** has a component in phase opposition to the series converter voltage $\mathbf{V}_{se}$. This implies that the UPFC series converter is delivering active power to the line. As already discussed earlier, the series converter must receive this active power from the shunt converter via the common DC link. The shunt converter, in turn, receives this active power from the sending end bus through its coupling transformer. This fact is reinforced from Figure 1.14c, in which the sending end bus voltage $\mathbf{V}_1$ is leading the shunt converter voltage $\mathbf{V}_{sh}$. This indicates that the sending end bus is delivering active power to the shunt converter for onward transfer to the line (through the series converter). Also, the shunt converter current $\mathbf{I}_{sh}$ has an in-phase component with the shunt converter voltage $\mathbf{V}_{sh}$, which shows that it is indeed absorbing power (from the sending end bus). The direction of the active power flow is shown by thick black (shaded) arrows (*P*) in Figure 1.15.

Next we consider the reactive powers. From Figure 1.14b, it can be observed that the line current **I** has a quadrature component leading the series converter voltage $\mathbf{V}_{se}$. This implies that the series converter is supplying reactive power to the line. The direction of the reactive power flow of the series converter is shown by a thick white (unshaded) arrow (Q_1) in Figure 1.15. Also, from Figure 1.14c, it can be observed that the shunt converter draws a current $\mathbf{I}_{sh}$, which has a quadrature component lagging the shunt converter output voltage $\mathbf{V}_{sh}$. This implies that the shunt converter is absorbing reactive power from the sending end bus. This is reinforced by

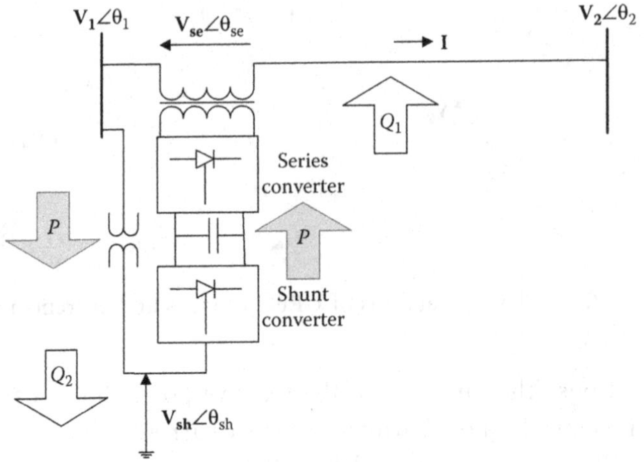

FIGURE 1.15 Directions of active and reactive powers of UPFC series and shunt converters.

the fact that the magnitude of the shunt converter voltage $\mathbf{V}_{sh}$ is less than the sending end bus voltage magnitude $\mathbf{V}_1$. The direction of the reactive power flow of the shunt converter is shown by a thick white (unshaded) arrow (Q_2) in Figure 1.15.

Case 4: UPFC series converter absorbs active power from the line and delivering reactive power to the line. The phasor diagrams for the bus and converter voltages and currents are shown in Figure 1.16a through c.

First, we consider the active power. From Figure 1.16b, it can be observed that the line current **I** has an in-phase component with the series converter voltage $\mathbf{V}_{se}$, which implies that the series converter is

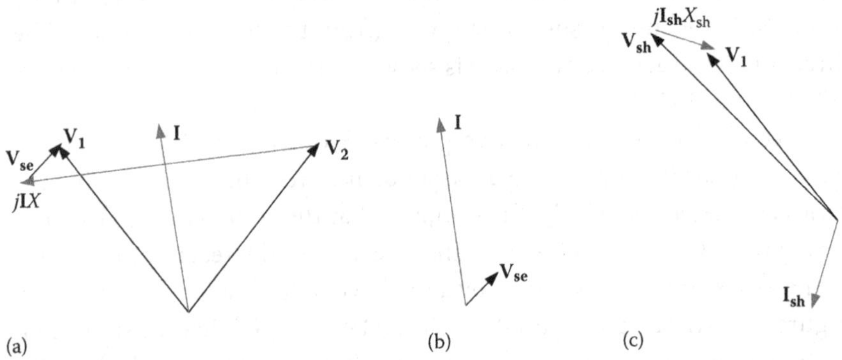

FIGURE 1.16 (a–c) Phasor diagrams of UPFC voltages and currents in general mode (case 4).

absorbing active power from the line. This active power must be transferred to the shunt converter via the common DC link for onward transmittal by the shunt converter to the sending end bus through the shunt coupling transformer. From Figure 1.16c, it can be observed that the shunt converter output voltage $\mathbf{V}_{sh}$ leads the sending end bus voltage $\mathbf{V}_1$. Also, the current drawn by the shunt converter $\mathbf{I}_{sh}$ has a component in phase opposition to the shunt converter output voltage $\mathbf{V}_{sh}$. Both of these imply that the shunt converter is delivering active power to the sending end bus. The direction of the active power flow is shown by thick black (shaded) arrows (P) in Figure 1.17.

Next, we consider the reactive powers. From Figure 1.16b, it can be observed that the line current **I** has a quadrature component that is leading the series converter voltage $\mathbf{V}_{se}$, which implies that the series converter is delivering reactive power to the line. The direction of the reactive power flow of the series converter is shown by a thick white (unshaded) arrow (Q_1) in Figure 1.17. Also, from Figure 1.16c, it can be observed that the magnitude of the shunt converter output voltage $\mathbf{V}_{sh}$ is more than the sending end bus voltage $\mathbf{V}_1$, implying that the shunt converter is delivering reactive power to the sending end bus. This is reiterated from the fact that in Figure 1.16c, the current drawn by the shunt converter $\mathbf{I}_{sh}$ has a quadrature component leading the shunt converter output voltage $\mathbf{V}_{sh}$. The direction of the reactive power flow of the shunt converter is shown by a thick white (unshaded) arrow (Q_2) in Figure 1.17.

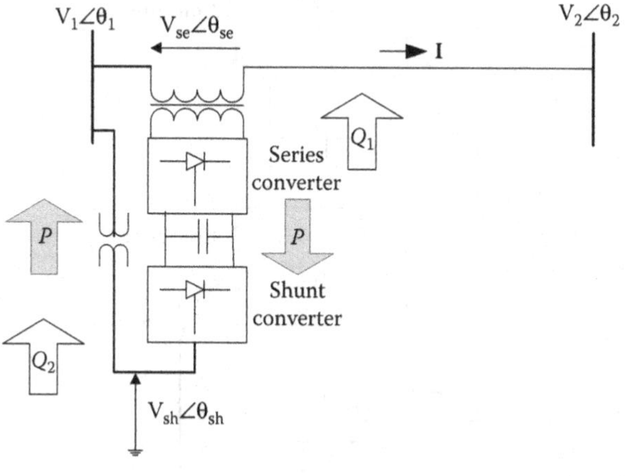

FIGURE 1.17 Directions of active and reactive powers of UPFC series and shunt converters.

1.5 THE IPFC AND THE GUPFC

Unlike the STATCOM, SSSC, or UPFC, which can control the power flow in a single transmission line only, the IPFC or GUPFC can address the problem of compensating a number of transmission lines simultaneously at a given substation. The IPFC employs a number of DC-to-AC converters linked together at their DC terminals, each providing series compensation for a different line. The converters are connected to the AC system through their series coupling transformers [22]. The most elementary two-converter IPFC has three degrees of freedom, being able to control independently, three electrical quantities of interest. On the other hand, the most elementary GUPFC possesses five degrees of freedom, being capable of controlling independently five electrical quantities of interest [23,24].

The schematic diagram of a simple two-converter IPFC is shown in Figure 1.18. The equivalent circuit of the diagram shown in the figure is shown in Figure 1.19.

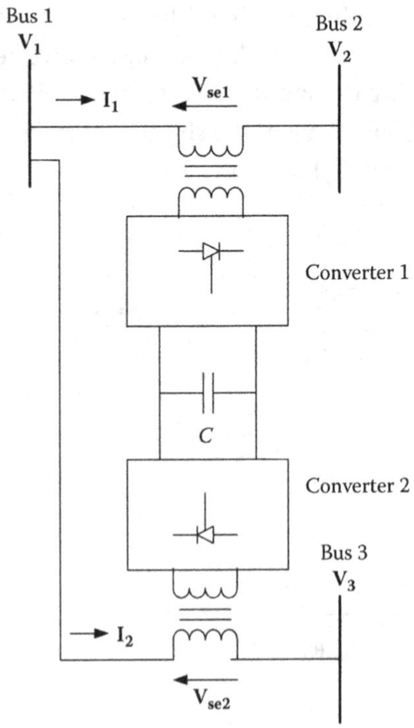

FIGURE 1.18 Schematic diagram of a IPFC with two series converters.

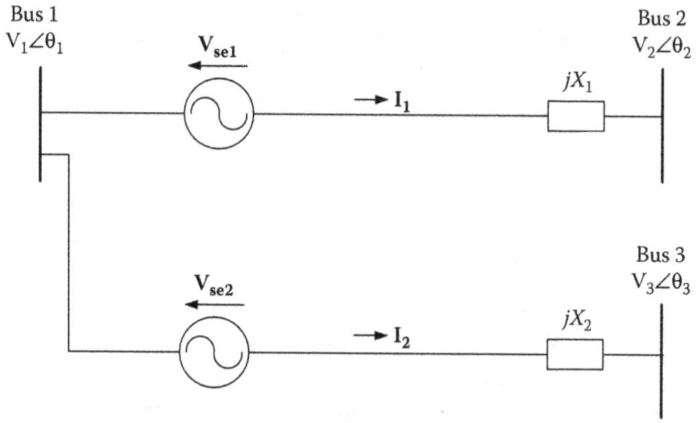

FIGURE 1.19 Equivalent circuit of the network shown in Figure 1.18.

Before we proceed to draw the phasor diagrams of the voltages and currents with this two-converter IPFC, it is important to make certain simplifying assumptions. These are given as follows:

1. The bus voltage magnitudes are taken to be same, that is, $|\mathbf{V}_1| = |\mathbf{V}_2| = |\mathbf{V}_3|$.
2. All resistances are negligible. Coupling transformer leakage reactance is combined with the transmission line reactances.
3. The reactances of both the lines are same, that is, $X_1 = X_2 = X$. This simplifies the analysis.
4. Before the IPFC was installed, active power was flowing from bus 1 to bus 2 in transmission line 1 and from bus 1 to bus 3 in transmission line 2. Thus, $\mathbf{V}_1$ is leading both $\mathbf{V}_2$ and $\mathbf{V}_3$.
5. Before the IPFC was installed, more active power was flowing in transmission line 1 than in transmission line 2. Since $X_1 = X_2 = X$, this implies that the phase angle difference between $\mathbf{V}_1$ and $\mathbf{V}_2$ is more than that between $\mathbf{V}_1$ and $\mathbf{V}_3$ i.e. $\mathbf{V}_3$ is leading $\mathbf{V}_2$.
6. Converter rating limits have been ignored.

The IPFC is used for simultaneously reducing the receiving end active power in transmission line 1 (connected between buses 1 and 2) and increasing the receiving end active power in transmission line 2 (connected between buses 1 and 3), in response to a simultaneous change of demand.

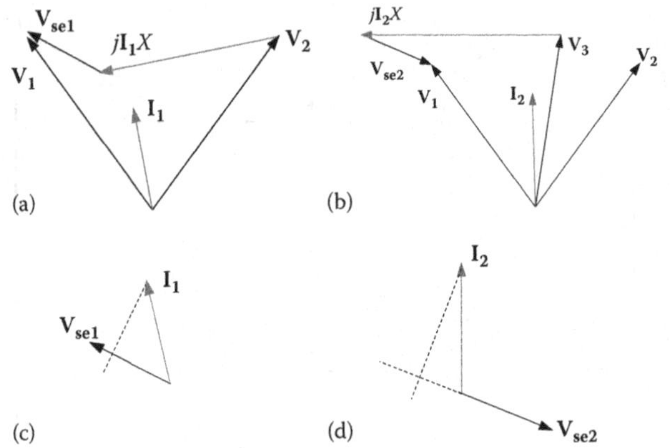

FIGURE 1.20 (a–d) Phasor diagrams of voltages and currents with IPFC.

Typical phasor diagrams to achieve this objective are shown in Figure 1.20a through d. It is important to note that the phasor diagrams shown in the figure correspond to only one out of a myriad of schemes possible to achieve this objective. Although the phase angles of the voltages injected by the series converters 1 and 2 are assumed to be arbitrary, we again follow the basic rule that the IPFC cannot deliver or absorb active power by itself, that is, $\mathrm{Re}\left[\mathbf{V}_{se1}(-\mathbf{I}_1^*)+\mathbf{V}_{se2}(-\mathbf{I}_2^*)\right]=0$.

First, we discuss the active power. From Figure 1.20a, it can be observed that due to the voltage $\mathbf{V}_{se1}$ injected by the series converter 1, the magnitudes of the line drop $\mathbf{I}_1X$ and the line current $\mathbf{I}_1$ decreases. Also, the receiving end power factor angle (angle between $\mathbf{V}_2$ and $\mathbf{I}_1$) of transmission line 1 increases compared to the case when there is no IPFC. This results in a reduction of active power at the receiving end of line 1. The phasor diagram showing the relationship between $\mathbf{V}_{se1}$ and $\mathbf{I}_1$ is shown in Figure 1.20c. From this figure, it can be observed that the current $\mathbf{I}_1$ has a current component in phase with $\mathbf{V}_{se1}$ (shown as the projection of $\mathbf{I}_1$ on $\mathbf{V}_{se1}$). This implies that the IPFC series converter 1 absorbs active power from the line through its coupling transformer. Because series converter 1 cannot consume this active power by itself, this must be transferred through the DC link over to the series converter 2. Because converter 2 cannot consume this active power either, it must deliver this active power to transmission line 2 through its coupling transformer. Because series converter 2 delivers active power to transmission line 2, the current $\mathbf{I}_2$ in transmission line 2 must have a component in phase opposition to the voltage $\mathbf{V}_{se2}$ injected by the series converter 2. This is observed from Figure 1.20b and also shown separately in Figure 1.20d. From Figure 1.20b, it is also observed that due to the voltage

$\mathbf{V}_{se2}$ injected by the series converter 2, the magnitudes of the line drop $\mathbf{I}_2X$ and the line current $\mathbf{I}_2$ increases. Also, the receiving end power factor angle (angle between $\mathbf{V}_3$ and $\mathbf{I}_2$) of transmission line 2 decreases compared to the case when there is no IPFC. This results in an increase of active power at the receiving end of line 2. Also, from Figure 1.20b and d, which shows separately the relationship between $\mathbf{V}_{se1}$ and $\mathbf{I}_1$ along with that between $\mathbf{V}_{se2}$ and $\mathbf{I}_2$, it can be observed that unlike the projection of $\mathbf{I}_1$ on $\mathbf{V}_{se1}$, the projection of $\mathbf{I}_2$ on $\mathbf{V}_{se2}$ is negative to fulfill the condition $\mathrm{Re}\left[\mathbf{V}_{se1}(-\mathbf{I}_1^*)+\mathbf{V}_{se2}(-\mathbf{I}_2^*)\right]=0$. The direction of the active power flow for series converters 1 and 2 is shown by thick black (shaded) arrows (P) in Figure 1.21.

Now, we discuss the reactive powers of the two series converters. From Figure 1.20c, it is observed that the current $\mathbf{I}_1$ has a quadrature component lagging the voltage $\mathbf{V}_{se1}$ injected by series converter 1. This implies that converter 1 absorbs reactive power from line 1 through its coupling transformer. The direction of the reactive power flow of the series converter 1 is shown by a thick white (unshaded) arrow (Q_1) in Figure 1.21. In a similar manner, it can be observed from Figure 1.20d that $\mathbf{I}_2$ has a quadrature

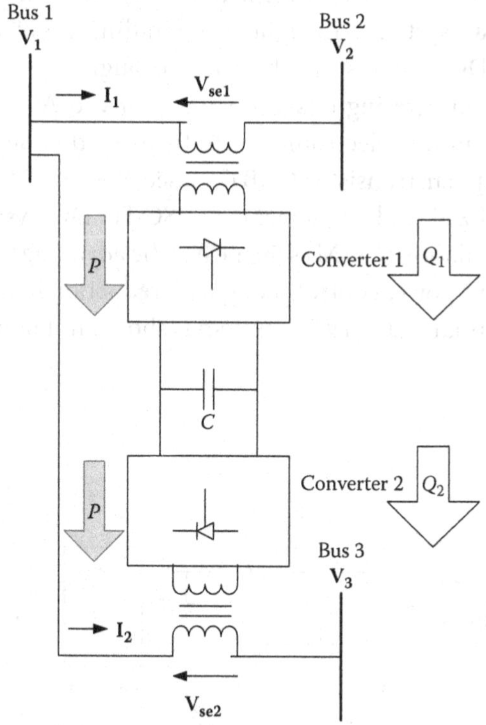

FIGURE 1.21 Directions of active and reactive powers of IPFC series converters.

component leading $\mathbf{V}_{se2}$ which implies that series converter 2 delivers reactive power to line 2 through its coupling transformer. The direction of the reactive power flow of the series converter 2 is shown by a thick white (unshaded) arrow (Q_2) in Figure 1.21. It is important to note that unlike the case of active power, both the converters can absorb or deliver reactive powers independently.

The GUPFC is an extension of the IPFC. Because the IPFC has already been discussed in some detail, the GUPFC is not given an exclusive introduction here. The interested reader is advised to refer to [23,24] for an introduction. We will directly discuss the power flow modeling of a GUPFC in Chapter 6.

1.6 VSC-HVDC SYSTEMS

In AC transmission, the length of transmission links are limited by stability considerations. No such limitation exists for DC transmission. In this context, a high-voltage DC (HVDC) link can be used to interconnect two AC substations that are separated by very long distances. A HVDC link can be used to improve system reliability by interconnecting two asynchronous AC systems. In light of dwindling fossil fuel resources, VSC-based HVDC systems can be used to augment power transmission capacity by integrating offshore wind farms to AC grids. With the advancement of power electronics, gate-turn-off thyristor (GTO)- and insulated gate bipolar transistor (IGBT)-based VSC-HVDC systems have been conceptualized and implemented. VSC-HVDC systems based on pulse width modulation (PWM) scheme has the advantage of independent active and reactive power control, along with reduction in filter size [38]. A typical two-terminal VSC-HVDC system is shown in Figure 1.22.

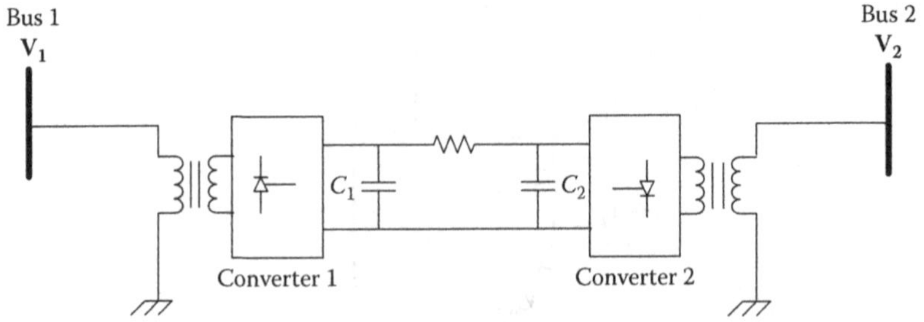

FIGURE 1.22 A typical two-terminal VSC-HVDC system.

Unlike a two-terminal VSC-HVDC interconnection, a multiterminal VSC-HVDC system is more versatile and better capable to utilize the economic and technical advantages of the VSC-HVDC technology.

In a multiterminal VSC-HVDC system, the converter stations can be located closely, in the same substation or remotely, at different locations. They are known to be in back-to-back (BTB) or point-to-point (PTP) configuration, respectively. Most of the VSC-HVDC systems installed over the world are in PTP configuration.

1.7 POWER FLOW MODELS OF FACTS CONTROLLERS AND VSC-HVDC SYSTEMS

For proper planning, design, and operation of power system networks incorporating these controllers, it is essential to evaluate the current system performance as well as to ascertain the effectiveness of alternative plans of system expansion to meet increased load demand with these FACTS controllers. This requires the power flow solution of the network(s) incorporating these FACTS controllers, and therefore, the development of suitable power flow models of these controllers is a fundamental requirement.

The earliest algorithms for power flow solution of networks were based on the Gauss–Seidel method. They suffered, however, from relatively poor convergence characteristics. Subsequently, the Newton–Raphson (NR) algorithm was developed. The underlying problem for the iterative Newton method is the solution of a matrix equation of large dimension. However, with the development of sparse matrix techniques and vast increases in low-cost computer memory and processor speed over the past several decades, the NR algorithm has emerged as the method of choice in commercial power flow packages. Some of the excellent references on the fundamentals of power flow applying the NR method are [25–31]. For computations involving power systems under the usual operating conditions, another simplification of the NR scheme is also possible. This modification is known as fast decoupled power flow. In this case, the matrices need not be updated at each iteration and the computational burden is greatly reduced [26,27,32–36].

Because of the need of suitable power flow models of the FACTS controllers and the adoption of NR algorithm as the *de facto* standard in the industry, many researchers have paid attention to the development of NR power flow models of these VSC-based FACTS devices [37–43].

Now, among the VSC-based FACTS devices, the STATCOM was the earliest to be conceived. Zhang et al. [44] have presented a comprehensive, multicontrol functional model of a STATCOM for power flow studies.

A robust strategy for the enforcement of multiple violated constraints associated with the STATCOM is also proposed in this work. A novel power injection model of a STATCOM for power flow and voltage stability studies, along with practical device limit constraints of the STATCOM, is addressed in [45]. Yang et al. [46] have accounted for the switching and conduction losses of the STATCOM in an improved STATCOM model. Inclusion of coupling transformer resistance in the power flow model of a STATCOM has been presented by Radman and Shultz [47]. Bhargava et al. [48] have demonstrated the incorporation of both STATCOM switching losses and coupling transformer losses using a simple two-step algorithm.

Unlike the STATCOM, the SSSC is a VSC-based series FACTS controller. The first comprehensive, multicontrol functional power flow model of the SSSC, inclusive of device limit constraints, was presented by Zhang [49]. This model can be used for steady-state control of any one of the line active or reactive power flows, voltage at a bus, or transmission line impedance. A multicontrol mode, power injection model of an SSSC for power flow analysis, inclusive of practical constraints, is addressed in [50]. Vinkovic and Mihalic [51] have presented a novel current-based model of the SSSC. Instead of the magnitude and phase angle of the injected voltage, this model uses the real and imaginary components of the SSSC current for its representation and is reported to exhibit faster convergence than the injection model [50]. SSSC models for three-phase power flow analysis have been presented by Zhang et al. [52]. The models have been implemented in a three-phase Newton power flow algorithm in rectangular coordinates. Zhao et al. [53] have demonstrated the realization of an SSSC model for power flow analysis through the user interface programmer of Power System Analysis Software Package.

Like the STATCOM and the SSSC, a lot of research work has been carried out in the area of power flow modeling of a UPFC. Sen [54] has addressed the theory and the modeling technique of a UPFC using an EMTP simulation package. Arabi and Kundur [55] have considered the UPFC as a coordinated and interconnected set of controllable shunt and series elements for comprehensive power flow and stability simulations. Iravani et al. [56] have presented mathematical models of the UPFC for steady-state, transient stability, and eigenvalue studies. For the steady-state model, the sending and receiving ends of the UPFC are decoupled. The sending end (shunt converter end) bus is converted to a voltage-controlled bus and the receiving end bus is converted to a load bus. Noroozian et al. [57] have described a UPFC injection model for power flow study and application of the UPFC for optimal power flow (OPF) control. Acha et al. [58] have presented a comprehensive UPFC model suitable for OPF solutions. An optimization-based method for steady-state analysis of power systems having UPFCs has been presented by Nguyen and Nguyen [59] with a high-level line optimization control (LOC). In this method, optimal reference inputs to UPFCs as required in LOC are determined using constrained optimization. A general UPFC power flow model capable of solving large power networks is presented in [60,61]. In this model, a set of analytical equations has been derived to provide good initial conditions for the UPFC parameters in order to retain the quadratic convergence of Newton method. Yan and Sekar [62] have presented a modified power flow model of the UPFC with line power flows and bus voltage magnitudes as independent variables. An approximate model for power flow studies that takes into consideration a lossless model of UPFC-embedded transmission lines has been presented by Alomoush [63]. Wei et al. [64] have used injected voltage sources to directly model a UPFC. Subsequently, the limits on the UPFC rating are imposed in an NR power flow algorithm in order to compute its maximum voltage stability-limited transfer capability. Sun et al. [65] have proposed a Newton power flow model of the UPFC in rectangular coordinates, in which the UPFC control functions are expressed by the equivalent loads of the related buses. A novel steady-state model of the UPFC that can be easily incorporated in existing power flow software and that provides an automatic adjustment of UPFC parameters and an imposition of UPFC operating limits has been presented by Santos et al. [66]. Dazhong et al. [67] have presented an approach for power flow analysis of a power system with UPFC using the equivalent power network given in the work. Eleven novel control modes of the UPFC have been successfully modeled in Newton power flow algorithm by Zhang and Godfrey [68]. A generalized power flow model of UPFC incorporating sparse techniques for formulation and computation of the Jacobian matrix is presented in [69]. Commercial software of power system analysis can be employed with this approach. The device limit constraints of the UPFC in power flow control are addressed in several publications [70–74]. Mhaskar et al. [75] have presented a methodology incorporating dual state and control variables for power flow solution with series FACTS controllers, including UPFC, in which an ingenious selection of variables renders decoupling of the power-flow problem. Efforts to reuse the original NR power flow codes with a UPFC model have been demonstrated by Nor et al. [76].

In respect of the IPFC, some excellent techniques for its power flow modeling are presented in [77,78]. Zhang et al. [78] have described a novel power injection model of the IPFC for power flow analysis, including its various limit constraints. The power injection model keeps intact the original structure and symmetry of the admittance matrix. Modeling of both UPFC and IPFC for power-flow, sensitivity, and dispatch analysis has been demonstrated by Wei et al. [79]. In this paper, a common modeling framework for the analysis and optimal dispatch of FACTS controllers is proposed, which also allows a consistent formulation of the sensitivity analysis of the VSC control variables. Modeling of IPFC rating limits in an NR power-flow algorithm for a maximum dispatch benefit strategy is demonstrated in [80]. In this paper, the power circulation between the two series converters is used as a parameter to optimize the voltage profile and power transfer.

Zhang [77] has presented a Newton power flow modeling of the GUPFC along with IPFC. In this paper, modeling of the IPFC and GUPFC with direct incorporation of their additional control constraints in Newton power flow is investigated. A mathematical model of the GUPFC suitable for power flow and OPF study is presented in [81]. In this work, analytical solutions for initial values of the GUPFC have also been derived for better convergence. Fardanesh [82] has demonstrated a generalized method utilizing an OPF-type formulation for modeling of UPFCs, IPFCs, and GUPFCs. Using this method, optimal dimensioning and sizing of these VSC-based controllers can be achieved. A simple approach for steady-state modeling of both IPFCs and GUPFCs based on the converters' power balance method is demonstrated in [83]. This approach uses the **d–q** orthogonal coordinates to present a direct solution for these controllers by solving a quadratic equation.

The power flow modeling of VSC-HVDC systems is discussed in Chapter 8.

1.8 ORGANIZATION OF THE BOOK

In this book, an attempt has been made to investigate systematically the modeling of VSC-based FACTS controllers and VSC-HVDC systems in an existing NR power flow algorithm. Newton power flow models of the SSSC, UPFC, IPFC, STATCOM, GUPFC, and finally VSC-HVDC systems have been developed.

The outline of the remaining chapters of the book is as follows:

Chapter 2 presents the NR method for solution of nonlinear algebraic equations. The application of the NR algorithm for solution of nonlinear equations in single and multiple variables is demonstrated.

Subsequently, the reader is introduced to the power flow problem and the application of the NR method for its solution. The power flow problem for an example six-bus system is considered for demonstration of the NR algorithm. The Jacobian matrix, its constituent subblocks, and the nature of the elements in these subblocks are described. Finally, the generalized form of the power flow problem is described.

Chapter 3 presents the development of a Newton power flow model of an SSSC. Similar to existing models, the proposed model [84,85] can handle multiple control functions of the SSSC, including control of bus voltage, line active power flow, line reactive power flow, and line reactance. The issue of practical device constraints of the SSSC has also been addressed in this chapter. For this purpose, two major device limit constraints have been considered: (1) the magnitude of the injected series converter voltage and (2) the magnitude of the line current through the converter. These device limit constraints have been accommodated by the principle that whenever a particular constraint limit is violated, it is kept at its specified limit, although a control objective is relaxed. Mathematically, this signifies the replacement of the control objectives by the corresponding limits violated during the formation of the Jacobian matrix. Furthermore, the proposed model can also accommodate the switching losses for a practical SSSC very easily. In fact, the proposed technique exhibits excellent convergence for any practical value of converter switching loss.

Chapter 4 addresses the development of a Newton power flow model [84,86] of a UPFC, which can also incorporate various practical device limit constraints. Four major device limit constraints of the UPFC have been considered: (1) the limits on the magnitude of the injected voltage of the series converter, (2) the magnitude of the line current through the series converter, (3) the DC link power transfer, and (4) the magnitude of the shunt converter current. These limit constraints have been enforced in three different ways: (1) limit violation of a single constraint, (2) limit violations of two separate constraints simultaneously, and (3) limit violations of all three separate constraints simultaneously.

Chapter 5 presents the development of a Newton power flow model [84,87] of an IPFC. Three major device limit constraints of the IPFC have been considered: (1) the limits on the magnitudes of the injected voltages of the series converters, (2) the magnitudes of the line currents through the series converters, and (3) the DC link power transfer. For demonstrating the enforcement of the three limits, a single elementary IPFC with two series converters installed in two different transmission lines has been considered. Single, double, or multiple (all three) device limit constraints of the IPFC have been enforced in different combinations.

Chapter 6 addresses the development of a Newton power flow model [84] of a GUPFC. The proposed model can accommodate five different practical limit constraints of the GUPFC: (1) the limits on the magnitudes of the injected voltages of the series converters, (2) the magnitudes of the line currents through the series converters, (3) the DC link power transfer, (4) the magnitude of the shunt converter current, and (5) the magnitudes of the bus voltages on the line side of the series converters. Enforcement of these five limit constraints have been considered in five different ways: (1) limit violation of a single constraint, (2) limit violations of two different constraints simultaneously, (3) limit violations of all three different constraints simultaneously, (4) limit violations of four different constraints simultaneously, and (5) limit violations of all five different constraints simultaneously. For demonstrating the enforcement of limits, a GUPFC with two series converters and a shunt converter (minimum possible configuration) has been considered.

Chapter 7 initially discusses the Newton power flow model [84,88] of a STATCOM, which also takes care of two practical device limit constraints: (1) the limits on the injected voltage of the shunt converter and (2) the limits on the shunt converter current. Subsequently, by applying decoupling techniques, a fast decoupled power-flow (FDLF) model of the STATCOM has also been developed [84,88]. It is to be noted that the FDLF model of the STATCOM is also capable of considering various practical device limit constraints. Finally, a number of comparative case studies of the Newton power flow and the decoupled power flow models of the STATCOM are also presented in this chapter.

Chapter 8 presents the development of a Newton power flow model [93] of multiterminal VSC-HVDC systems with PWM control schemes. The model is applicable to the more generalized and widely installed point-to-point VSC-HVDC configurations. Both the converter modulation indices and the converter DC side voltages along with the phase angle of the converter AC side voltage appear as unknowns in the model.

Feasibility studies of all the models proposed in Chapters 3 through 8 have been carried out on the IEEE 118- and/or 300-bus test systems to validate the convergence characteristics of the proposed techniques.

Derivations of all the difficult formulae in various chapters are detailed in the Appendix.

1.9 SOLVED PROBLEMS

PROBLEM 1.1

Figure 1.23 shows the equivalent circuit of a STATCOM connected to a load bus. The bus voltage is $\mathbf{V_{bus}} = 0.97\angle 10°$ p.u. The coupling transformer reactance is $X_{sh} = 0.2$ p.u.

1. Assuming a lossless STATCOM, find the magnitude (fundamental) of the STATCOM output voltage V_{sh}, if the STATCOM delivers a reactive power of 0.1 p.u to the load bus.
2. If the STATCOM has losses and it receives an active power of 0.02 p.u. from the load bus, find the reactive power delivered by the STATCOM, if the magnitude of the STATCOM output voltage $V_{sh} = 1.02$ p.u.

Solution

1. The expression for the active power transfer from the load bus to the STATCOM is given by $P_{sh} = \left((V_{bus}V_{sh})/X_{sh}\right)\sin(\theta_{bus} - \theta_{sh})$. For a lossless STATCOM, $P_{sh} = 0$. This implies $\theta_{sh} = \theta_{bus}$. Hence, $\theta_{sh} = 10°$.

   Now, the expression for the reactive power absorbed by the STATCOM from the load bus is given by

$$Q_{sh} = \mathrm{Im}[\mathbf{V_{sh}}\mathbf{I_{sh}^*}] = \mathrm{Im}\left[\mathbf{V_{sh}}\left\{\frac{\mathbf{V_{bus}} - \mathbf{V_{sh}}}{jX_{sh}}\right\}^*\right] = \frac{V_{sh}V_{bus}}{X_{sh}}\cos(\theta_{sh} - \theta_{bus}) - \frac{V_{sh}^2}{X_{sh}}.$$

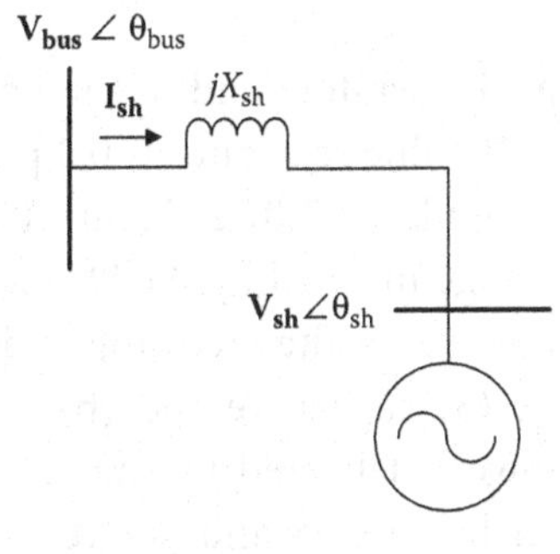

FIGURE 1.23 Equivalent circuit of a STATCOM connected to a load bus.

Thus,

$$-0.1 = \frac{0.97\,V_{sh}}{0.2}\cos(10° - 10°) - \frac{V_{sh}^2}{0.2}$$

gives $V_{sh} = 0.9902$ p.u. or $V_{sh} = -0.0202$ p.u. (It may be noted that the sign of Q_{sh} is taken negative as the STATCOM is not absorbing, but delivering a reactive power of 0.1 p.u.).

Since magnitude of a quantity can not be negative, $V_{sh} = 0.9902$ p.u.

It is important to note that had there been two positive solutions, say $V_{sh} = 0.9902$ p.u. or $V_{sh} = 0.95$ p.u., one might have been in a dilemma to choose the correct solution. But it should be noted that as the STATCOM is delivering reactive power, magnitude V_{sh} must exceed V_{bus} (as reactive power flows from a bus with a higher voltage magnitude to a one with a lower voltage magnitude, in transmission systems). Hence, in that case too, $V_{sh} = 0.9902$ p.u.

2. If the STATCOM receives an active power of 0.02 p.u. from the load bus (to supply its losses)

$$P_{sh} = \frac{V_{bus}V_{sh}}{X_{sh}}\sin(\theta_{bus} - \theta_{sh}) = \frac{0.97 \times 1.02}{0.2}\sin(10° - \theta_{sh}) = 0.02 \text{ p.u.}$$

Thus, $\theta_{sh} = 9.77°$.

Thus, the reactive power delivered by the STATCOM is given by

$$Q_{del} = -Q_{sh} = \frac{V_{sh}^2}{X_{sh}} - \frac{V_{sh}V_{bus}}{X_{sh}}\cos(\theta_{sh} - \theta_{bus})$$

$$= \frac{1.02^2}{0.2} - \frac{1.02 \times 0.97}{0.2}\cos(10° - 9.77°) = 0.255 \text{p.u.}$$

PROBLEM 1.2

An SSSC is connected at the sending end of the line between two buses as shown in Figure 1.24. The line reactance is 0.3 p.u. The bus voltages at the sending and receiving ends are $1.0\angle 22°$ and $1.0\angle 2°$ p.u., respectively. The magnitude of the voltage injected by the SSSC converter is $\mathbf{V}_{se} = 0.2$ p.u. Compute the active power at the receiving end of the line when the SSSC is operating in the (a) capacitive and (b) inductive modes. Also compute the reactive powers delivered by the SSSC converter for both modes. Neglect converter harmonics and line resistance. Assume a loss-less converter.

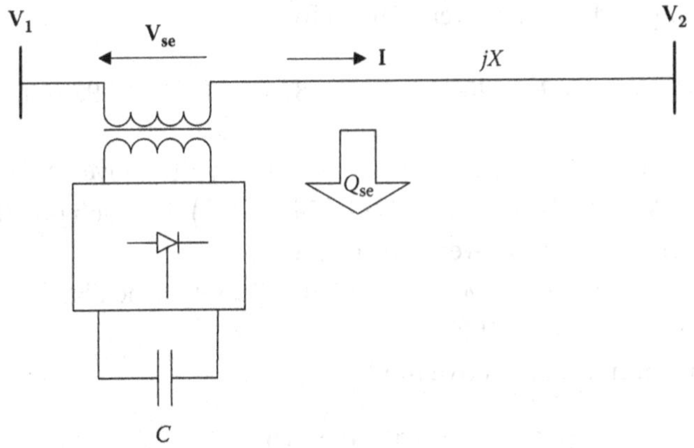

FIGURE 1.24 SSSC connected at the sending end of a transmission line.

Solution

The phasor diagram of the voltages is shown in Figure 1.25.

Now, applying KVL to the line, $\mathbf{V}_1 = \mathbf{V}_2 + \mathbf{V}_{se} + jX\mathbf{I}$.

Hence, $\mathbf{I} = \left(\mathbf{V}_1 - \mathbf{V}_2 - \mathbf{V}_{se}\right)/jX$.

However, the phase angle of $\mathbf{V}_{se}$ needs to be known before we can compute the current phasor **I**. From the phasor diagram in Figure 1.25, it can be observed that the current phasor **I** bisects the sending and receiving end voltage phasors $\mathbf{V}_1$ and $\mathbf{V}_2$, respectively. Thus, the phase angle of **I** is $2° + \left[(22° - 2°)/2\right] = 12°$.

Assuming a lossless converter in the capacitive mode (Figure 1.25a) of the SSSC, $\mathbf{V}_{se}$ lags **I** by 90°. Hence, $V_{se} = 0.2\angle -78°$.

Thus, the current in the capactive mode is

$$\mathbf{I} = \frac{1.0\angle 22° - 1.0\angle 2° - 0.2\angle -78°}{j0.3} = 1.824\angle 12°$$

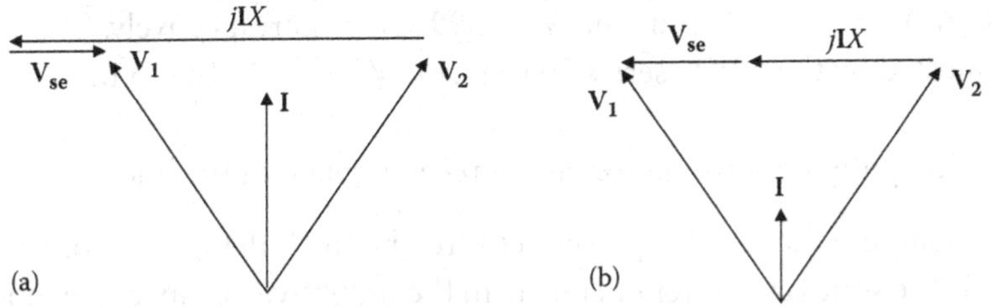

FIGURE 1.25 Phasor diagrams of voltages in (a) the capacitive and (b) inductive modes.

The receiving end active power is given by

$$P_{RE} = \text{Re}\left(\mathbf{V}_2\mathbf{I}^*\right) = \text{Re}\left(1.0\angle 2^\circ \times 1.824\angle -12^\circ\right) = 1.797 \text{ p.u.}$$

From Figure 1.24, the reactive power absorbed by the converter is given by $Q_{SSSC} = \text{Im}\left(\mathbf{V}_{se}\mathbf{I}^*\right) = \text{Im}\left(0.2\angle -78^\circ \times 1.824\angle -12^\circ\right) = -0.365$ p.u. Thus, the SSSC delivers a reactive power of 0.365 p.u.

Similarly, in the inductive mode (Figure 1.25b) of the SSSC, $\mathbf{V}_{se}$ leads **I** by 90°. Hence, $\mathbf{V}_{se} = 0.2\angle 102^\circ$.

The current in the inductive mode is

$$\mathbf{I} = \frac{1.0\angle 22^\circ - 1.0\angle 2^\circ - 0.2\angle 102^\circ}{j0.3} = 0.491\angle 12^\circ$$

The receiving end active power is given by

$$P_{RE} = \text{Re}\left(\mathbf{V}_2\mathbf{I}^*\right) = \text{Re}\left(1.0\angle 2^\circ \times 0.491\angle -12^\circ\right) = 0.484 \text{ p.u.}$$

The reactive power absorbed by the converter is given by $Q_{SSSC} = \text{Im}\left(\mathbf{V}_{se}\mathbf{I}^*\right) = \text{Im}\left(0.2\angle 102^\circ \times 0.491\angle -12^\circ\right) = 0.098$ p.u. Thus, the SSSC delivers a reactive power of −0.098 p.u. which is shown by the thick white (unshaded) arrow (Q_{se}) in Figure 1.24.

It may be noted that because the converter output voltage phasor $\mathbf{V}_{se}$ is perpendicular to the current phasor **I**, the reactive powers can be computed from the product of V_{se} and I in both the cases.

PROBLEM 1.3

A UPFC is connected at the sending end of the line between two buses as shown in Figure 1.26. The combined reactance of the line and the series coupling transformer is 0.3 p.u. The leakage reactance of the shunt coupling transformer is 0.2 p.u. The bus voltages at the sending and receiving ends are $\mathbf{V}_1 = 0.99\angle 21^\circ$ p.u. and $\mathbf{V}_2 = 0.99\angle 3^\circ$ p.u., respectively. The voltage injected by the UPFC series converter is $\mathbf{V}_{se} = 0.15\angle 195^\circ$ p.u.

1. Compute the active power at the receiving end of the line.
2. Compute the reactive power delivered by the UPFC series converter. Is the series converter operating in the inductive or capacitive mode?
3. Compute the reactive power delivered by the UPFC shunt converter if the magnitude of the shunt converter output voltage is $\mathbf{V}_{sh} = 1.029$ p.u.

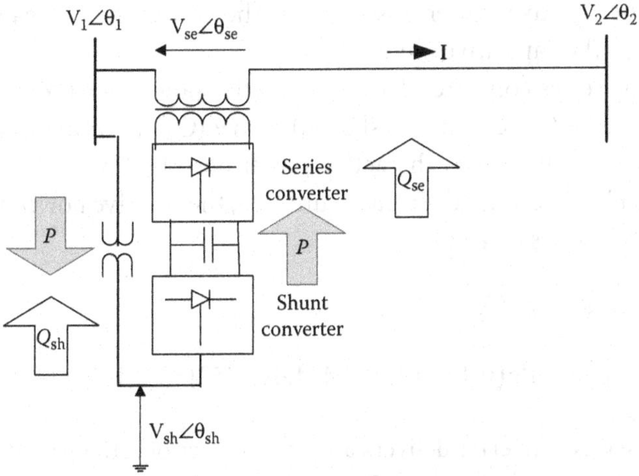

FIGURE 1.26 UPFC connected in a transmission line between two buses.

Neglect converter harmonics and all resistances. Also assume lossless converters.

Solution

Applying KVL to the transmission line, $\mathbf{V}_1 = \mathbf{V}_2 + \mathbf{V}_{se} + jX\mathbf{I}$.

$$\mathbf{I} = \frac{\left(\mathbf{V}_1 - \mathbf{V}_2 - \mathbf{V}_{se}\right)}{jX}$$

Hence,

$$= \frac{\left(0.99\angle 21° - 0.99\angle 3° - 0.15\angle 195°\right)}{j0.3}$$

$$= 1.171\angle -13.25° \text{p.u.}$$

The receiving end active power is given by

$$P_{RE} = \text{Re}\left(\mathbf{V}_2\mathbf{I}^*\right)$$

$$= \text{Re}\left(0.99\angle 3° \times 1.171\angle 13.25°\right) = 1.113 \text{ p.u.}$$

From Figure 1.26, the reactive power absorbed by the UPFC series converter is

$$Q_{se} = \text{Im}\left(\mathbf{V}_{se}\mathbf{I}^*\right)$$

$$= \text{Im}\left(0.15\angle 195° \times 1.171\angle 13.25°\right) = -0.083 \text{ p.u.}$$

Because the reactive power absorbed by the converter is negative, it is operating in the capacitive mode.

Thus, the series converter delivers a reactive power of 0.083 p.u. which is shown by the thick white (unshaded) arrow (Q_{se}) pointing upwards in Figure 1.26. At this stage, the active power absorbed by the series converter must be determined. From Figure 1.26, the reactive power absorbed by the UPFC series converter is

$$P_{SSSC} = \text{Re}\left(\mathbf{V}_{se}\mathbf{I}^*\right)$$

$$= \text{Re}\left(0.15\angle 195° \times 1.171\angle 13.25°\right) = -0.155 \text{ p.u.}$$

Thus, the series converter delivers an active power of 0.155 p.u. This active power must come from the sending end bus through the shunt coupling transformer and the shunt converter, as both the converters are lossless. The direction of this active power is shown as *P* inside thick black (shaded) arrows in Figure 1.26 (downward from the sending end bus through the coupling transformer and the shunt converter, and then upward through the series converter to the series coupling transformer).

Because all resistances are neglected, the active power transfer from the sending end bus (through the shunt coupling transformer) to the UPFC shunt converter is

$$P = \frac{V_1 V_{sh}}{X_{sh}} \sin(\theta_1 - \theta_{sh})$$

$$= \frac{0.99 \times 1.029}{0.2} \sin(21° - \theta_{sh})$$

This must be equal to 0.155 p.u. It is the active power going toward the series converter.

Thus, $\sin(21° - \theta_{sh}) = (0.155 \times 0.2)/(0.99 \times 1.029)$, or $\theta_{sh} = 19.3°$.

The reactive power delivered by the shunt converter is given by

$$Q_{sh} = \frac{V_{sh}^2}{X_{sh}} - \frac{V_{sh} V_1}{X_{sh}} \cos(\theta_{sh} - \theta_1)$$

$$= \frac{1.029^2}{0.2} - \frac{1.029 \times 0.99}{0.2} \cos(21° - 19.3°) = 0.2 \text{ p.u.}$$

This is shown by the thick white (unshaded) arrow (Q_{sh}) in Figure 1.26, also pointing upwards.

CHAPTER 2

Introduction to the Newton–Raphson Method and the Power Flow Problem

2.1 INTRODUCTION

In this chapter, the reader is introduced to the Newton–Raphson method used for solving nonlinear algebraic equations. He or she is also introduced to the power flow problem and its solution using the Newton–Raphson algorithm [25–36].

2.2 THE NEWTON–RAPHSON METHOD

Let us consider the solution of the following equation:

$$x^3 - 3x^2 + 3x = 2 \tag{2.1}$$

The above equation can be written as

$$f(x) = f^{SP} \tag{2.2}$$

where:

$$f(x) = x^3 - 3x^2 + 3x \tag{2.3}$$

and

$$f^{SP} = 2 \tag{2.4}$$

One can start with a value of $x_0 = 3$. Because a wild guess has been made, it is extremely unlikely to be the correct solution. However, if it is assumed that the actual solution lies at some small distance Δx from $x_0 = 3$, and somehow Δx can be computed, then the actual solution can be determined to be $x_0 + \Delta x$, that is, $3 + \Delta x$. Let us try this.

Because the solution is assumed to lie at a distance Δx from $x_0 = 3$, Equation 2.2 can be written as

$$f(3 + \Delta x) = 2 = f^{SP} \tag{2.5}$$

Now, by Taylor's series expansion,

$$f(3 + \Delta x) = f(3) + f'(3)\Delta x + \frac{f''(3)}{2!}(\Delta x)^2 + \cdots \tag{2.6}$$

If the initial guess is close to the actual solution, Δx will be small and $(\Delta x)^2$ will be even smaller. If we neglect the term involving $(\Delta x)^2$ and the subsequent ones, we can write Equation 2.6 as

$$f(3 + \Delta x) \approx f(3) + f'(3)\Delta x \tag{2.7}$$

Substituting the value of $f(3 + \Delta x)$ from Equation 2.5 in Equation 2.7, we get

$$f^{SP} = 2 = f(3) + f'(3)\Delta x \quad \text{or} \quad \Delta x = \frac{2 - f(3)}{f'(3)} = \frac{f^{SP} - f(3)}{f'(3)} \tag{2.8}$$

Now, $f(3) = (3)^3 - 3(3)^2 + 3(3) = 9$ and $f^{SP} = 2$. Also, $f'(x) = 3x^2 - 6x + 3$, making $f'(3) = 3(3)^2 - 6(3) + 3 = 12$.

Thus, from Equation 2.8, $\Delta x = (2 - 9)/12 = -7/12 = -0.5833$. The solution after one iteration is $3 + \Delta x = 2.4167$.

The starting point has moved from 3 (initial guess) to 2.4167. Hence, in the next step, it is assumed that the actual solution lies at a distance of Δx from 2.4167, that is,

$$f(2.4167 + \Delta x) = 2 = f^{SP} \quad \text{or,} \quad f(2.4167) + f'(2.4167)\Delta x = 2 = f^{SP}$$

Hence,

$$\Delta x = \frac{f^{SP} - f(2.4167)}{f'(2.4167)} = \frac{2 - 3.8434}{6.0211} = -0.3062$$

Thus, the solution after the second iteration is $2.4167+\Delta x=2.1105$. After the third iteration,

$$\Delta x=\frac{2-f(2.1105)}{f'(2.1105)}=\frac{2-2.3695}{3.6996}=-0.0999$$

making the solution after the third equation as $2.1105+\Delta x=2.0106$. After the fourth iteration,

$$\Delta x=\frac{2-f(2.0106)}{f'(2.0106)}=\frac{2-2.0321}{3.0639}=-0.0105$$

making the solution after the fourth iteration as 2.0001. The actual solution is 2.0.

What would happen if an initial guess of 1.5 is made?

$$\Delta x=\frac{2-f(1.5)}{f'(1.5)}=\frac{2-1.125}{0.75}=1.1667$$

Δx is positive now. The solution after the first iteration is $1.5+\Delta x=2.6667$. For the next iteration,

$$\Delta x=\frac{2-f(2.6667)}{f'(2.6667)}=-0.4356$$

And the solution is $2.6667+\Delta x=2.2311$, and so on.

It is interesting to note that when a value greater than the actual solution (2.0) is taken up as the initial guess, Δx is negative. Similarly, when a value lesser than 2.0 is taken up, Δx is positive. Thus, whatever be the initial guess, the subsequent points always move in the correct direction, that is, toward the solution. It is important to note that this is not a coincidence. In fact, this is what makes the algorithm convergent.

Now, a more difficult problem can be considered. Let us consider the solution of the nonlinear system of equations:

$$x^2-2x-y=-0.5 \tag{2.9}$$

$$x^2+4y^2=4 \tag{2.10}$$

Because two variables x and y are involved, the above equations can be written as

$$f(x,y)=f^{SP} \tag{2.11}$$

$$g(x,y)=g^{SP} \tag{2.12}$$

where:

$$f(x,y)=x^2-2x-y \quad \text{and} \quad g(x,y)=x^2+4y^2 \tag{2.13}$$

$$f^{SP}=-0.5 \quad \text{and} \quad g^{SP}=4 \tag{2.14}$$

Again, an initial guess $x_0=1,\ y_0=1$ is made. Being a wild guess, it is extremely unlikely to be the correct solution. However, like the previous case, it is assumed that the actual solution lies at $(x_0+\Delta x,\ y_0+\Delta y)$, that is, $(1+\Delta x,1+\Delta y)$.

Then Equations 2.11 and 2.12 can be written as

$$f(1+\Delta x,1+\Delta y) = -0.5=f^{SP} \tag{2.15}$$

$$g(1+\Delta x,1+\Delta y) = 4=g^{SP} \tag{2.16}$$

Now, by Taylor's series expansion in two variables,

$$f(x_0+\Delta x,y_0+\Delta y)=f(x_0,y_0)+\frac{\partial f(x_0,y_0)}{\partial x}\Delta x+\frac{\partial f(x_0,y_0)}{\partial y}\Delta y+\cdots \tag{2.17}$$

In a similar manner,

$$g(x_0+\Delta x,y_0+\Delta y)=g(x_0,y_0)+\frac{\partial g(x_0,y_0)}{\partial x}\Delta x+\frac{\partial g(x_0,y_0)}{\partial y}\Delta y+\cdots \tag{2.18}$$

Neglecting the terms involving $(\Delta x)^2$, $(\Delta y)^2$, and the subsequent higher ones in Equations 2.17 and 2.18 and combining with Equations 2.15 and 2.16, we get (because $x_0=1,\ y_0=1$),

$$f(1+\Delta x,1+\Delta y)=-0.5=f^{SP}\approx f(1,1)+\frac{\partial f(1,1)}{\partial x}\Delta x+\frac{\partial f(1,1)}{\partial y}\Delta y \tag{2.19}$$

$$g(1+\Delta x, 1+\Delta y) = 4 = g^{SP} \approx g(1,1) + \frac{\partial g(1,1)}{\partial x}\Delta x + \frac{\partial g(1,1)}{\partial y}\Delta y \quad (2.20)$$

The above equations can be rewritten as

$$\begin{bmatrix} \dfrac{\partial f(1,1)}{\partial x} & \dfrac{\partial f(1,1)}{\partial y} \\ \dfrac{\partial g(1,1)}{\partial x} & \dfrac{\partial g(1,1)}{\partial y} \end{bmatrix} \begin{bmatrix} \Delta x \\ \Delta y \end{bmatrix} = \begin{bmatrix} f^{SP} - f(1,1) \\ g^{SP} - g(1,1) \end{bmatrix} = \begin{bmatrix} -0.5 - f(1,1) \\ 4 - g(1,1) \end{bmatrix} \quad (2.21)$$

or

$$[J]\begin{bmatrix} \Delta x \\ \Delta y \end{bmatrix} = \begin{bmatrix} f^{SP} - f(1,1) \\ g^{SP} - g(1,1) \end{bmatrix} = \begin{bmatrix} -0.5 - f(1,1) \\ 4 - g(1,1) \end{bmatrix} \quad (2.22)$$

where $\boldsymbol{J}$ is known as the Jacobian matrix.

Now, let us compute Δx and Δy. First we compute $\boldsymbol{J}$. From Equation 2.13, $\partial f/\partial x = 2x - 2$, $\partial f/\partial y = -1$, $\partial g/\partial x = 2x$, and $\partial g/\partial y = 8y$. Evaluated at $x_0 = 1$, $y_0 = 1$, we get

$$\boldsymbol{J} = \begin{bmatrix} \dfrac{\partial f(1,1)}{\partial x} & \dfrac{\partial f(1,1)}{\partial y} \\ \dfrac{\partial g(1,1)}{\partial x} & \dfrac{\partial g(1,1)}{\partial y} \end{bmatrix} = \begin{bmatrix} 0 & -1 \\ 2 & 8 \end{bmatrix}$$

Similarly, the matrix on the right-hand side (RHS) of Equation 2.22 becomes

$$\begin{bmatrix} f^{SP} - f(1,1) \\ g^{SP} - g(1,1) \end{bmatrix} = \begin{bmatrix} -0.5 - f(1,1) \\ 4 - g(1,1) \end{bmatrix} = \begin{bmatrix} 1.5 \\ -1 \end{bmatrix}$$

Thus, Equation 2.22 becomes

$$\begin{bmatrix} 0 & -1 \\ 2 & 8 \end{bmatrix}\begin{bmatrix} \Delta x \\ \Delta y \end{bmatrix} = \begin{bmatrix} 1.5 \\ -1 \end{bmatrix}$$

which gives

$$\begin{bmatrix} \Delta x \\ \Delta y \end{bmatrix} = \begin{bmatrix} 5.5 \\ -1.5 \end{bmatrix}$$

Thus, the solution after the first iteration is $(1+\Delta x, 1+\Delta y)=(6.5,-0.5)$. We have moved from the intial guess of (1,1) to (6.5,–0.5). Thus, in the next step, the actual solution is assumed to lie at $(6.5+\Delta x, -0.5+\Delta y)$.

Again the Jacobian matrix is evaluated at (6.5,–0.5).

$$\mathbf{J}=\begin{bmatrix} \dfrac{\partial f(6.5,-0.5)}{\partial x} & \dfrac{\partial f(6.5,-0.5)}{\partial y} \\ \dfrac{\partial g(6.5,-0.5)}{\partial x} & \dfrac{\partial g(6.5,-0.5)}{\partial y} \end{bmatrix}=\begin{bmatrix} 11 & -1 \\ 13 & -4 \end{bmatrix}$$

The matrix on the RHS of Equation 2.22 becomes

$$\begin{bmatrix} -0.5-f(6.5,-0.5) \\ 4-g(6.5,-0.5) \end{bmatrix}=\begin{bmatrix} -30.25 \\ -39.25 \end{bmatrix}$$

Thus,

$$\begin{bmatrix} 11 & -1 \\ 13 & -4 \end{bmatrix}\begin{bmatrix} \Delta x \\ \Delta y \end{bmatrix}=\begin{bmatrix} -30.25 \\ -39.25 \end{bmatrix} \text{ or, } \begin{bmatrix} \Delta x \\ \Delta y \end{bmatrix}=\begin{bmatrix} -2.6371 \\ 1.2419 \end{bmatrix}$$

making our next starting point as $(6.5+\Delta x, -0.5+\Delta y)=(3.8629, 0.7419)$. Subsequently, we repeat the steps to compute

$$\mathbf{J}(3.8629,0.7419)=\begin{bmatrix} 5.7258 & -1 \\ 7.7258 & 5.9352 \end{bmatrix}$$

and

$$\begin{bmatrix} -0.5-f(3.8629,0.7419) \\ 4-g(3.8629,0.7419) \end{bmatrix}=\begin{bmatrix} -6.9543 \\ -13.1237 \end{bmatrix}$$

yielding

$$\begin{bmatrix} \Delta x \\ \Delta y \end{bmatrix}=\begin{bmatrix} -1.3042 \\ -0.5135 \end{bmatrix}$$

and making the next starting point $(3.8629+\Delta x, 0.7419+\Delta y)=(2.5587, 0.2284)$.

In the next iteration,

$$\begin{bmatrix} \Delta x \\ \Delta y \end{bmatrix} = \begin{bmatrix} \frac{\partial f(2.5587,0.2284)}{\partial x} & \frac{\partial f(2.5587,0.2284)}{\partial y} \\ \frac{\partial g(2.5587,0.2284)}{\partial x} & \frac{\partial g(2.5587,0.2284)}{\partial y} \end{bmatrix}^{-1}$$

$$\begin{bmatrix} -0.5 - f(2.5587,0.2284) \\ 4 - g(2.5587,0.2284) \end{bmatrix}$$

$$= \begin{bmatrix} -0.5423 \\ 0.0106 \end{bmatrix}$$

making the next starting point as $(2.5587+\Delta x, 0.2284+\Delta y) = (2.0164, 0.239)$. If continued further, the next three starting points are obtained as (1.9082,0.3132), (1.9007,0.3112), and (1.9007,0.3112). As the last two iterations yield the same solution (no update in values), the computed solution can be taken to be equal to the actual solution, for all practical purposes.

We are now ready to face the power flow problem.

2.3 THE POWER FLOW PROBLEM

The power system network consists of synchronous generators (sources) and loads (sinks) interconnected through transmission lines (impedances). The entire system is modeled as a set of nodes (buses) interconnected by impedances. At different nodes, sources (generators) injecting complex powers and/or sinks (loads) absorbing complex powers may be connected. The generators produce complex powers that flow through the transmission lines for consumption by the loads. A small fraction of the complex power produced by the generators is also absorbed by the transmission lines as line losses (real loss) and reactive drops in the lines. As discussed elsewhere [25–28], power flow is used to compute analytically the voltage magnitudes and phase angles at the buses (nodes) of the transmission network under balanced three-phase steady-state conditions. The analytical computation of all the bus voltage magnitudes and phase angles is known in the technical jargon as *power flow analysis*. From this basic information, additional electrical quantities of interest can be further computed, for example, the active and reactive power flows in any transmission line, the losses in the line, and the reactive power supplied by the generators. After the power flow computation, it may be found out that some system buses may be having voltage magnitudes which are outside the acceptable limits (too low or too large). Too low a voltage may cause lamp flickers, fans running at low speeds, and pumps unable to build up their required heads of pressure. Similarly, too large a voltage may damage the equipment. In addition, one or more transmission lines may be overloaded (close to their thermal limits), or a few of them may be close to their stability limits. Here lies the importance of power flow analysis. There are some excellent references [25–28] in the area of power flow analysis for the interested reader.

Now, to compute certain quantities analytically that are usually termed as *unknowns*, we must have some quantities that are usually given or termed as *knowns*. In this respect, it is important to note that the active and reactive powers consumed by a load are assumed to be known constants. These are usually given or known from the load's consumption history. In a similar manner, the active powers supplied by the generators are also specified, as they can be controlled by varying their respective turbine powers. Thus, to proceed, we have to establish the relationships between the knowns and the unknowns. These relationships are termed as *power flow equations*. As we have already assumed a three-phase balanced system, we can carry out per-phase analysis.

The input data for the power flow problem consist of transmission line data, transformer data, and bus data. The transmission line data consist of series impedances and shunt admittances of the equivalent pi circuits of the transmission lines in per unit, along with the bus numbers between which the lines are connected. The transformer data comprise resistances and leakage reactances of the equivalent circuits of the transformers in per-unit and transformer tap ratios, along with the bus numbers between which the windings are connected. The bus data comprise the active and reactive powers consumed by the loads at the load buses along with the active powers supplied by the generators and the generator reactive power limits at the generator buses. From the transmission line and transformer data, the bus admittance matrix of the system is computed.

2.4 POWER FLOW EQUATIONS

As discussed in Section 2.3, a power system is modeled as a set of nodes (buses) interconnected by impedances (transmission lines). At different nodes, generators and loads are connected, which inject and absorb complex powers.

To start our analysis, let us consider a small power system network. It is a six-bus system with three generators, which is represented by its single

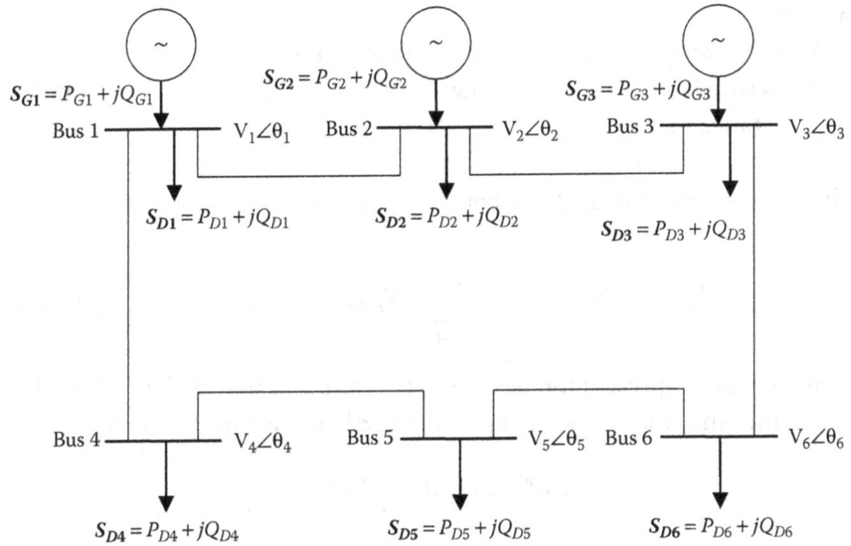

FIGURE 2.1 A six-bus power system.

line diagram as shown in Figure 2.1. For establishment of the power flow equations at any system bus i, we will be following certain conventions [25–28], which are elaborated as follows:

1. By virtue of a common convention, at any bus, the complex power produced by the generator per phase is denoted as $\mathbf{S}_{Gi} = P_{Gi} + jQ_{Gi}$ and the complex power absorbed by the load per phase is denoted by $\mathbf{S}_{Di} = P_{Di} + jQ_{Di}$.

2. P_{Gi} and Q_{Gi} are the real and imaginary parts of $\mathbf{S}_{Gi}$. Similarly, P_{Di} and Q_{Di} are the real and imaginary parts of $\mathbf{S}_{Di}$. Usually, all the P_{Di}'s and Q_{Di}'s are assumed to be constants known from the load's consumption history.

3. After taking care of the per-phase complex power $\mathbf{S}_{Di}$ absorbed by the load at bus i, $\mathbf{S}_{Gi} - \mathbf{S}_{Di} = \mathbf{S}_i$ is the per-phase complex power that is remaining of $\mathbf{S}_{Gi}$. This complex power $\mathbf{S}_i$ goes into the transmission system or the network and is known as *net injected (per-phase) complex power at bus* i.

Also, the net complex current injection [28] at any bus i is given by

$$\mathbf{I}_i = \sum_{k=1}^{n} \mathbf{Y}_{ik} \mathbf{V}_k \qquad i = 1,2,\ldots 6 \tag{2.23}$$

where:

- $\mathbf{V}_i$ is the complex (rms) bus voltage phasor at bus i
- $\mathbf{Y}_{ab}$ is the element in the ath row and the bth column of the bus admittance matrix $\mathbf{Y}$

Thus, using Equation 2.23, the complex power injected at bus i is

$$\mathbf{S}_i = P_i + jQ_i = \mathbf{V}_i\mathbf{I}_i^* = \sum_{k=1}^{n} \mathbf{V}_i\mathbf{V}_k^*\mathbf{Y}_{ik}^* \qquad i=1,2,\ldots,6 \tag{2.24}$$

If we use polar representations for voltage phasors as well as elements of the bus admittance matrix, we can proceed further. For this, let us consider

$$\mathbf{V}_i = V_i e^{j\theta_i} = V_i\angle\theta_i \qquad i=1,2,\ldots,6 \tag{2.25}$$

$$\mathbf{Y}_{ik} = \mathrm{Y}_{ik}e^{j\varphi_{ki}} = \mathrm{Y}_{ik}\angle\varphi_{ik} = \mathrm{G}_{ik} + j\mathrm{B}_{ik} \qquad i,k=1,2,\ldots,6 \tag{2.26}$$

Substituting the above equations in Equation 2.24,

$$\mathbf{S}_i = \sum_{k=1}^{n} \mathrm{V}_i\mathrm{V}_k Y_{ik}\angle(\theta_i - \theta_k - \varphi_{ik}) \qquad i,k=1,2,\ldots 6 \tag{2.27}$$

$$\text{or } \mathbf{S}_i = \sum_{k=1}^{n} \mathrm{V}_i\mathrm{V}_k\left[\cos(\theta_i-\theta_k) + j\sin(\theta_i-\theta_k)\right]Y_{ik}\left(\cos\varphi_{ik} - j\sin\varphi_{ik}\right)$$

$$\text{or } \mathbf{S}_i = \sum_{k=1}^{n} \mathrm{V}_i\mathrm{V}_k\left[\cos(\theta_i-\theta_k) + j\sin(\theta_i-\theta_k)\right]\left(G_{ik} - jB_{ik}\right) \tag{2.28}$$

From Equation 2.27, equating the real and imaginary parts gives us the polar form of the power flow equations as

$$P_i = \sum_{k=1}^{n} \mathrm{V}_i\mathrm{V}_k Y_{ik}\cos(\theta_i - \theta_k - \varphi_{ik}) \quad i,k=1,2,\ldots,6 \tag{2.29}$$

$$Q_i = \sum_{k=1}^{n} \mathrm{V}_i\mathrm{V}_k Y_{ik}\sin(\theta_i - \theta_k - \varphi_{ik}) \quad i,k=1,2,\ldots,6 \tag{2.30}$$

In a similar manner, Equation 2.28 gives us the rectangular form of the power flow equations as

$$P_i = \sum_{k=1}^{n} V_i V_k \left[G_{ik}\cos(\theta_i - \theta_k) + B_{ik}\sin(\theta_i - \theta_k) \right] \quad i,k = 1,2,\ldots 6 \qquad (2.31)$$

$$Q_i = \sum_{k=1}^{n} V_i V_k \left[G_{ik}\sin(\theta_i - \theta_k) - B_{ik}\cos(\theta_i - \theta_k) \right] \quad i,k = 1,2,\ldots,6 \qquad (2.32)$$

2.5 THE CLASSIFICATION OF BUSES

Once the power flow equations have been written, it is now time to understand the known and unknown quantities. We will proceed set by step, starting with the unknowns.

1. The motive of *power flow* is to compute the magnitudes and phase angles of the bus voltages. Hence, it appears that the unknowns for the six-bus system shown in Figure 2.1 should be

$$\theta_1, V_1, \theta_2, V_2, \theta_3, V_3, \theta_4, V_4, \theta_5, V_5, \theta_6, \text{and } V_6 \qquad (2.33)$$

2. At buses 1, 2, and 3, generators are connected. As a usual practice, the automatic voltage regulator (AVR) of each generator keeps its terminal voltage at the desired value by control of the generator's field current. V_1, V_2, and V_3 should be controlled, that is, specified or known quantities. Hence, the apparent list of unknowns gets modified to

$$\theta_1, \theta_2, \theta_3, \theta_4, V_4, \theta_5, V_5, \theta_6, \text{and } V_6 \qquad (2.34)$$

3. Now, let us take the stock of the known quantities. As discussed earlier, the active (real) and reactive powers consumed by the load are assumed to be known from their *consumption history*, and hence, $P_{D1}, Q_{D1}, P_{D2}, Q_{D2}, P_{D3}, Q_{D3}, P_{D4}, Q_{D4}, P_{D5}, Q_{D5}, P_{D6}$, and Q_{D6} are all known quantities. Also, at buses 4, 5, and 6, there are no *sources* of active or reactive power, that is, the active and reactive powers generated are zero (loads do not generate—but absorb active and reactive powers). Thus, $P_{G4} = P_{G5} = P_{G6} = 0$ and $Q_{G4} = Q_{G5} = Q_{G6} = 0$. From the

definition of net injected active and reactive powers, it follows that at bus 4, $P_4 = P_{G4} - P_{D4} = -P_{D4}$ and $Q_4 = Q_{G4} - Q_{D4} = -Q_{D4}$ are known quantities. In a similar manner, at buses 5 and 6, P_5, P_6, Q_5, and Q_6 are also known quantities. Thus, corresponding to the apparent list of unknowns given in Equation 2.34, the list of knowns appear to be

$$P_4, Q_4, P_5, Q_5, P_6, \text{and } Q_6 \tag{2.35}$$

4. Now, at any bus, four quantities are involved: the voltage magnitude, the voltage phase angle, the net injected active power, and the net injected reactive power. One can solve for two quantities (say, the bus voltage magnitude and the phase angle) if the other two are known or specified. At a generator bus, the voltage magnitude is specified (the AVR maintains the terminal voltage of a generator at a specified value). In a similar manner, at a generator bus, the active power injected by the generator is specified (as the turbine power is controllable by steam/gate valve control). Thus, it appears that the known quantities are

$$P_{G1}, P_{G2}, \text{and } P_{G3} \tag{2.36}$$

5. By the law of conservation of complex power,

$$\text{Active power generated} = \text{active power absorbed}$$

Hence,

$$P_{G1} + P_{G2} + P_{G3} = P_{D1} + P_{D2} + P_{D3} + P_{D4} + P_{D5} + P_{D6} + \text{line losses} \tag{2.37}$$

The line losses are the sum of the I^2R losses in the individual transmission lines. The expression for the I^2R loss in any transmission line can be obtained very easily. As an example, the line loss in the transmission line connecting buses 2 and 3 is

$$P_{\text{LOSS}\,23} = \left|\mathbf{I}_{23}\right|^2 R_{23} = \text{I}_{23}^2 R_{23} \tag{2.38}$$

where:

$$\mathbf{I}_{23} = \mathbf{y}_{23}(\mathbf{V}_2 - \mathbf{V}_3) = \mathbf{Y}_{23}(\mathbf{V}_3 - \mathbf{V}_2) = Y_{23} \angle \varphi_{y23}(\text{V}_3\angle\theta_3 - \text{V}_2\angle\theta_2) \tag{2.39}$$

In the above equation, Y_{23} and φ_{y23} are known quantities, because they are related to an element of the bus admittance matrix. V_2 and V_3 are known quantities as already explained earlier. However, θ_2 and θ_3 are unknowns (Equation 2.34). Hence, from Equations 2.38 and 2.39, it can be observed that $\mathbf{I}_{23}$ and hence $P_{\mathrm{LOSS}\,23}$ cannot be determined beforehand. In a similar manner, it can be shown that the current in any of the individual transmission lines cannot be computed beforehand, making the line losses an unknown quantity.

6. In Equation 2.37, all the quantities P_{D1}, P_{D2}, P_{D3}, P_{D4}, P_{D5}, and P_{D6} are known quantities, being active powers consumed by the loads. However, because the line losses are unknown, the RHS of Equation 2.37 is an unknown quantity. Thus, in the left-hand side (LHS), all the three quantities P_{G1}, P_{G2}, and P_{G3} cannot be simultaneously known. This contradicts Equation 2.36.
7. The above situation is tackled by assuming that one of the generator active powers is not known beforehand. By virtue of convention, P_{G1} is kept as an unknown to balance the active powers and take up the slack in determining the line losses. Usually, in any system, the generator with the largest capacity is allotted this function.
8. Hence, P_{G2} and P_{G3} are known quantities along with P_{D1}, P_{D2}, P_{D3}, P_{D4}, P_{D5}, and P_{D6} (being active powers consumed by the load). It thus follows that $P_2 = P_{G2} - P_{D2}$ and $P_3 = P_{G3} - P_{D3}$ are known quantities. Hence, using Equations 2.35 and 2.36 and the discussion in point 6 (above), the final list of known quantities is

$$P_2, P_3, P_4, Q_4, P_5, Q_5, P_6, \text{and } Q_6 \tag{2.40}$$

9. From Equation 2.34, the apparent list of unknowns are

$$\theta_1, \theta_2, \theta_3, \theta_4, V_4, \theta_5, V_5, \theta_6, \text{and } V_6 \tag{2.41}$$

The number of unknowns is nine and exceeds the number of known quantities (eight in Equation 2.40). This problem can be tackled easily.

10. From Equations 2.29 through 2.32, the nature of expression of any of the P_i or Q_i shows that we can solve only for the differences in phase angles $(\theta_i - \theta_k)$ and not individual phase angles. Because of this, we take $\theta_1 = 0$ (reference) and compute all the other unknown

phase angles with reference to θ_1. This removes θ_1 from the list of unknowns. Thus, the final power flow problem boils down to

Solve $\theta_2, \theta_3, \theta_4, V_4, \theta_5, V_5, \theta_6, \text{and } V_6$

Given the following:

$$P_2, P_3, P_4, Q_4, P_5, Q_5, P_6, \text{and } Q_6 \tag{2.42}$$

Now, before we proceed to solve the above expression in Section 2.6, we conclude this section by classifying the system buses depending on the known and unknown quantities at these buses. The usual practice is to classify the system buses into three types:

1. *Load buses or* PQ *buses*: These are the buses to which generators are not connected. Thus, $P_{Gi} = Q_{Gi} = 0$ and P_{Di} and Q_{Di} are known. For the three-bus system considered (Figure 2.1), buses 4, 5, and 6 are load buses.
2. *Voltage-controlled buses*: These are the buses at which the injected active power and the bus voltage magnitude are known or specified. Thus, $P_i = P_{Gi} - P_{Di}$ and V_i is known. Usually, these can be buses to which either generators are connected (and operating within their reactive power limits) or sources of reactive powers are connected (and operating within their reactive power limits), which are used to maintain the bus voltage magnitude to the specified value.
3. *Slack or swing bus*: This is a bus that is connected to a generator. Usually, in any system, the bus that is connected to the generator with the largest capacity is the slack bus. By virtue of convention, bus 1 is usually denoted as the slack bus. The magnitude as well as the phase angle of the bus voltage is specified.

2.6 SOLUTION OF THE POWER FLOW PROBLEM

We again take up the small six-bus power system as shown in Figure 2.1. We recall the power flow problem from Section 2.5 as follows.

Solve $\theta_2, \theta_3, \theta_4, V_4, \theta_5, V_5, \theta_6$, and V_6.

Given $P_2, P_3, P_4, Q_4, P_5, Q_5, P_6$, and Q_6.

Usually, the quantities known or specified are denoted with a superscript *SP* such as P_2^{SP} or Q_5^{SP}. Thus, the active and reactive powers can be rewritten in the form

$$\begin{aligned} P_2 &= P_2^{SP} \\ P_3 &= P_3^{SP} \\ P_4 &= P_4^{SP} \\ P_5 &= P_5^{SP} \\ P_6 &= P_6^{SP} \\ Q_4 &= Q_4^{SP} \\ Q_5 &= Q_5^{SP} \\ Q_6 &= Q_6^{SP} \end{aligned} \tag{2.43}$$

From Equations 2.29 and 2.30, it can be observed that the LHSs are governed by equations; the RHSs are specified constants. Now, as an example, let us pick any one from Equation 2.43, say, the second one, that is,

$$P_3 = P_3^{SP} \tag{2.44}$$

Now, for the small six-bus system considered in Figure 2.1, using Equation 2.29,

$$P_3 = \sum_{k=1}^{6} V_3 V_k Y_{3k} \cos(\theta_3 - \theta_k - \varphi_{3k}) \tag{2.45}$$

Expanding, we get

$$\begin{aligned} P_3 &= V_3 V_1 Y_{31} \cos(\theta_3 - \theta_1 - \varphi_{31}) + V_3 V_2 Y_{32} \cos(\theta_3 - \theta_2 - \varphi_{32}) \\ &\quad + V_3^2 Y_{33} \cos(\varphi_{33}) + V_3 V_4 Y_{34} \cos(\theta_3 - \theta_4 - \varphi_{34}) \\ &\quad + V_3 V_5 Y_{35} \cos(\theta_3 - \theta_5 - \varphi_{35}) + V_3 V_6 Y_{36} \cos(\theta_3 - \theta_6 - \varphi_{36}) \end{aligned} \tag{2.46}$$

From Equation 2.45, it can be observed that P_3 is a function of $\theta_1, V_1, \theta_2, V_2, \theta_3, V_3, \theta_4, V_4, \theta_5, V_5, \theta_6,$ and V_6. However, we should recall that $\theta_1, V_1, V_2, \text{and } V_3$ are known constants ($\theta_1 = 0$ is the chosen reference, whereas $V_1, V_2, \text{and } V_3$ are controlled to a specified value by the AVRs of the respective generators). Hence, P_3 is a function of the variables $\theta_2, \theta_3, \theta_4, V_4, \theta_5, V_5, \theta_6, \text{and } V_6$. Thus, we can write P_3 as

$$P_3(\theta_2, \theta_3, \theta_4, V_4, \theta_5, V_5, \theta_6, V_6)$$

Thus, Equation 2.44 can now be rewritten as

$$P_3(\theta_2, \theta_3, \theta_4, V_4, \theta_5, V_5, \theta_6, V_6) = P_3^{SP} \tag{2.47}$$

In a similar manner, from the system of equations given in Equation 2.43, it can be shown that $P_2, P_4, Q_4, P_5, Q_5, P_6, \text{and } Q_6$ are all functions of $\theta_2, \theta_3, \theta_4, V_4, \theta_5, V_5, \theta_6, \text{and } V_6$. Thus, Equation 2.43 can be summarized as

$$\begin{aligned}
P_2(\theta_2, \theta_3, \theta_4, V_4, \theta_5, V_5, \theta_6, V_6) &= P_2^{SP} \\
P_3(\theta_2, \theta_3, \theta_4, V_4, \theta_5, V_5, \theta_6, V_6) &= P_3^{SP} \\
P_4(\theta_2, \theta_3, \theta_4, V_4, \theta_5, V_5, \theta_6, V_6) &= P_4^{SP} \\
Q_4(\theta_2, \theta_3, \theta_4, V_4, \theta_5, V_5, \theta_6, V_6) &= Q_4^{SP} \\
P_5(\theta_2, \theta_3, \theta_4, V_4, \theta_5, V_5, \theta_6, V_6) &= P_5^{SP} \\
Q_5(\theta_2, \theta_3, \theta_4, V_4, \theta_5, V_5, \theta_6, V_6) &= Q_5^{SP} \\
P_6(\theta_2, \theta_3, \theta_4, V_4, \theta_5, V_5, \theta_6, V_6) &= P_6^{SP} \\
Q_6(\theta_2, \theta_3, \theta_4, V_4, \theta_5, V_5, \theta_6, V_6) &= Q_6^{SP}
\end{aligned} \tag{2.48}$$

We recall that in Section 2.2, a nonlinear system of equations comprising two functions *f*() and *g*() was taken up. Both functions *f*() and *g*() involved two independent variables *x* and *y*. Also, the Taylor series expansion of the two functions *f*() and *g*() was shown in Equations 2.17 and 2.18, respectively.

In a similar manner, the Taylor series expansion of the nonlinear system of equations given in Equation 2.48 can be written very easily. We start with the first equation of Equation 2.48, that is,

$$P_2(\theta_2, \theta_3, \theta_4, V_4, \theta_5, V_5, \theta_6, V_6) = P_2^{SP} \tag{2.49}$$

As already done in previous cases, we start with our initial guess of $(\theta_2^0, \theta_3^0, \theta_4^0, V_4^0, \theta_5^0, V_5^0, \theta_6^0, V_6^0)$ in the neighborhood of the actual solution and assume that the actual solution lies at an infinitesimal distance away from our initial guess. This is same as assuming that the actual solution is at $(\theta_2^0 + \Delta\theta_2, \theta_3^0 + \Delta\theta_3, \theta_4^0 + \Delta\theta_4, V_4^0 + \Delta V_4, \theta_5^0 + \Delta\theta_5, V_5^0 + \Delta V_5, \theta_6^0 + \Delta\theta_6, V_6^0 + \Delta V_6)$. This means that

$$P_2(\theta_2^0 + \Delta\theta_2, \theta_3^0 + \Delta\theta_3, \theta_4^0 + \Delta\theta_4, V_4^0 + \Delta V_4, \theta_5^0 + \Delta\theta_5, V_5^0$$

$$+ \Delta V_5, \theta_6^0 + \Delta\theta_6, V_6^0 + \Delta V_6) = P_2^{SP}$$

Subsequently, the Taylor series expansion of the function $P_2(\,)$ around the initial values $\theta_2^0, \theta_3^0, \theta_4^0, V_4^0, \theta_5^0, V_5^0, \theta_6^0, \text{and}\, V_6^0$ with terms such as $(\Delta\theta_2)^2$ and $(\Delta\theta_3)^2$ and the subsequent higher ones neglected yields

$$P_2\left(\theta_2^0, \theta_3^0, \theta_4^0, V_4^0, \theta_5^0, V_5^0, \theta_6^0, V_6^0\right) + \frac{\partial P_2}{\partial \theta_2}\Delta\theta_2 + \frac{\partial P_2}{\partial \theta_3}\Delta\theta_3$$

$$+ \frac{\partial P_2}{\partial \theta_4}\Delta\theta_4 + \frac{\partial P_2}{\partial V_4}\Delta V_4 + \frac{\partial P_2}{\partial \theta_5}\Delta\theta_5 + \frac{\partial P_2}{\partial V_5}\Delta V_5 + \frac{\partial P_2}{\partial \theta_6}\Delta\theta_6 + \frac{\partial P_2}{\partial V_6}\Delta V_6 = P_2^{SP} \tag{2.50}$$

or

$$\frac{\partial P_2}{\partial \theta_2}\Delta\theta_2 + \frac{\partial P_2}{\partial \theta_3}\Delta\theta_3 + \frac{\partial P_2}{\partial \theta_4}\Delta\theta_4 + \frac{\partial P_2}{\partial V_4}\Delta V_4 + \frac{\partial P_2}{\partial \theta_5}\Delta\theta_5$$

$$+ \frac{\partial P_2}{\partial V_5}\Delta V_5 + \frac{\partial P_2}{\partial \theta_6}\Delta\theta_6 + \frac{\partial P_2}{\partial V_6}\Delta V_6 \tag{2.51}$$

$$= P_2^{SP} - P_2\left(\theta_2^0, \theta_3^0, \theta_4^0, V_4^0, \theta_5^0, V_5^0, \theta_6^0, V_6^0\right)$$

$$= P_2^{SP} - P_2^x$$

where:

$P_2^x = P_2\left(\theta_2^0, \theta_3^0, \theta_4^0, V_4^0, \theta_5^0, V_5^0, \theta_6^0, V_6^0\right)$ denotes the value of the function $P_2 = \sum_{k=1}^{6} V_2 V_k Y_{2k} \cos(\theta_2 - \theta_k - \varphi_{2k})$ computed at the initial guess $(\theta_2^0, \theta_3^0, \theta_4^0, V_4^0, \theta_5^0, V_5^0, \theta_6^0, V_6^0)$

P_2^{SP} is a given value, a constant

From now on, in all subsequent analyses, we will use the notation P_i^x (or Q_i^x) to denote computed values of the function P_i (or Q_i), respectively.

In a similar manner, for the equation corresponding to P_3 (in Equation 2.48),

$$\frac{\partial P_3}{\partial \theta_2}\Delta\theta_2 + \frac{\partial P_3}{\partial \theta_3}\Delta\theta_3 + \frac{\partial P_3}{\partial \theta_4}\Delta\theta_4 + \frac{\partial P_3}{\partial V_4}\Delta V_4 + \frac{\partial P_3}{\partial \theta_5}\Delta\theta_5$$

$$+\frac{\partial P_3}{\partial V_5}\Delta V_5 + \frac{\partial P_3}{\partial \theta_6}\Delta\theta_6 + \frac{\partial P_3}{\partial V_6}\Delta V_6 \tag{2.52}$$

$$= P_3^{SP} - P_3\left(\theta_2^0, \theta_3^0, \theta_4^0, V_4^0, \theta_5^0, V_5^0, \theta_6^0, V_6^0\right)$$

$$= P_3^{SP} - P_3^x$$

and so on. Thus, the system of equations given in Equation 2.48 can be summarized in matrix form as

$$\begin{bmatrix}
\frac{\partial P_2}{\partial \theta_2} & \frac{\partial P_2}{\partial \theta_3} & \frac{\partial P_2}{\partial \theta_4} & \frac{\partial P_2}{\partial \theta_5} & \frac{\partial P_2}{\partial \theta_6} & \frac{\partial P_2}{\partial V_4} & \frac{\partial P_2}{\partial V_5} & \frac{\partial P_2}{\partial V_6} \\
\frac{\partial P_3}{\partial \theta_2} & \frac{\partial P_3}{\partial \theta_3} & \frac{\partial P_3}{\partial \theta_4} & \frac{\partial P_3}{\partial \theta_5} & \frac{\partial P_3}{\partial \theta_6} & \frac{\partial P_3}{\partial V_4} & \frac{\partial P_3}{\partial V_5} & \frac{\partial P_3}{\partial V_6} \\
\frac{\partial P_4}{\partial \theta_2} & \frac{\partial P_4}{\partial \theta_3} & \frac{\partial P_4}{\partial \theta_4} & \frac{\partial P_4}{\partial \theta_5} & \frac{\partial P_4}{\partial \theta_6} & \frac{\partial P_4}{\partial V_4} & \frac{\partial P_4}{\partial V_5} & \frac{\partial P_4}{\partial V_6} \\
\frac{\partial P_5}{\partial \theta_2} & \frac{\partial P_5}{\partial \theta_3} & \frac{\partial P_5}{\partial \theta_4} & \frac{\partial P_5}{\partial \theta_5} & \frac{\partial P_5}{\partial \theta_6} & \frac{\partial P_5}{\partial V_4} & \frac{\partial P_5}{\partial V_5} & \frac{\partial P_5}{\partial V_6} \\
\frac{\partial P_6}{\partial \theta_2} & \frac{\partial P_6}{\partial \theta_3} & \frac{\partial P_6}{\partial \theta_4} & \frac{\partial P_6}{\partial \theta_5} & \frac{\partial P_6}{\partial \theta_6} & \frac{\partial P_6}{\partial V_4} & \frac{\partial P_6}{\partial V_5} & \frac{\partial P_6}{\partial V_6} \\
\frac{\partial Q_4}{\partial \theta_2} & \frac{\partial Q_4}{\partial \theta_3} & \frac{\partial Q_4}{\partial \theta_4} & \frac{\partial Q_4}{\partial \theta_5} & \frac{\partial Q_4}{\partial \theta_6} & \frac{\partial Q_4}{\partial V_4} & \frac{\partial Q_4}{\partial V_5} & \frac{\partial Q_4}{\partial V_6} \\
\frac{\partial Q_5}{\partial \theta_2} & \frac{\partial Q_5}{\partial \theta_3} & \frac{\partial Q_5}{\partial \theta_4} & \frac{\partial Q_5}{\partial \theta_5} & \frac{\partial Q_5}{\partial \theta_6} & \frac{\partial Q_5}{\partial V_4} & \frac{\partial Q_5}{\partial V_5} & \frac{\partial Q_5}{\partial V_6} \\
\frac{\partial Q_6}{\partial \theta_2} & \frac{\partial Q_6}{\partial \theta_3} & \frac{\partial Q_6}{\partial \theta_4} & \frac{\partial Q_6}{\partial \theta_5} & \frac{\partial Q_6}{\partial \theta_6} & \frac{\partial Q_6}{\partial V_4} & \frac{\partial Q_6}{\partial V_5} & \frac{\partial Q_6}{\partial V_6}
\end{bmatrix}
\begin{bmatrix} \Delta\theta_2 \\ \Delta\theta_3 \\ \Delta\theta_4 \\ \Delta\theta_5 \\ \Delta\theta_6 \\ \Delta V_4 \\ \Delta V_5 \\ \Delta V_6 \end{bmatrix} \tag{2.53}$$

$$= \begin{bmatrix} P_2^{SP} - P_2^x \\ P_3^{SP} - P_3^x \\ P_4^{SP} - P_4^x \\ P_5^{SP} - P_5^x \\ P_6^{SP} - P_6^x \\ Q_4^{SP} - Q_4^x \\ Q_5^{SP} - Q_5^x \\ Q_6^{SP} - Q_6^x \end{bmatrix} = \begin{bmatrix} \Delta P_2 \\ \Delta P_3 \\ \Delta P_4 \\ \Delta P_5 \\ \Delta P_6 \\ \Delta Q_4 \\ \Delta Q_5 \\ \Delta Q_6 \end{bmatrix}$$

It is important to note that in the above equation, the computed values of the functions ($P_2^x, P_3^x, P_4^x, P_5^x, P_6^x, Q_4^x, Q_5^x, Q_6^x$) on the RHS are computed at the initial guess ($\theta_2^0, \theta_3^0, \theta_4^0, V_4^0, \theta_5^0, V_5^0, \theta_6^0, V_6^0$). Similarly, all the partial derivatives on the LHS are also evaluated at ($\theta_2^0, \theta_3^0, \theta_4^0, V_4^0, \theta_5^0, V_5^0, \theta_6^0, V_6^0$).

Now, because our initial guess is ($\theta_2^0, \theta_3^0, \theta_4^0, V_4^0, \theta_5^0, V_5^0, \theta_6^0, V_6^0$) and we have assumed that the solution lies at ($\theta_2^0 + \Delta\theta_2, \theta_3^0 + \Delta\theta_3, \theta_4^0 + \Delta\theta_4, V_4^0 + \Delta V_4, \theta_5^0 + \Delta\theta_5, V_5^0 + \Delta V_5, \theta_6^0 + \Delta\theta_6, V_6^0 + \Delta V_6$), we need to compute ($\Delta\theta_2, \Delta\theta_3, \Delta\theta_4, \Delta\theta_5, \Delta\theta_6, \Delta V_4, \Delta V_5, \Delta V_6$) to find the solution. From Equation 2.53,

$$\begin{bmatrix} \Delta\theta_2 \\ \Delta\theta_3 \\ \Delta\theta_4 \\ \Delta\theta_5 \\ \Delta\theta_6 \\ \Delta V_4 \\ \Delta V_5 \\ \Delta V_6 \end{bmatrix} = \begin{bmatrix} \frac{\partial P_2}{\partial \theta_2} & \frac{\partial P_2}{\partial \theta_3} & \frac{\partial P_2}{\partial \theta_4} & \frac{\partial P_2}{\partial \theta_5} & \frac{\partial P_2}{\partial \theta_6} & \frac{\partial P_2}{\partial V_4} & \frac{\partial P_2}{\partial V_5} & \frac{\partial P_2}{\partial V_6} \\ \frac{\partial P_3}{\partial \theta_2} & \frac{\partial P_3}{\partial \theta_3} & \frac{\partial P_3}{\partial \theta_4} & \frac{\partial P_3}{\partial \theta_5} & \frac{\partial P_3}{\partial \theta_6} & \frac{\partial P_3}{\partial V_4} & \frac{\partial P_3}{\partial V_5} & \frac{\partial P_3}{\partial V_6} \\ \frac{\partial P_4}{\partial \theta_2} & \frac{\partial P_4}{\partial \theta_3} & \frac{\partial P_4}{\partial \theta_4} & \frac{\partial P_4}{\partial \theta_5} & \frac{\partial P_4}{\partial \theta_6} & \frac{\partial P_4}{\partial V_4} & \frac{\partial P_4}{\partial V_5} & \frac{\partial P_4}{\partial V_6} \\ \frac{\partial P_5}{\partial \theta_2} & \frac{\partial P_5}{\partial \theta_3} & \frac{\partial P_5}{\partial \theta_4} & \frac{\partial P_5}{\partial \theta_5} & \frac{\partial P_5}{\partial \theta_6} & \frac{\partial P_5}{\partial V_4} & \frac{\partial P_5}{\partial V_5} & \frac{\partial P_5}{\partial V_6} \\ \frac{\partial P_6}{\partial \theta_2} & \frac{\partial P_6}{\partial \theta_3} & \frac{\partial P_6}{\partial \theta_4} & \frac{\partial P_6}{\partial \theta_5} & \frac{\partial P_6}{\partial \theta_6} & \frac{\partial P_6}{\partial V_4} & \frac{\partial P_6}{\partial V_5} & \frac{\partial P_6}{\partial V_6} \\ \frac{\partial Q_4}{\partial \theta_2} & \frac{\partial Q_4}{\partial \theta_3} & \frac{\partial Q_4}{\partial \theta_4} & \frac{\partial Q_4}{\partial \theta_5} & \frac{\partial Q_4}{\partial \theta_6} & \frac{\partial Q_4}{\partial V_4} & \frac{\partial Q_4}{\partial V_5} & \frac{\partial Q_4}{\partial V_6} \\ \frac{\partial Q_5}{\partial \theta_2} & \frac{\partial Q_5}{\partial \theta_3} & \frac{\partial Q_5}{\partial \theta_4} & \frac{\partial Q_5}{\partial \theta_5} & \frac{\partial Q_5}{\partial \theta_6} & \frac{\partial Q_5}{\partial V_4} & \frac{\partial Q_5}{\partial V_5} & \frac{\partial Q_5}{\partial V_6} \\ \frac{\partial Q_6}{\partial \theta_2} & \frac{\partial Q_6}{\partial \theta_3} & \frac{\partial Q_6}{\partial \theta_4} & \frac{\partial Q_6}{\partial \theta_5} & \frac{\partial Q_6}{\partial \theta_6} & \frac{\partial Q_6}{\partial V_4} & \frac{\partial Q_6}{\partial V_5} & \frac{\partial Q_6}{\partial V_6} \end{bmatrix}^{-1} \begin{bmatrix} P_2^{SP} - P_2^x \\ P_3^{SP} - P_3^x \\ P_4^{SP} - P_4^x \\ P_5^{SP} - P_5^x \\ P_6^{SP} - P_6^x \\ Q_4^{SP} - Q_4^x \\ Q_5^{SP} - Q_5^x \\ Q_6^{SP} - Q_6^x \end{bmatrix} \quad (2.54)$$

Hence, the solution lies at

$$\begin{bmatrix} \theta_2^1 \\ \theta_3^1 \\ \theta_4^1 \\ \theta_5^1 \\ \theta_6^1 \\ V_4^1 \\ V_5^1 \\ V_6^1 \end{bmatrix} = \begin{bmatrix} \theta_2^0 + \Delta\theta_2 \\ \theta_3^0 + \Delta\theta_3 \\ \theta_4^0 + \Delta\theta_4 \\ \theta_5^0 + \Delta\theta_5 \\ \theta_6^0 + \Delta\theta_6 \\ V_4^0 + \Delta V_4 \\ V_5^0 + \Delta V_5 \\ V_6^0 + \Delta V_6 \end{bmatrix} = \begin{bmatrix} \theta_2^0 \\ \theta_3^0 \\ \theta_4^0 \\ \theta_5^0 \\ \theta_6^0 \\ V_4^0 \\ V_5^0 \\ V_6^0 \end{bmatrix}$$

$$+ \begin{bmatrix} \frac{\partial P_2}{\partial \theta_2} & \frac{\partial P_2}{\partial \theta_3} & \frac{\partial P_2}{\partial \theta_4} & \frac{\partial P_2}{\partial \theta_5} & \frac{\partial P_2}{\partial \theta_6} & \frac{\partial P_2}{\partial V_4} & \frac{\partial P_2}{\partial V_5} & \frac{\partial P_2}{\partial V_6} \\ \frac{\partial P_3}{\partial \theta_2} & \frac{\partial P_3}{\partial \theta_3} & \frac{\partial P_3}{\partial \theta_4} & \frac{\partial P_3}{\partial \theta_5} & \frac{\partial P_3}{\partial \theta_6} & \frac{\partial P_3}{\partial V_4} & \frac{\partial P_3}{\partial V_5} & \frac{\partial P_3}{\partial V_6} \\ \frac{\partial P_4}{\partial \theta_2} & \frac{\partial P_4}{\partial \theta_3} & \frac{\partial P_4}{\partial \theta_4} & \frac{\partial P_4}{\partial \theta_5} & \frac{\partial P_4}{\partial \theta_6} & \frac{\partial P_4}{\partial V_4} & \frac{\partial P_4}{\partial V_5} & \frac{\partial P_4}{\partial V_6} \\ \frac{\partial P_5}{\partial \theta_2} & \frac{\partial P_5}{\partial \theta_3} & \frac{\partial P_5}{\partial \theta_4} & \frac{\partial P_5}{\partial \theta_5} & \frac{\partial P_5}{\partial \theta_6} & \frac{\partial P_5}{\partial V_4} & \frac{\partial P_5}{\partial V_5} & \frac{\partial P_5}{\partial V_6} \\ \frac{\partial P_6}{\partial \theta_2} & \frac{\partial P_6}{\partial \theta_3} & \frac{\partial P_6}{\partial \theta_4} & \frac{\partial P_6}{\partial \theta_5} & \frac{\partial P_6}{\partial \theta_6} & \frac{\partial P_6}{\partial V_4} & \frac{\partial P_6}{\partial V_5} & \frac{\partial P_6}{\partial V_6} \\ \frac{\partial Q_4}{\partial \theta_2} & \frac{\partial Q_4}{\partial \theta_3} & \frac{\partial Q_4}{\partial \theta_4} & \frac{\partial Q_4}{\partial \theta_5} & \frac{\partial Q_4}{\partial \theta_6} & \frac{\partial Q_4}{\partial V_4} & \frac{\partial Q_4}{\partial V_5} & \frac{\partial Q_4}{\partial V_6} \\ \frac{\partial Q_5}{\partial \theta_2} & \frac{\partial Q_5}{\partial \theta_3} & \frac{\partial Q_5}{\partial \theta_4} & \frac{\partial Q_5}{\partial \theta_5} & \frac{\partial Q_5}{\partial \theta_6} & \frac{\partial Q_5}{\partial V_4} & \frac{\partial Q_5}{\partial V_5} & \frac{\partial Q_5}{\partial V_6} \\ \frac{\partial Q_6}{\partial \theta_2} & \frac{\partial Q_6}{\partial \theta_3} & \frac{\partial Q_6}{\partial \theta_4} & \frac{\partial Q_6}{\partial \theta_5} & \frac{\partial Q_6}{\partial \theta_6} & \frac{\partial Q_6}{\partial V_4} & \frac{\partial Q_6}{\partial V_5} & \frac{\partial Q_6}{\partial V_6} \end{bmatrix}^{-1} \begin{bmatrix} P_2^{SP} - P_2^x \\ P_3^{SP} - P_3^x \\ P_4^{SP} - P_4^x \\ P_5^{SP} - P_5^x \\ P_6^{SP} - P_6^x \\ Q_4^{SP} - Q_4^x \\ Q_5^{SP} - Q_5^x \\ Q_6^{SP} - Q_6^x \end{bmatrix} \quad (2.55)$$

This becomes our new starting point for the next iteration. We again compute the value of the functions $P_2^x, P_3^x, P_4^x, P_5^x, P_6^x, Q_4^x, Q_5^x, Q_6^x$ and the partial derivatives corresponding to the new starting point $(\theta_2^1, \theta_3^1, \theta_4^1,$

$\theta_5^1, \theta_6^1, V_4^1, V_5^1, V_6^1$) and use Equation 2.55 to get the next starting point and so on. If the process is convergent, a stage comes when the computed values ($P_2^x, P_3^x, P_4^x, P_5^x, P_6^x, Q_4^x, Q_5^x, Q_6^x$) appear within a given tolerance of the specified values ($P_2^{SP}, P_3^{SP}, P_4^{SP}, P_5^{SP}, P_6^{SP}, Q_4^{SP}, Q_5^{SP}, Q_6^{SP}$). Usually, for power flow applications, this tolerance is on the order of 10^{-4} p.u.

2.7 THE JACOBIAN MATRIX

The square matrix in Equation 2.53

$$\mathbf{J} = \begin{bmatrix} \frac{\partial P_2}{\partial \theta_2} & \frac{\partial P_2}{\partial \theta_3} & \frac{\partial P_2}{\partial \theta_4} & \frac{\partial P_2}{\partial \theta_5} & \frac{\partial P_2}{\partial \theta_6} & \frac{\partial P_2}{\partial V_4} & \frac{\partial P_2}{\partial V_5} & \frac{\partial P_2}{\partial V_6} \\ \frac{\partial P_3}{\partial \theta_2} & \frac{\partial P_3}{\partial \theta_3} & \frac{\partial P_3}{\partial \theta_4} & \frac{\partial P_3}{\partial \theta_5} & \frac{\partial P_3}{\partial \theta_6} & \frac{\partial P_3}{\partial V_4} & \frac{\partial P_3}{\partial V_5} & \frac{\partial P_3}{\partial V_6} \\ \frac{\partial P_4}{\partial \theta_2} & \frac{\partial P_4}{\partial \theta_3} & \frac{\partial P_4}{\partial \theta_4} & \frac{\partial P_4}{\partial \theta_5} & \frac{\partial P_4}{\partial \theta_6} & \frac{\partial P_4}{\partial V_4} & \frac{\partial P_4}{\partial V_5} & \frac{\partial P_4}{\partial V_6} \\ \frac{\partial P_5}{\partial \theta_2} & \frac{\partial P_5}{\partial \theta_3} & \frac{\partial P_5}{\partial \theta_4} & \frac{\partial P_5}{\partial \theta_5} & \frac{\partial P_5}{\partial \theta_6} & \frac{\partial P_5}{\partial V_4} & \frac{\partial P_5}{\partial V_5} & \frac{\partial P_5}{\partial V_6} \\ \frac{\partial P_6}{\partial \theta_2} & \frac{\partial P_6}{\partial \theta_3} & \frac{\partial P_6}{\partial \theta_4} & \frac{\partial P_6}{\partial \theta_5} & \frac{\partial P_6}{\partial \theta_6} & \frac{\partial P_6}{\partial V_4} & \frac{\partial P_6}{\partial V_5} & \frac{\partial P_6}{\partial V_6} \\ \frac{\partial Q_4}{\partial \theta_2} & \frac{\partial Q_4}{\partial \theta_3} & \frac{\partial Q_4}{\partial \theta_4} & \frac{\partial Q_4}{\partial \theta_5} & \frac{\partial Q_4}{\partial \theta_6} & \frac{\partial Q_4}{\partial V_4} & \frac{\partial Q_4}{\partial V_5} & \frac{\partial Q_4}{\partial V_6} \\ \frac{\partial Q_5}{\partial \theta_2} & \frac{\partial Q_5}{\partial \theta_3} & \frac{\partial Q_5}{\partial \theta_4} & \frac{\partial Q_5}{\partial \theta_5} & \frac{\partial Q_5}{\partial \theta_6} & \frac{\partial Q_5}{\partial V_4} & \frac{\partial Q_5}{\partial V_5} & \frac{\partial Q_5}{\partial V_6} \\ \frac{\partial Q_6}{\partial \theta_2} & \frac{\partial Q_6}{\partial \theta_3} & \frac{\partial Q_6}{\partial \theta_4} & \frac{\partial Q_6}{\partial \theta_5} & \frac{\partial Q_6}{\partial \theta_6} & \frac{\partial Q_6}{\partial V_4} & \frac{\partial Q_6}{\partial V_5} & \frac{\partial Q_6}{\partial V_6} \end{bmatrix}$$

is known as the Jacobian matrix. The elements of this matrix comprise four major blocks

$$\mathbf{J} = \left[\begin{array}{c|c} \frac{\partial \mathbf{P}}{\partial \boldsymbol{\theta}} & \frac{\partial \mathbf{P}}{\partial \mathbf{V}} \\ \hline \frac{\partial \mathbf{Q}}{\partial \boldsymbol{\theta}} & \frac{\partial \mathbf{Q}}{\partial \mathbf{V}} \end{array}\right]$$

where:

$$\frac{\partial \mathbf{P}}{\partial \boldsymbol{\theta}}=\begin{bmatrix} \frac{\partial P_2}{\partial \theta_2} & \frac{\partial P_2}{\partial \theta_3} & \frac{\partial P_2}{\partial \theta_4} & \frac{\partial P_2}{\partial \theta_5} & \frac{\partial P_2}{\partial \theta_6} \\ \frac{\partial P_3}{\partial \theta_2} & \frac{\partial P_3}{\partial \theta_3} & \frac{\partial P_3}{\partial \theta_4} & \frac{\partial P_3}{\partial \theta_5} & \frac{\partial P_3}{\partial \theta_6} \\ \frac{\partial P_4}{\partial \theta_2} & \frac{\partial P_4}{\partial \theta_3} & \frac{\partial P_4}{\partial \theta_4} & \frac{\partial P_4}{\partial \theta_5} & \frac{\partial P_4}{\partial \theta_6} \\ \frac{\partial P_5}{\partial \theta_2} & \frac{\partial P_5}{\partial \theta_3} & \frac{\partial P_5}{\partial \theta_4} & \frac{\partial P_5}{\partial \theta_5} & \frac{\partial P_5}{\partial \theta_6} \\ \frac{\partial P_6}{\partial \theta_2} & \frac{\partial P_6}{\partial \theta_3} & \frac{\partial P_6}{\partial \theta_4} & \frac{\partial P_6}{\partial \theta_5} & \frac{\partial P_6}{\partial \theta_6} \end{bmatrix} \frac{\partial \mathbf{P}}{\partial \mathbf{V}}=\begin{bmatrix} \frac{\partial P_2}{\partial V_4} & \frac{\partial P_2}{\partial V_5} & \frac{\partial P_2}{\partial V_6} \\ \frac{\partial P_3}{\partial V_4} & \frac{\partial P_3}{\partial V_5} & \frac{\partial P_3}{\partial V_6} \\ \frac{\partial P_4}{\partial V_4} & \frac{\partial P_4}{\partial V_5} & \frac{\partial P_4}{\partial V_6} \\ \frac{\partial P_5}{\partial V_4} & \frac{\partial P_5}{\partial V_5} & \frac{\partial P_5}{\partial V_6} \\ \frac{\partial P_6}{\partial V_4} & \frac{\partial P_6}{\partial V_5} & \frac{\partial P_6}{\partial V_6} \end{bmatrix} \quad (2.56)$$

$$\frac{\partial \mathbf{Q}}{\partial \boldsymbol{\theta}}=\begin{bmatrix} \frac{\partial Q_4}{\partial \theta_2} & \frac{\partial Q_4}{\partial \theta_3} & \frac{\partial Q_4}{\partial \theta_4} & \frac{\partial Q_4}{\partial \theta_5} & \frac{\partial Q_4}{\partial \theta_6} \\ \frac{\partial Q_5}{\partial \theta_2} & \frac{\partial Q_5}{\partial \theta_3} & \frac{\partial Q_5}{\partial \theta_4} & \frac{\partial Q_5}{\partial \theta_5} & \frac{\partial Q_5}{\partial \theta_6} \\ \frac{\partial Q_6}{\partial \theta_2} & \frac{\partial Q_6}{\partial \theta_3} & \frac{\partial Q_6}{\partial \theta_4} & \frac{\partial Q_6}{\partial \theta_5} & \frac{\partial Q_6}{\partial \theta_6} \end{bmatrix}$$

$$\frac{\partial \mathbf{Q}}{\partial \mathbf{V}}=\begin{bmatrix} \frac{\partial Q_4}{\partial V_4} & \frac{\partial Q_4}{\partial V_5} & \frac{\partial Q_4}{\partial V_6} \\ \frac{\partial Q_5}{\partial V_4} & \frac{\partial Q_5}{\partial V_5} & \frac{\partial Q_5}{\partial V_6} \\ \frac{\partial Q_6}{\partial V_4} & \frac{\partial Q_6}{\partial V_5} & \frac{\partial Q_6}{\partial V_6} \end{bmatrix} \quad (2.57)$$

Let us begin with the first submatrix $\partial \mathbf{P}/\partial \boldsymbol{\theta}$ of the Jacobian matrix $\mathbf{J}$. It can be observed that a typical element of the submatrix has the form $\partial P_a/\partial \theta_b$, where $b = a$ or $b \neq a$. Because these elements have to be computed (first at the initial guess point and subsequently at the updated starting points obtained after every iteration), we now try to find the expression for each such element.

From Equation 2.29,

$$\begin{aligned} P_i &= \sum_{k=1}^{6} V_i V_k Y_{ik} \cos(\theta_i - \theta_k - \varphi_{ik}) \\ &= \sum_{k=1,\ k\neq i}^{6} V_i V_k Y_{ik} \cos(\theta_i - \theta_k - \varphi_{ik}) + V_i^2 Y_{ii} \cos\varphi_{ii} \end{aligned} \tag{2.58}$$

In a similar manner, from Equation 2.30,

$$\begin{aligned} Q_i &= \sum_{k=1}^{6} V_i V_k Y_{ik} \sin(\theta_i - \theta_k - \varphi_{ik}) \\ &= \sum_{k=1,\ k\neq i}^{6} V_i V_k Y_{ik} \sin(\theta_i - \theta_k - \varphi_{ik}) - V_i^2 Y_{ii} \sin\varphi_{ii} \end{aligned} \tag{2.59}$$

Thus, from Equations 2.58 and 2.59,

$$P_a = \sum_{k=1,\ k\neq i}^{6} V_a V_k Y_{ak} \cos(\theta_a - \theta_k - \varphi_{ak}) + V_a^2 Y_{aa} \cos\varphi_{aa} \tag{2.60}$$

$$Q_a = \sum_{k=1,\ k\neq i}^{6} V_a V_k Y_{ak} \sin(\theta_a - \theta_k - \varphi_{ak}) - V_a^2 Y_{aa} \sin\varphi_{aa} \tag{2.61}$$

From Equation 2.60, we have

$$\frac{\partial P_a}{\partial \theta_b} = V_a V_b Y_{ab} \sin(\theta_a - \theta_b - \varphi_{ab}) \tag{2.62}$$

Also,

$$\frac{\partial P_a}{\partial \theta_a} = -\sum_{k=1,\ k\neq i}^{6} V_a V_k Y_{ak} \sin(\theta_a - \theta_k - \varphi_{ak}) \tag{2.63}$$

From Equation 2.61, we can write

$$-\sum_{k=1,\,k\neq i}^{6} V_a V_k Y_{ak}\sin(\theta_a-\theta_k-\varphi_{ak}) = -Q_a - V_a^2 Y_{aa}\sin\varphi_{aa} \tag{2.64}$$

From Equations 2.63 and 2.64, we get

$$\frac{\partial P_a}{\partial \theta_a} = -\sum_{k=1,\,k\neq i}^{6} V_a V_k Y_{ak}\sin(\theta_a-\theta_k-\varphi_{ak}) = -Q_a - V_a^2 Y_{aa}\sin\varphi_{aa} \tag{2.65}$$

Thus, from Equations 2.62 and 2.65, we can write

$$\begin{aligned}\frac{\partial P_a}{\partial \theta_b} &= -Q_a - V_a^2 Y_{aa}\sin\varphi_{aa}, \quad \text{if } b=a \\ &= V_a V_b Y_{ab}\sin(\theta_a-\theta_b-\varphi_{ab}), \quad \text{if } b\neq a\end{aligned} \tag{2.66}$$

In a similar manner, we compute the elements for the second submatrix $\partial \mathbf{P}/\partial \mathbf{V}$ of $\mathbf{J}$.

From Equation 2.60, we get

$$\frac{\partial P_a}{\partial V_b} = V_a Y_{ab}\cos(\theta_a-\theta_b-\varphi_{ab}) \tag{2.67}$$

and

$$\frac{\partial P_a}{\partial V_a} = \sum_{k=1,\,k\neq i}^{6} V_k Y_{ak}\cos(\theta_a-\theta_k-\varphi_{ak}) + 2V_a Y_{aa}\cos\varphi_{aa} \tag{2.68}$$

However, from Equation 2.60, if all the terms are divided by V_a, then we have

$$\frac{P_a}{V_a} - V_a Y_{aa}\cos\varphi_{aa} = \sum_{k=1,\,k\neq i}^{6} V_k Y_{ak}\cos(\theta_a-\theta_k-\varphi_{ak}) \tag{2.69}$$

Substituting the summation terms of Equation 2.68 by Equation 2.69 gives

$$\frac{\partial P_a}{\partial V_a} = \frac{P_a}{V_a} + V_a Y_{aa}\cos\varphi_{aa} \tag{2.70}$$

Thus, from Equations 2.67 and 2.70, we can write

$$\frac{\partial P_a}{\partial V_b} = V_a Y_{ab} \cos(\theta_a - \theta_b - \varphi_{ab}), \quad \text{if } b \neq a$$
$$= \frac{P_a}{V_a} + V_a Y_{aa} \cos\varphi_{aa}, \quad \text{if } b = a \tag{2.71}$$

We are now left with only two of the Jacobian matrix subblocks, namely, $\partial\mathbf{Q}/\partial\boldsymbol{\theta}$ and $\partial\mathbf{Q}/\partial\mathbf{V}$.

From Equation 2.61,

$$\frac{\partial Q_a}{\partial \theta_b} = -V_a V_b Y_{ab} \cos(\theta_a - \theta_b - \varphi_{ab}) \tag{2.72}$$

and

$$\frac{\partial Q_a}{\partial \theta_a} = \sum_{k=1,\, k\neq i}^{6} V_a V_k Y_{ak} \cos(\theta_a - \theta_k - \varphi_{ak}) \tag{2.73}$$

From Equation 2.60, it can be observed that

$$P_a - V_a^2 Y_{aa} \cos\varphi_{aa} = \sum_{k=1,\, k\neq i}^{6} V_a V_k Y_{ak} \cos(\theta_a - \theta_k - \varphi_{ak}) \tag{2.74}$$

Substituting the summation terms of Equation 2.73 by Equation 2.74 gives

$$\frac{\partial Q_a}{\partial \theta_a} = P_a - V_a^2 Y_{aa} \cos\varphi_{aa} \tag{2.75}$$

Hence, from Equations 2.72 and 2.75,

$$\frac{\partial Q_a}{\partial \theta_b} = -V_a V_b Y_{ab} \cos(\theta_a - \theta_b - \varphi_{ab}), \quad \text{if } b \neq a$$
$$= P_a - V_a^2 Y_{aa} \cos\varphi_{aa}, \quad \text{if } b = a \tag{2.76}$$

We are now left with only the subblock $\partial\mathbf{Q}/\partial\mathbf{V}$.

From Equation 2.61,

$$\frac{\partial Q_a}{\partial V_b} = V_a Y_{ab} \sin(\theta_a - \theta_b - \varphi_{ab}) \tag{2.77}$$

and

$$\frac{\partial Q_a}{\partial V_a} = \sum_{k=1,\ k\neq i}^{6} V_k Y_{ak} \sin(\theta_a - \theta_k - \varphi_{ak}) - 2V_a Y_{aa} \sin\varphi_{aa} \tag{2.78}$$

Now, from Equation 2.61, if all the terms are divided by V_a, then we have

$$\frac{Q_a}{V_a} + V_a Y_{aa} \sin\varphi_{aa} = \sum_{k=1,\ k\neq i}^{6} V_k Y_{ak} \sin(\theta_a - \theta_k - \varphi_{ak}) \tag{2.79}$$

Substituting the summation terms of Equation 2.78 by Equation 2.79 gives

$$\frac{\partial Q_a}{\partial V_a} = \frac{Q_a}{V_a} - V_a Y_{aa} \sin\varphi_{aa} \tag{2.80}$$

Hence, from Equations 2.77 and 2.80,

$$\begin{aligned} \frac{\partial Q_a}{\partial V_b} &= V_a Y_{ab} \sin(\theta_a - \theta_b - \varphi_{ab}), \quad \text{if } b \neq a \\ &= \frac{Q_a}{V_a} - V_a Y_{aa} \sin\varphi_{aa}, \quad \text{if } b = a \end{aligned} \tag{2.81}$$

Thus, the typical elements of the four Jacobian subblocks can be summarized as shown in Table 2.1.

TABLE 2.1 Generalized Expression of Jacobian Elements

Jacobian Subblock	Typical Element	Expression
$\frac{\partial \mathbf{P}}{\partial \theta}$	$\frac{\partial P_a}{\partial \theta_b}$	$-Q_a - V_a^2 Y_{aa} \sin\phi_{aa}$ if $b = a$
		$V_a V_b Y_{ab} \sin(\theta_a - \theta_b - \phi_{ab})$ if $b \neq a$
$\frac{\partial \mathbf{P}}{\partial \mathbf{V}}$	$\frac{\partial P_a}{\partial V_b}$	$V_a Y_{ab} \cos(\theta_a - \theta_b - \phi_{ab})$ if $b \neq a$
		$\frac{P_a}{V_a} + V_a Y_{aa} \cos\phi_{aa}$ if $b = a$
$\frac{\partial \mathbf{Q}}{\partial \theta}$	$\frac{\partial Q_a}{\partial \theta_b}$	$-V_a V_b Y_{ab} \cos(\theta_a - \theta_b - \phi_{ab})$ if $b \neq a$
		$P_a - V_a^2 Y_{aa} \cos\phi_{aa}$ if $b = a$
$\frac{\partial \mathbf{Q}}{\partial \mathbf{V}}$	$\frac{\partial Q_a}{\partial V_b}$	$V_a Y_{ab} \sin(\theta_a - \theta_b - \phi_{ab})$ if $b \neq a$
		$\frac{Q_a}{V_a} - V_a Y_{aa} \sin\phi_{aa}$ if $b = a$

2.8 POWER FLOW SOLUTION: THE GENERALIZED FORM

Until Section 2.7, we based our analysis for a very small six-bus system. In reality, power systems are enormous, interconnecting hundreds of generators, transformers, and transmission lines, and thousands of buses. It would be correct at this stage to formulate the power flow equations for an n-bus system consisting of m generators ($m \le n$). Without any loss of generality, it is also assumed that there are m generators connected at the first m buses of this system. By convention then, the first bus (bus 1) will be assumed to be the slack bus, the next (m - 1) buses from bus 2 to bus m will be voltage-controlled (PV) buses, and the remaining ($n - m$) buses from bus (m + 1) to bus n will all be load (PQ) buses. Both the unknown variables and the specified quantities can be tabulated as shown in Table 2.2.

The power flow problem for the above n-bus system can be summarized as

Compute $\theta_2, \theta_3, \ldots, \theta_n, V_{m+1}, V_{m+2}, \ldots, V_n$
Given $P_2, P_3, \ldots, P_n, Q_{m+1}, Q_{m+2}, \ldots, Q_n$

TABLE 2.2 Types of Buses and Unknown/Specified Quantities

Bus No.	Type of Bus	Unknown Quantities (Bus Voltage Magnitudes and Angles)	Specified (Known) Quantities (Net Active and Reactive Power Injections at Buses)	Remarks
1	Slack or swing	–	– (I^2R losses are not known. Therefore, P_{D1} is specified but not P_{G1}. Also, Q_{D1} is specified but not Q_{G1})	V_1 is controlled (by AVR) to a specified value $\theta_1 = 0$ is taken as reference.
2 to m	Voltage-controlled (PV) buses	$\theta_2, \theta_3, \ldots, \theta_m$	$P_2, P_3, \ldots, P_m$ (both P_{Gi} and P_{Di} are specified. Q_{Di} is specified but not Q_{Gi}, $i = 2,3,\ldots,m$)	$V_2, V_3, \ldots, V_m$ are controlled (by AVR) to specified values.
(m + 1) to n	Load (PQ) buses	$\theta_{m+1}, \theta_{m+2}, \ldots, \theta_n$ $V_{m+1}, V_{m+2}, \ldots, V_n$	$P_{m+1}, P_{m+2}, \ldots, P_n$ $Q_{m+1}, Q_{m+2}, \ldots, Q_n$	No generation, so $P_{Gi} = 0$, $Q_{Gi} = 0$. However, P_{Di} and Q_{Di} ($i = 2,3,\ldots,m$) are specified.

The Newton power flow equation for the above system can be written as

$$\begin{bmatrix} \theta_2^{k+1} \\ \theta_3^{k+1} \\ \vdots \\ \theta_n^{k+1} \\ V_{m+1}^{k+1} \\ V_{m+2}^{k+1} \\ \vdots \\ V_n^{k+1} \end{bmatrix} = \begin{bmatrix} \theta_2^{k} \\ \theta_3^{k} \\ \vdots \\ \theta_n^{k} \\ V_{m+1}^{k} \\ V_{m+2}^{k} \\ \vdots \\ V_n^{k} \end{bmatrix} + \begin{bmatrix} \frac{\partial P_2}{\partial \theta_2} & \frac{\partial P_2}{\partial \theta_3} & \cdots & \frac{\partial P_2}{\partial \theta_n} & \frac{\partial P_2}{\partial V_{m+1}} & \frac{\partial P_2}{\partial V_{m+2}} & \cdots & \frac{\partial P_2}{\partial V_n} \\ \frac{\partial P_3}{\partial \theta_2} & \frac{\partial P_3}{\partial \theta_3} & \cdots & \frac{\partial P_3}{\partial \theta_n} & \frac{\partial P_3}{\partial V_{m+1}} & \frac{\partial P_3}{\partial V_{m+2}} & \cdots & \frac{\partial P_3}{\partial V_n} \\ \vdots & \vdots & \vdots & \vdots & \vdots & \vdots & \vdots & \vdots \\ \frac{\partial P_n}{\partial \theta_2} & \frac{\partial P_n}{\partial \theta_3} & \cdots & \frac{\partial P_n}{\partial \theta_n} & \frac{\partial P_n}{\partial V_{m+1}} & \frac{\partial P_n}{\partial V_{m+2}} & \cdots & \frac{\partial P_n}{\partial V_n} \\ \frac{\partial Q_{m+1}}{\partial \theta_2} & \frac{\partial Q_{m+1}}{\partial \theta_3} & \cdots & \frac{\partial Q_{m+1}}{\partial \theta_n} & \frac{\partial Q_{m+1}}{\partial V_{m+1}} & \frac{\partial Q_{m+1}}{\partial V_{m+2}} & \cdots & \frac{\partial Q_{m+1}}{\partial V_n} \\ \frac{\partial Q_{m+2}}{\partial \theta_2} & \frac{\partial Q_{m+2}}{\partial \theta_3} & \cdots & \frac{\partial Q_{m+2}}{\partial \theta_n} & \frac{\partial Q_{m+2}}{\partial V_{m+1}} & \frac{\partial Q_{m+2}}{\partial V_{m+2}} & \cdots & \frac{\partial Q_{m+2}}{\partial V_n} \\ \vdots & \vdots & \vdots & \vdots & \vdots & \vdots & \vdots & \vdots \\ \frac{\partial Q_n}{\partial \theta_2} & \frac{\partial Q_n}{\partial \theta_3} & \cdots & \frac{\partial Q_n}{\partial \theta_n} & \frac{\partial Q_n}{\partial V_{m+1}} & \frac{\partial Q_n}{\partial V_{m+2}} & \cdots & \frac{\partial Q_n}{\partial V_n} \end{bmatrix}_k^{-1} \times \begin{bmatrix} P_2^{SP} - P_2^x \\ P_3^{SP} - P_3^x \\ \vdots \\ P_n^{SP} - P_n^x \\ Q_{m+1}^{SP} - Q_{m+1}^x \\ Q_{m+2}^{SP} - Q_{m+2}^x \\ \vdots \\ Q_n^{SP} - Q_n^x \end{bmatrix}_k \tag{2.82}$$

In the above equation, the supercript k in the unknown quantities denotes their values in the kth computational iteration. Similarly, the subscript k in the Jacobian matrix indicates that its elements are computed corresponding to the values of the variables at the kth iteration. In a similar manner, the mismatch vector on the rightmost side is also computed corresponding to the values of the variables at the kth iteration.

In practice, typical power systems are enormous, comprising thousands of buses. For such systems, the size of the Jacobian matrix is very large. Its storage and inversion at every iterative step will call for huge memory requirement, computational effort, and time. In this regard, sparsity of the Jacobian matrix is utilized to reduce the storage requirements and computational effort drastically. Although sparsity techniques are beyond the scope of this book, the interested reader may refer to [29,31] for details.

2.9 SUMMARY

In this chapter, the Newton–Raphson algorithm is introduced. The application of this powerful method for solving nonlinear algebraic equations in single and multiple variables is demonstrated. Subsequently, the reader is introduced to the power flow problem and the application of the Newton–Raphson method for its solution. The Jacobian matrix and its different blocks have been described with the generalized form of its elements. Finally, the generalized form of the power flow problem is described.

In the subsequent chapters, the Newton power flow models of the various voltage-sourced converter (VSC)-based flexible AC transmission system controllers and the VSC-HVDC are discussed. The Newton power flow model of the SSSC is developed in Chapter 3.

CHAPTER 3

Newton Power Flow Model of the Static Synchronous Series Compensator

3.1 INTRODUCTION

A lot of research work has been carried out in the literature for developing efficient power flow algorithms for the static synchronous series compensator (SSSC) [49–51]. However, in all these works, it is observed that the incorporation of an SSSC in an existing Newton–Raphson algorithm greatly enhances the complexity of software codes. The voltage source representing the SSSC contributes new terms to the power flow equations of the sending end (SE) and receiving end (RE) buses of the line incorporating the SSSC. Moreover, a completely new equation for the real power handled by the SSSC comes into the picture. All this requires modification of the existing Newton–Raphson codes as well as the development of fresh codes. The same is also true of the Jacobian matrix, in which entirely new Jacobian subblocks related to the SSSC come into the picture. For each of these subblocks, fresh codes need to be written. The problem of development of new codes increases manifold when the number of SSSCs in a system increases.

Existing works on power flow modeling of an SSSC do not directly address this problem. To address this issue, a novel modeling approach for the SSSC [84,85] is proposed in this chapter. By this modeling approach, an existing power system installed with SSSCs is transformed into an equivalent augmented network without any SSSC. This results in a substantial reduction in the programming complexity because of the following reasons:

1. In the absence of any SSSC, the expressions for the power injections at the buses associated with the SSSC no longer contain any terms contributed from the SSSCs. Thus, all bus power injections can be computed in the proposed model using the existing Newton–Raphson codes.
2. The equations for the SSSC real powers do not exist anymore. The SSSCs are transformed to additional power-flow buses, power flow equations of which can be computed using existing codes.
3. Only three Jacobian subblocks need to be evaluated in the proposed model. Two of these subblocks can be computed using existing Jacobian codes directly, whereas the third one can be computed with very minor modifications of the existing Jacobian codes.

Furthermore, similar to the other models already reported in the literature, the proposed model can also handle multiple control functions of the SSSCs such as control of bus voltage, line active power, line reactive power, and line reactance. This proposed model can also account for various device limit constraints of the SSSC. The developed model is described in Sections 3.2 through 3.5 in detail. It is to be noted that in this chapter as well as in the subsequent chapters, boldfaced quantities are used to denote complex variables and equivalent pi models are used to represent transmission lines. Also, all the quantities shown in this chapter as well as in the subsequent chapters carry their usual meanings.

3.2 SSSC MODEL FOR NEWTON POWER FLOW ANALYSIS

Figure 3.1 shows an SSSC connected in the branch i–j between buses i and j of an existing n-bus power system network. The series impedance of the branch i–j alone is $\mathbf{Z}_{ij}$ (not shown in the figure). The equivalent circuit of the above network is shown in Figure 3.2 with the simplification that the two half-line charging shunt admittances are shown connected at the two

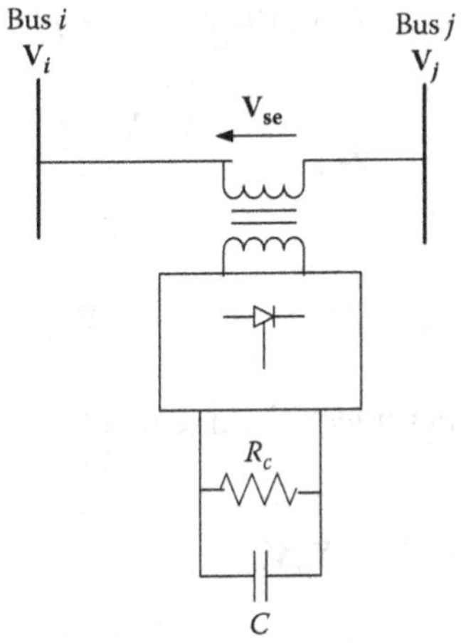

FIGURE 3.1 SSSC connected between buses i and j of an n-bus power system.

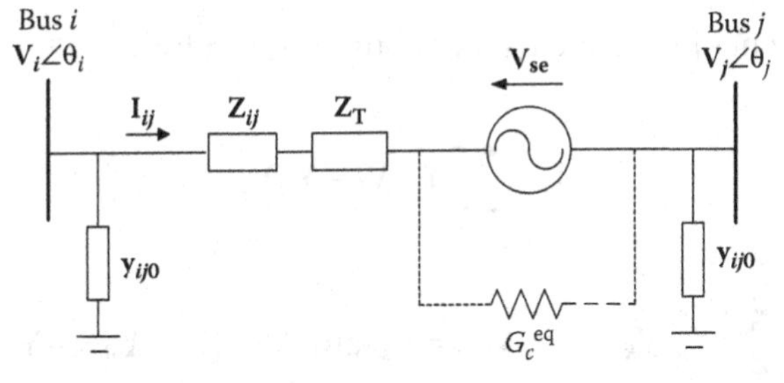

FIGURE 3.2 Equivalent circuit of the line incorporating SSSC.

ends of the line—SSSC combination. This little simplification makes the computation of the Y-bus matrix quite simple, as this case is equivalent to augmenting the corresponding line series impedance by the coupling transformer impedance.

In Figure 3.2, the voltages at buses i and j are represented as $\mathbf{V}_i$ and $\mathbf{V}_j$, respectively. The SSSC is represented by a voltage source $\mathbf{V}_{se}$ in series with the coupling transformer impedance $\mathbf{Z}_{\mathrm{T}}$. In the figure, let us define $\mathbf{Z}_{ij} = R_{ij} + jX_{ij}$, $\mathbf{Z}_{\mathrm{T}} = R_{\mathrm{T}} + jX_{\mathrm{T}}$, $\mathbf{Zc}_{ij} = \mathbf{Z}_{ij} + \mathbf{Z}_{\mathrm{T}}$, $\mathbf{y}_{ij} = 1/\mathbf{Zc}_{ij} = \mathbf{y}_{ji}$, and $\mathbf{Y}_{ij} = \mathbf{Y}_{ji} = -\mathbf{y}_{ij}$. As shown in the figure, the current $\mathbf{I}_{ij}$ flowing in the branch i–j is the same as the current drawn by the SSSC, which will be represented as $\mathbf{I}_{se}$. This current is in quadrature with $\mathbf{V}_{se}$. It is to be noted that an SSSC cannot supply any active power (as it does not have any active power source) but can only supply reactive power.

From Figure 3.2, the current in the branch i–j is given by

$$\mathbf{I}_{ij} = \mathbf{I}_{se} = \frac{\mathbf{V}_i - \mathbf{V}_j - \mathbf{V}_{se}}{\mathbf{Zc}_{ij}} = \mathbf{Y}_{ij}(\mathbf{V}_j - \mathbf{V}_i) + \mathbf{Y}_{ij}\mathbf{V}_{se} \tag{3.1}$$

Similarly,

$$\mathbf{I}_{ji} = \mathbf{Y}_{ij}(\mathbf{V}_i - \mathbf{V}_j) - \mathbf{Y}_{ij}\mathbf{V}_{se} \tag{3.2}$$

From Equation 3.1, the net injected current at bus i is

$$\mathbf{I}_i = \sum_{k=1}^{n} \mathbf{Y}_{ik}\,\mathbf{V}_k + \mathbf{Y}_{ij}\mathbf{V}_{se} \tag{3.3}$$

$$= \sum_{k=1}^{n+1} \mathbf{Y}_{ik}\,\mathbf{V}_k \quad \text{if} \quad \mathbf{V}_{n+1} = \mathbf{V}_{se} \text{ and } \mathbf{Y}_{i(n+1)} = \mathbf{Y}_{ij} \tag{3.4}$$

Similarly, the net injected current at bus j is given by

$$\mathbf{I}_j = \sum_{k=1}^{n} \mathbf{Y}_{jk}\,\mathbf{V}_k - \mathbf{Y}_{ji}\mathbf{V}_{se} \tag{3.5}$$

$$= \sum_{k=1}^{n+1} \mathbf{Y}_{jk}\,\mathbf{V}_k \quad \text{if} \quad \mathbf{V}_{n+1} = \mathbf{V}_{se} \text{ and } \mathbf{Y}_{j(n+1)} = -\mathbf{Y}_{ji} = -\mathbf{Y}_{ij} \tag{3.6}$$

Thus, from Equations 3.4 and 3.6, it is observed that the effect of the SSSC series converter is equivalent to an additional bus in the existing network. The magnitude and angle of the voltage of this $(n + 1)$th fictitious bus are equal to the magnitude and angle of the representative series voltage source $\mathbf{V}_{se}$ of the SSSC.

Now, from Figure 3.2, the net injected current at this fictitious $(n + 1)$th bus equals the current flowing into the transmission system from this bus and is given by

$$\mathbf{I}_{n+1} = -\mathbf{I}_{se} = \mathbf{Y}_{ij}(\mathbf{V}_i - \mathbf{V}_j) - \mathbf{Y}_{ij}\mathbf{V}_{se} = \sum_{k=1}^{n+1} \mathbf{Y}_{(n+1)k}\,\mathbf{V}_k \tag{3.7}$$

or

$$\mathbf{I}_{n+1} = \sum_{k=1}^{n+1} \mathbf{Y}_{(n+1)k}\,\mathbf{V}_k \tag{3.8}$$

$$\text{if } \mathbf{Y}_{(n+1)i} = \mathbf{Y}_{ij},\ \mathbf{Y}_{(n+1)j} = -\mathbf{Y}_{ij},\ \mathbf{Y}_{(n+1)(n+1)} = -\mathbf{Y}_{ij} \text{ as } \mathbf{V}_{n+1} = \mathbf{V}_{se} \tag{3.9}$$

Thus, it is observed that in the proposed model, the existing n bus power system network with the SSSC is transformed into an $(n + 1)$ bus network without any SSSC. In this transformed system, the net injected current at any bus g (g can be any sending or receiving end bus or a bus representing the SSSC) can be expressed as

$$\mathbf{I}_g = \sum_{k=1}^{n+1} \mathbf{Y}_{gk}\,\mathbf{V}_k,\quad 1 \le g \le (n+1) \tag{3.10}$$

In general, in the proposed model, an existing n bus network containing p SSSCs is transformed into an $(n + p)$ bus network without any SSSC and the expression for the net injected current at any bus h of the network would be

$$\mathbf{I}_h = \sum_{k=1}^{n+p} \mathbf{Y}_{hk}\,\mathbf{V}_k,\quad 1 \le h \le (n+p) \tag{3.11}$$

In particular, the mth $\{1 \le m \le p\}$ SSSC connected between buses u (sending end (SE)) and v (receiving end (RE)) is transformed to the $(n + m)$th bus and the net injected current at this mth fictitious bus is given by

$$\mathbf{I}_{n+m} = \sum_{k=1}^{n+p} \mathbf{Y}_{(n+m)k}\,\mathbf{V}_k \tag{3.12}$$

where

$$\mathbf{Y}_{(n+m)u} = \mathbf{Y}_{uv},\ \mathbf{Y}_{(n+m)v} = -\mathbf{Y}_{uv},\ \mathbf{Y}_{(n+m)(n+m)} = -\mathbf{Y}_{uv} \text{ and } \mathbf{V}_{n+m} = \mathbf{V}_{sem}$$

3.3 POWER FLOW EQUATIONS IN THE PROPOSED SSSC MODEL

With existing SSSC models in the literature, the net active power injection at any bus a $\{1 \le a \le (n+p)\}$ in an n bus system with p SSSCs can be written from Equations 3.3 and 3.5 as

$$\begin{aligned} P_a = \mathrm{Re}\{\mathbf{V}_a\mathbf{I}_a^*\} = \sum_{k=1}^{n} V_a V_k Y_{ak} \cos\{\theta_a - \theta_k - \varphi_{yak}\} \\ + V_a V_{sec} Y_{ab} \cos\{\theta_a - \theta_{sec} - \varphi_{yab}\} \end{aligned} \tag{3.13}$$

if SSSC c is connected between buses a and b with a as SE bus

$$=\sum_{k=1}^{n} V_a V_k Y_{ak} \cos\{\theta_a - \theta_k - \varphi_{yak}\} - V_a V_{sec} Y_{ab} \cos\{\theta_a - \theta_{sec} - \varphi_{yab}\} \quad (3.14)$$

if SSSC c is connected between buses a and b with a as RE bus

$$=\sum_{k=1}^{n} V_a V_k Y_{ak} \cos\{\theta_a - \theta_k - \varphi_{yak}\} \quad (3.15)$$

if no SSSC is connected to bus a

Thus, from the above equation it is observed that with existing SSSC models, contributions from the voltage source representing the SSSC necessitate modifications in the bus power injection equations, which call for consequent modifications in the existing Newton–Raphson software codes.

In the proposed model, the n bus system with p SSSCs is transformed to a $(n + p)$ bus network without any SSSC and the net active power injection at any bus a $\{1 \leq a \leq (n+p)\}$ can be written using Equations 3.4 and 3.6 as

$$P_a = \mathrm{Re}\{\mathbf{V}_a \mathbf{I}_a^*\} = \mathrm{Re}\{\mathbf{V}_a \sum_{k=1}^{n+p} \mathbf{Y}_{ak}^* \mathbf{V}_k^*\} = \sum_{k=1}^{n+p} V_a V_k Y_{ak} \cos\{\theta_a - \theta_k - \varphi_{yak}) \quad (3.16)$$

In a similar way, it can be shown that

$$Q_a = \sum_{k=1}^{n+p} V_a V_k Y_{ak} \sin\{\theta_a - \theta_k - \varphi_{yak}) \quad (3.17)$$

Thus, in the proposed model, it is observed that both active and reactive power injection equations at any bus can be computed using the existing Newton–Raphson codes.

Now, with existing SSSC models, the expression for the real power delivered by any SSSC c $\{1 \leq c \leq p\}$, connected between any two buses a and b (with a as SE bus) can be written as

$$\begin{aligned} P_{sec} &= \mathrm{Re}[\mathbf{V}_{sec}\{-\mathbf{I}_{sec}^*\}] \\ &= V_{sec} V_a Y_{ab} \cos\{\theta_{sec} - \theta_a - \varphi_{yab}\} - V_{sec} V_b Y_{ab} \cos\{\theta_{sec} - \theta_b - \varphi_{yab}\} \\ &\quad - V_{sec}^2 Y_{ab} \cos\varphi_{yab} \end{aligned} \quad (3.18)$$

From Equation 3.18 it is observed that entirely new terms come into the picture in the expression of $P_{se\,c}$ because of the SSSC, and consequently fresh codes are required for its evaluation. In the proposed model, the SSSC c is transformed to a fictitious power-flow bus ($n + c$), and using Equation 3.7, the expression for the real power delivered by SSSC c is

$$P_{se\,c} = \text{Re}[\mathbf{V}_{\mathbf{se}\,c}\{-\mathbf{I}^*_{\mathbf{se}\,c}\}]$$

$$= \text{Re}[\mathbf{V}_{n+c}\{\mathbf{I}^*_{n+c}\}] = P_{n+c} = \sum_{k=1}^{n+p} V_{n+c}V_kY_{(n+c)k}\cos\{\theta_{n+c} - \theta_k - \varphi_{y(n+c)k}\} \tag{3.19}$$

From Equation 3.19, it can be observed that in the proposed model, the active power injection at the (n + c)th fictitious power-flow bus equals the real power of the cth SSSC, which can be computed using the existing Newton–Raphson power flow codes.

Now, since the SSSC accommodates multiple control functions such as control of bus voltage, line active power, line reactive power, and line reactance, the question of additional complexities of software codes for their implementation also needs to be addressed.

From Figure 3.2, the expressions for the active and reactive power flows in any line containing SSSC can be written for both existing and the proposed SSSC models. With the existing models, the expressions for line active and reactive power flow with any SSSC c connected in a line between any two buses a (SE) and b (RE) can be written using Equation 3.1 as

$$P_{ab} = \text{Re}\{\mathbf{V}_a\mathbf{I}^*_{ab}\} = \text{Re}\{\mathbf{V}_a\mathbf{I}^*_{\mathbf{se}\,c}\}$$

or

$$P_{ab} = V_aV_bY_{ab}\cos\{\theta_a - \theta_b - \varphi_{yab}\} - V_a^2Y_{ab}\cos\varphi_{yab} + V_aV_{se\,c}Y_{ab}\cos\{\theta_a - \theta_{se\,c} - \varphi_{yab}\} \tag{3.20}$$

Similarly,

$$Q_{ab} = V_aV_bY_{ab}\sin\{\theta_a - \theta_b - \varphi_{yab}\} + V_a^2Y_{ab}\sin\varphi_{yab} + V_aV_{se\,c}Y_{ab}\sin\{\theta_a - \theta_{se\,c} - \varphi_{yab}\} \tag{3.21}$$

Both Equations 3.20 and 3.21 have a contribution term from the series voltage source representing the SSSC series converter, and therefore modifications in the existing software codes are required for their implementation. In the proposed model, these expressions become

$$P_{ab} = \mathrm{Re}\{\mathbf{V}_a \mathbf{I}_{ab}^*\} = \mathrm{Re}\{\mathbf{V}_a \mathbf{I}_{sec}^*\} = \mathrm{Re}[\mathbf{V}_a\{-\mathbf{I}_{n+c}^*\}]$$

or

$$P_{ab} = -\sum_{k=1}^{n+p} V_a V_k Y_{(n+c)k} \cos\{\theta_a - \theta_k - \varphi_{y(n+c)k}\} \tag{3.22}$$

as $\mathbf{I}_{n+c} = -\mathbf{I}_{sec} = \sum_{k=1}^{n+p} \mathbf{Y}_{(n+c)k}\, \mathbf{V}_k$ in the proposed model (using Equation 3.7)

Similarly

$$Q_{ab} = -\sum_{k=1}^{n+p} V_a V_k Y_{(n+c)k} \sin\{\theta_a - \theta_k - \varphi_{y(n+c)k}\} \tag{3.23}$$

From Equations 3.22 and 3.23, it is observed that both the line active and reactive power flow can be evaluated in the proposed model using very minor modifications in the existing power flow codes.

3.4 IMPLEMENTATION IN NEWTON POWER FLOW ANALYSIS

If the number of voltage controlled buses is (m-1), the power-flow problem for a n bus system with p SSSCs can be formulated as follows.

Solve,

$$\begin{gathered}\theta = [\theta_2 \; \ldots\ldots \; \theta_n]^T, \mathbf{V} = [V_{m+1} \; \ldots\ldots \; V_n]^T, \boldsymbol{\theta}_{\mathbf{se}} = [\theta_{se1} \; \ldots\ldots \; \theta_{se\,p}]^T, \\ \mathbf{V}_{\mathbf{se}} = [V_{se1} \; \ldots\ldots \; V_{se\,p}]^T\end{gathered} \tag{3.24}$$

Specified,

$$\begin{gathered}\mathbf{P} = [P_2 \; \ldots\ldots P_n]^T, \mathbf{Q} = [Q_{m+1} \; \ldots\ldots \; Q_n]^T, \mathbf{P}_{\mathbf{se}} = [P_{se1} \; \ldots\ldots \; P_{se\,p}]^T, \\ \mathbf{R} = [R_1 \; \ldots\ldots \; R_P]^T\end{gathered} \tag{3.25}$$

where $\boldsymbol{P}_{se}$ and $\boldsymbol{R}$ represent the vectors for the specified real powers and the control modes or device limit constraint specifications of the p SSSCs, respectively. For this formulation, it has been assumed that the m generators are connected at the first m buses of the system with bus 1 being the slack bus. Thus, the basic power-flow Equation for the Newton power flow solution would be represented as

$$\begin{bmatrix} \frac{\partial P}{\partial \theta} & \frac{\partial P}{\partial V} & \frac{\partial P}{\partial \theta_{se}} & \frac{\partial P}{\partial V_{se}} \\ \frac{\partial Q}{\partial \theta} & \frac{\partial Q}{\partial V} & \frac{\partial Q}{\partial \theta_{se}} & \frac{\partial Q}{\partial V_{se}} \\ \frac{\partial P_{se}}{\partial \theta} & \frac{\partial P_{se}}{\partial V} & \frac{\partial P_{se}}{\partial \theta_{se}} & \frac{\partial P_{se}}{\partial V_{se}} \\ \frac{\partial R}{\partial \theta} & \frac{\partial R}{\partial V} & \frac{\partial R}{\partial \theta_{se}} & \frac{\partial R}{\partial V_{se}} \end{bmatrix} \begin{bmatrix} \Delta\theta \\ \Delta V \\ \Delta\theta_{se} \\ \Delta V_{se} \end{bmatrix} = \begin{bmatrix} \Delta P \\ \Delta Q \\ \Delta P_{se} \\ \Delta R \end{bmatrix} \tag{3.26}$$

or

$$\begin{bmatrix} \multicolumn{2}{c}{\multirow{2}{*}{J_{old}}} & J_{se1} & J_{se3} \\ & & J_{se2} & J_{se4} \\ J_{se5} & J_{se6} & J_{se7} & J_{se8} \\ J_{se9} & J_{se10} & J_{se11} & J_{se12} \end{bmatrix} \begin{bmatrix} \Delta\theta \\ \Delta V \\ \Delta\theta_{se} \\ \Delta V_{se} \end{bmatrix} = \begin{bmatrix} \Delta P \\ \Delta Q \\ \Delta P_{se} \\ \Delta R \end{bmatrix} \tag{3.27}$$

In Equation 3.27,

$$J_{old} = \begin{bmatrix} \frac{\partial P}{\partial \theta} & \frac{\partial P}{\partial V} \\ \frac{\partial Q}{\partial \theta} & \frac{\partial Q}{\partial V} \end{bmatrix}$$

is the conventional power-flow Jacobian subblock corresponding to the angle and voltage magnitude variables of the n buses. The other Jacobian submatrices can be identified easily from Equations 3.26 and 3.27.

Now, in the proposed model, there would be $(n + p)$ buses. Thus, the quantities to be solved for power-flow are θ^{new} and $\mathbf{V}^{new}$, where

$$\theta^{new} = [\theta_2 \; ..... \; \theta_{n+p}]^T \text{ and } \mathbf{V}^{new} = [V_{m+1} \; ..... \; V_{n+p}]^T \tag{3.28}$$

Thus, Equation 3.26 is transformed in the proposed model as

$$\begin{bmatrix} \mathbf{JX1} \\ --- \\ \mathbf{JX2} \\ --- \\ \mathbf{JX3} \end{bmatrix} \begin{bmatrix} \Delta\boldsymbol{\theta}^{\text{new}} \\ ----- \\ \Delta\mathbf{V}^{\text{new}} \end{bmatrix} = \begin{bmatrix} \Delta\mathbf{PQ} \\ ---- \\ \Delta\mathbf{P}_{\text{se}} \\ ---- \\ \Delta\mathbf{R} \end{bmatrix} \tag{3.29}$$

where

$$\Delta\mathbf{PQ} = [\,\Delta\mathbf{P}^{\text{T}}\ \Delta\mathbf{Q}^{\text{T}}\,]^{\text{T}} \tag{3.30}$$

Also, matrices **JX1**, **JX2**, and **JX3** are identified easily from Equation 3.29.

Now, it can be shown that in Equation 3.29,

1. The matrices **JX1** and **JX2** can be computed using the existing Jacobian codes.
2. The matrix **JX3** can be computed using very minor modifications of the existing Jacobian codes.

The justification of the above two statements are shown as follows.

For computation of matrices **JX1** and **JX2** in the proposed model with $(n + p)$ buses, let

$$\mathbf{P}^{\text{new}} = [P_2\ .....\ P_{n+p}]^T = [\mathbf{P}^T\ \ \mathbf{P}_{\text{se}}^T]^T \tag{3.31}$$

Equation 3.31 has been written by noting that from Equations 3.19 and 3.25, $\mathbf{P}_{\text{se}} = [P_{\text{se}1}\ .....\ P_{sep}]^T = [P_{n+1}\ .....\ P_{n+p}]^T$

Subsequently, a new Jacobian matrix is computed as

$$\mathbf{J}^{\text{new}} = \begin{bmatrix} \dfrac{\partial\mathbf{P}^{\text{new}}}{\partial\boldsymbol{\theta}^{\text{new}}} & \dfrac{\partial\mathbf{P}^{\text{new}}}{\partial\mathbf{V}^{\text{new}}} \\ ----- & ----- \\ \dfrac{\partial\mathbf{Q}}{\partial\boldsymbol{\theta}^{\text{new}}} & \dfrac{\partial\mathbf{Q}}{\partial\mathbf{V}^{\text{new}}} \end{bmatrix} = \begin{bmatrix} \mathbf{J}^{\text{A}} \\ -- \\ \mathbf{J}^{\text{B}} \end{bmatrix} \tag{3.32}$$

It can be observed (from Equations 3.16 and 3.17) that in the proposed model, the expressions for P_i $(2 \le i \le n+p)$ and Q_i $(m+1 \le i \le n)$ in $\mathbf{P}^{\text{new}}$ and $\mathbf{Q}$ respectively (of Equation 3.32), can be computed using the existing power flow codes. Hence, the matrix $\mathbf{J}^{\text{new}}$ can be computed with the existing codes for the Jacobian.

Now, from Equations 3.28 and 3.31,

$$\frac{\partial\mathbf{P}^{\text{new}}}{\partial\boldsymbol{\theta}^{\text{new}}} = \begin{bmatrix} \dfrac{\partial\mathbf{P}}{\partial\boldsymbol{\theta}^{\text{new}}} \\ ----- \\ \dfrac{\partial\mathbf{P}_{\text{se}}}{\partial\boldsymbol{\theta}^{\text{new}}} \end{bmatrix} \tag{3.33}$$

From Equation 3.33, it can be observed that the matrix $\partial \mathbf{P}/\partial \boldsymbol{\theta}^{new}$ is contained within $\partial \mathbf{P}^{new}/\partial \boldsymbol{\theta}^{new}$. In a similar way, it can be shown that $\partial \mathbf{P}/\partial \mathbf{V}^{new}$ is contained within $\partial \mathbf{P}^{new}/\partial \mathbf{V}^{new}$. Hence, once the matrix $\mathbf{J}^{new}$ is computed, it can be shown that the matrix

$$\mathbf{JX1} = \begin{bmatrix} \dfrac{\partial \mathbf{P}}{\partial \boldsymbol{\theta}^{new}} & \dfrac{\partial \mathbf{P}}{\partial \mathbf{V}^{new}} \\ \hdashline \multicolumn{2}{c}{\mathbf{J}^{B}} \end{bmatrix}$$

can be very easily extracted from the matrix $\mathbf{J}^{new}$ using elementary matrix extraction codes only. Hence, no fresh codes need to be written for computing **JX1**.

Again, it is observed from Equation 3.33 that $\partial \mathbf{P}_{se}/\partial \boldsymbol{\theta}^{new}$ is contained within $\partial \mathbf{P}^{new}/\partial \boldsymbol{\theta}^{new}$. In a similar way, it can be shown that $\partial \mathbf{P}_{se}/\partial \mathbf{V}^{new}$ is contained within $\partial \mathbf{P}^{new}/\partial \mathbf{V}^{new}$. Thus, the matrix

$$\mathbf{JX2} = \begin{bmatrix} \dfrac{\partial \mathbf{P}_{se}}{\partial \boldsymbol{\theta}^{new}} & \dfrac{\partial \mathbf{P}_{se}}{\partial \mathbf{V}^{new}} \end{bmatrix}$$

can also be extracted from the matrix $\mathbf{J}^{new}$. Thus, both matrices **JX1** and **JX2** need not be computed and require only extraction from the matrix $\mathbf{J}^{new}$ using simple codes for matrix extraction.

For computation of the matrix **JX3**, the vector **R** (in Equation 3.29) needs to be specified. Now it is to be noted that for computing the voltage magnitude and angle of any SSSC, two quantities need to be specified. For the cth SSSC, if it is lossless, its real power P_{sec} is specified to be zero, thereby leaving the SSSC with only one degree of freedom. Thus, the quantity R_C {for the cth SSSC} signifies the other quantity that needs to be specified. Depending on the operating condition, there exist two distinct possibilities:

1. The SSSC is operating within its operational constraints.
2. One or more device limit(s) of the SSSC is (are) violated.

These two cases are now discussed in detail below.

3.4.1 SSSC Is Operating within Its Operational Constraints

Under this condition, no device constraint limits are violated, and the SSSC can be used to control the bus voltage, the line active power, the line reactive power, or the reactance of a line. The control mode specification R_C (in Equation 3.29) for all the above cases, pertaining to a SSSC connected between SE and RE buses i and j respectively, are given below:

1. *Line active power control*

   In this case,

$$P_{ij} = P_{ij}^{SP} \text{ or } P_{ji} = P_{ji}^{SP} \tag{3.34}$$

   For this control mode, the matrix **JX3** can be computed using very minor modifications of the existing Jacobian codes. This is shown below:

   From Equation 3.22, the real power flow in the line between buses i and j incorporating SSSC k is given by,

$$P_{ij} = -\sum_{q=1}^{n+p} V_i V_q Y_{(n+k)q} \cos\{\theta_i - \theta_q - \phi_{y(n+k)q}\} \tag{3.35}$$

   Similarly,

$$P_{ji} = \sum_{q=1}^{n+p} V_j V_q Y_{(n+k)q} \cos\{\theta_j - \theta_q - \phi_{y(n+k)q}\} \tag{3.36}$$

   Equations 3.35 and 3.36 show that the line active power flows can be computed using very minor modifications of the existing power flow codes. Consequently, the matrix **JX3,** which constitutes the partial derivatives of Equations 3.35 or 3.36 with respect to the relevant variables, can also be computed with very minor modifications of the existing Jacobian codes.

2. *Line reactive power control*

   In this case,

$$Q_{ij} = Q_{ij}^{SP} \text{ or } Q_{ji} = Q_{ji}^{SP} \tag{3.37}$$

For this control mode also, the matrix **JX3** can be computed using very minor modifications of the existing Jacobian codes as shown below:

From Equation 3.23, the reactive power flow in the line between buses i and j incorporating SSSC k is given by

$$Q_{ij} = -\sum_{q=1}^{n+p} V_i V_q Y_{(n+k)q} \sin\{\theta_i - \theta_q - \varphi_{y(n+k)q}\} \tag{3.38}$$

Similarly

$$Q_{ji} = \sum_{q=1}^{n+p} V_j V_q Y_{(n+k)q} \sin\{\theta_j - \theta_q - \varphi_{y(n+k)q}\} \tag{3.39}$$

Thus, similar to the case of the line active power flows, the line reactive power flows and the associated jacobian matrix (matrix **JX3**) can be computed using very minor modifications of the existing power flow and Jacobian codes, respectively.

3. *Bus voltage control*

In this case,

$$V_i = V_i^{SP} \text{ or } V_j = V_j^{SP} \tag{3.40}$$

For this control mode, the vector **R** constitutes the voltage magnitudes of p sending or receiving end buses. Consequently, the elements of the matrix **JX3** are either unity or zero, depending on whether an element of the vector **R** is also an element of the vector $\mathbf{V}^{\text{new}}$ or not. Therefore, in this case, the matrix **JX3** is a constant matrix, which is known a priori.

4. *Line reactance control*

In this control mode, the effective reactance of the SSSC is maintained at a specified value, that is,

$$X_{se} = X_{se}^{SP} \tag{3.41}$$

where

$$X_{se} = \text{Im}\{\mathbf{Z}_{se}\} = \text{Im}\left\{\frac{\mathbf{V}_{se}}{\mathbf{I}_{se}}\right\} = \frac{c_1}{c_2} \tag{3.42}$$

and

$$c_1 = V_i V_{se}\sin\{\theta_i - \theta_{se} + \varphi_{yij}\} - V_j V_{se}\sin\{\theta_j - \theta_{se} + \phi_{yij}\} - V_{se}^2 \sin\varphi_{yij}$$

$$c_2 = Y_{ij}[V_i^2 + V_j^2 + V_{se}^2 - 2V_i V_j \cos\{\theta_i - \theta_j\} - 2V_i V_{se}\cos\{\theta_i - \theta_{se}\} + 2V_j V_{se}\cos\{\theta_j - \theta_{se}\}] \quad (3.43)$$

Elements of the matrix **JX3** are computed accordingly. The derivation of Equation 3.43 is given in the Appendix.

3.4.2 Device Limit Constraints of the SSSC Are Violated

In this case, one or more practical device limit constraint(s) of the SSSC is (are) violated. The two major device limit constraints of the SSSC considered in this chapter are as follows:

1. The injected series (converter) voltage magnitude (V_{se}^{Lim})
2. The magnitude of the line current through the converter (I_{se}^{Lim})

The above two device limit constraints have been taken into account following [49]. The device limit constraints have been accommodated by the principle [70] that whenever a particular constraint limit is violated, it is kept at its specified limit (for the rest of the computation process), while a control objective of the SSSC is relaxed. Mathematically, this signifies replacing the control objectives (V_{bus}^{SP}, Q_{Line}^{SP}, X_{se}^{SP} or P_{Line}^{SP}) by the corresponding limits violated (V_{se}^{Lim} or I_{se}^{Lim}) during the formulation of the Jacobian matrix. The vector $\Delta\mathbf{R}$ (in Equation 3.29) now constitutes the device constraint limit mismatch (es). The control strategies to incorporate the above two limits are detailed below:

1. *Limit on* V_{se}

   In this case,

$$V_{se} = V_{se}^{Lim} \quad (3.44)$$

If V_{se}^{Lim} is violated for the mth SSSC, V_{n+m} is fixed at the corresponding limit and the SSSC control objective (line active or reactive power flow, the voltage at a bus or the line reactance) is relaxed. The relaxed control objective mismatch is replaced by $\Delta V_{n+m} = V_{sem}^{Lim} - V_{n+m}$ and the mth row of the matrix **JX3** is rendered constant with all elements known a priori – all of them are equal to zero except the entry corresponding to V_{n+m}

which is unity. If V_{se}^{Lim} is not violated, the control objective for the mth SSSC is retained and the elements corresponding to the mth row of **JX3** are computed using existing codes, as shown earlier. A violation of V_{se}^{Lim} for all the p SSSCs would render **JX3** a predetermined, constant matrix

2. *Limit on I_{se}*

In this case,

$$I_{se} = I_{se}^{Lim} \tag{3.45}$$

where I_{se} can be computed from Equation 3.1 as

$$\begin{aligned} I_{se} = Y_{ij}[V_i^2 + V_j^2 + V_{se}^2 - 2V_i V_j \cos\{\theta_i - \theta_j\} - 2V_i V_{se} \cos\{\theta_i - \theta_{se}\} \\ + 2V_j V_{se} \cos\{\theta_j - \theta_{se}\}]^{1/2} \end{aligned} \tag{3.46}$$

If I_{se}^{Lim} is violated for the mth SSSC, I_{n+m} is fixed at the corresponding limit and the SSSC control objective is relaxed. The relaxed control objective mismatch is replaced by $\Delta I_{n+m} = I_{sem}^{Lim} - I_{n+m}$. If I_{se}^{Lim} is not violated, the control objective for the mth SSSC is retained and the elements corresponding to the mth row of **JX3** are computed using existing codes, as shown earlier.

3.5 INCLUSION OF SSSC SWITCHING LOSSES

In the previous sections, only ideal SSSCs, that is, SSSCs without any losses have been considered. However, in the proposed model, switching losses for practical SSSCs can also be very easily accommodated as shown below.

The SSSC converter switching loss is accounted for by the resistance R_C shown across the capacitor in Figure 3.1. In Figure 3.2, G_C^{eq} (in dashed lines) is the effective conductance due to the switching loss reflected on the line side of the coupling transformer. This conductance is connected in parallel with the series voltage source $\mathbf{V_{se}}$ representing the SSSC. The relation between G_C^{eq} and R_C can be established as follows.

It is known that $V_{se} = k_{se} V_{DC}$ [8], where V_{DC} is the DC side voltage (across the capacitor in Figure 3.1) and k_{se} is a constant that accounts for the type of converter. The switching loss is given as

$$P_{SW} = \frac{V_{DC}^2}{R_C} = \frac{1}{R_C}\left[\frac{V_{se}^2}{k_{se}^2}\right] = G_C^{eq} V_{se}^2 \tag{3.47}$$

where:

$$G_C^{eq} = \frac{1}{R_C k_{se}^2}$$

If any ideal (lossless) SSSC c $\{1 \leq c \leq p\}$ is connected between any two buses a and b (with a as SE bus), then from Equation 3.18, $P_{sec} = \mathrm{Re}[\mathbf{V}_{sec}\{-\mathbf{I}_{sec}^*\}] = 0$, as the SSSC can only supply reactive power. On the other hand, if switching losses are considered, the SSSC absorbs real power from the system to replenish its losses. Hence, from Equations 3.18 and 3.47,

$$\begin{aligned} -G_C^{eq} V_{sec}^2 = &-V_{sec}^2 Y_{ab} \cos\varphi_{yab} - V_{sec} V_b Y_{ab} \cos\{\theta_{sec} - \theta_b - \varphi_{yab}\} \\ &+ V_{sec} V_a Y_{ab} \cos\{\theta_{sec} - \theta_a - \varphi_{yab}\} \end{aligned} \tag{3.48}$$

In the proposed model, Equation 3.48 becomes (from Equation 3.19),

$$-G_C^{eq} V_{n+c}^2 = \sum_{k=1}^{n+p} V_{n+c} V_k Y_{(n+c)k} \cos\{\theta_{n+c} - \theta_k - \varphi_{y(n+c)k}\}$$

or

$$\begin{aligned} -G_C^{eq} V_{n+c}^2 = &\, V_{n+c}^2 Y_{(n+c)(n+c)} \cos\varphi_{y(n+c)(n+c)} \\ &+ \sum_{k=1, k \neq n+c}^{n+p} V_{n+c} V_k Y_{(n+c)k} \cos\{\theta_{n+c} - \theta_k - \varphi_{y(n+c)k}\} \end{aligned} \tag{3.49}$$

Thus, accommodation of switching losses of the SSSC is trivial. For any SSSC c connected between any two buses a and b (with a as SE bus), only the value of the self-admittance of the $(n + c)$th fictitious power-flow bus is modified (using Equation 3.12) to

$$\mathbf{Y}_{(n+c)(n+c)} = -\mathbf{Y}_{ab} + G_C^{eq} \tag{3.50}$$

It is also to be noted that only the real part of the self-admittance is affected.

3.6 CASE STUDIES AND RESULTS

The validity of the proposed method was tested on both IEEE 118- and 300-bus systems. In each of these test systems, multiple SSSCs with different control functions were included and studies were carried out for three cases: (1) ideal SSSCs (no switching loss) without any device limit constraints, (2) practical SSSCs (switching loss incorporated) without device limit constraints, and (3) practical SSSCs with device limit constraints. For representing the SSSC switching loss, a value of 0.02 p.u. has been chosen for G_C^{eq}. Moreover, the convergence property of the proposed methodology has also been validated with different values of G_C^{eq}. In all these case studies, a convergence tolerance of 10^{-12} p.u. has been chosen. The initial values of the voltage magnitude and angle of the fictitious power-flow bus (representing the SSSC) were chosen as $V_{se}^0 \angle \theta_{se}^0 = 0.1 \angle -(\pi/2)$ in accordance with [49]. Although a large number of case studies confirmed the validity of the model, a few sets of representative results are presented in this chapter.

3.6.1 Studies with Ideal SSSCs without Any Device Limit Constraints

3.6.1.1 IEEE 118-Bus System

In this system, three SSSCs have been considered on branches among buses 75-118 (SSSC-1), 5-11 (SSSC-2), and 95-96 (SSSC-3) for control of bus voltage, reactive power flow, and active power flow, respectively. The control references chosen are $V_{118}^{SP} = 0.96$ p.u., $Q_{5\text{-}11}^{SP} = 5$ MVAR, and $P_{96\text{-}95}^{SP} = 20$ MW. The results are shown in Table 3.1. The converged final values of the control objectives (COs) are shown in bold cases. In this table (and all subsequent tables in this chapter), the symbols NI and γ_{se} denote the number of iterations taken by the algorithm and the phase angle difference between $\mathbf{I}_{se}$ and $\mathbf{V}_{se}$, respectively. It is observed from Table 3.1 that without switching loss, the value of γ_{se} is exactly 90° (as the SSSC current is totally reactive in nature). In this system, it has been found that for the base case power-flow (without any SSSC), the convergence tolerance of 10^{-12} p.u. is achieved with six iterations.

From Table 3.1, it is also observed that in the presence of SSSCs, the line active and reactive power flow levels are enhanced compared to their corresponding values in the base case. This is also true for the voltage at

TABLE 3.1 Study of IEEE 118-Bus System with Ideal SSSC (No Device Limit Constraints)

Solution of Base Case Power Flow (without Any SSSC)			
V_{118}	$Q_{5\text{–}11}$	$P_{96\text{–}95}$	**NI**
0.9493	3.44 MVAR	14.67 MW	6
Unconstrained Solution of SSSC Quantities Neglecting Switching Losses ($G_C^{eq} = 0$)			
Quantity	**SSSC-1**	**SSSC-2**	**SSSC-3**
V_{se} (p.u.)	0.1144	0.0562	0.0289
θ_{se}	−130.84°	−110.35°	−102.87°
I_{se} (p.u.)	0.6959	0.6773	0.2021
γ_{se}	90°	90°	90°
CO	V_{118} = **0.96 p.u.**	$Q_{5\text{-}11}$ = **5 MVAR**	$P_{96\text{-}95}$ = **20 MW**
NI	7		

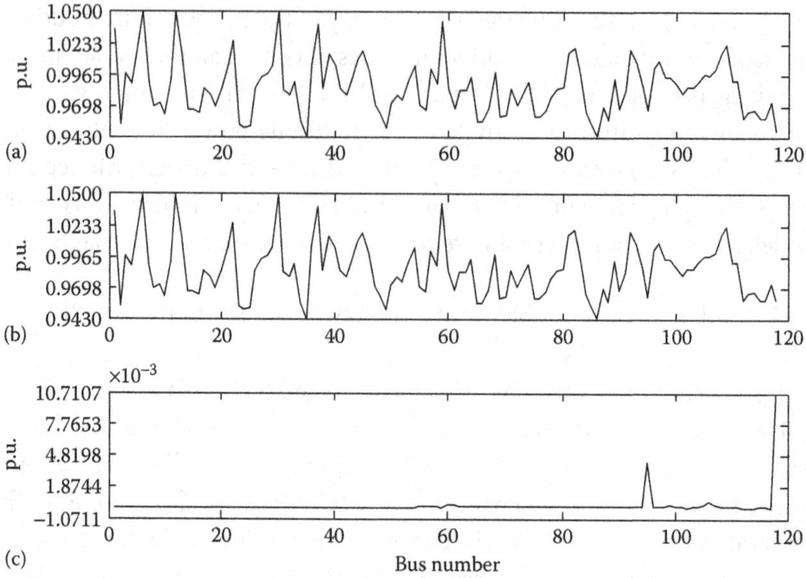

FIGURE 3.3 Bus voltage profile corresponding to the case study of Table 3.1. (a) Bus voltage magnitude without SSSC; (b) bus voltage magnitude with SSSC; (c) voltage magnitude difference.

bus 118. Furthermore, the number of iterations needed to obtain convergence marginally increases in the presence of SSSC. The bus voltage profiles without and with SSSCs are shown in Figure 3.3a and b, respectively. Further, the difference between the bus voltage magnitudes with and without SSSC is shown in Figure 3.3c. From this figure, it is observed that in the presence of SSSC, there is very little change in the bus voltage profile.

3.6.1.2 IEEE 300-Bus System

In this system, four SSSCs have been considered on branches among buses 58-237 (SSSC-1), 71-83 (SSSC-2), 190-191 (SSSC-3), and 5-9 (SSSC-4) for control of bus voltage, reactive power flow, active power flow, and line reactance (impedance), respectively. The control references chosen are $V_{237}^{SP} = 0.97$ p.u., $Q_{83\text{-}71}^{SP} = 25$ MVAR, $P_{191\text{-}190}^{SP} = 150$ MW, and $X_{5\text{-}9}^{SP} = -0.1$ p.u. It is to be noted that the negative value of $X_{5\text{-}9}^{SP}$ denotes the capacitive nature of the reactance. The results are shown in Table 3.2. The converged final values of the COs are again shown in bold cases. It is again observed from the table that without switching loss, the value of γ_{se} is exactly 90°.

It is observed from Table 3.2 that similar to the case study of the 118-bus system, the values of the COs can be enhanced with SSSCs. Also, similar

TABLE 3.2 Study of IEEE 300-Bus System with Ideal SSSC (No Device Limit Constraints)

Solution of Base Case Power Flow (without Any SSSC)				
V_{237}	$Q_{83\text{-}71}$	$P_{191\text{-}190}$	$X_{5\text{-}9}$	NI
0.9693	19.93 MVAR	127.25 MW	0.03 p.u.	7
Unconstrained Solution of SSSC Quantities Neglecting Switching Losses ($G_C^{eq} = 0$)				
Quantity	**SSSC-1**	**SSSC-2**	**SSSC-3**	**SSSC-4**
V_{se} (p.u.)	0.0349	0.0611	0.2076	0.1218
θ_{se}	−89.66°	−73.84°	−141.15°	−99.04°
I_{se} (p.u.)	0.3668	0.4944	1.6918	1.2179
γ_{se}	90°	90°	90°	90°
CO	V_{237} = **0.97 p.u.**	$Q_{83\text{-}71}$ = **25 MVAR**	$P_{191\text{-}190}$ = **150 MW**	$X_{5\text{-}9}$ = **−0.1 p.u.**
NI	10			

to the 118-bus system, the number of iterations to obtain convergence increases compared to their corresponding values in the base case. The bus voltage profiles are shown in Figure 3.4 for this case. From this figure also, it is observed that in the presence of SSSC, the bus voltage profile does not change very much.

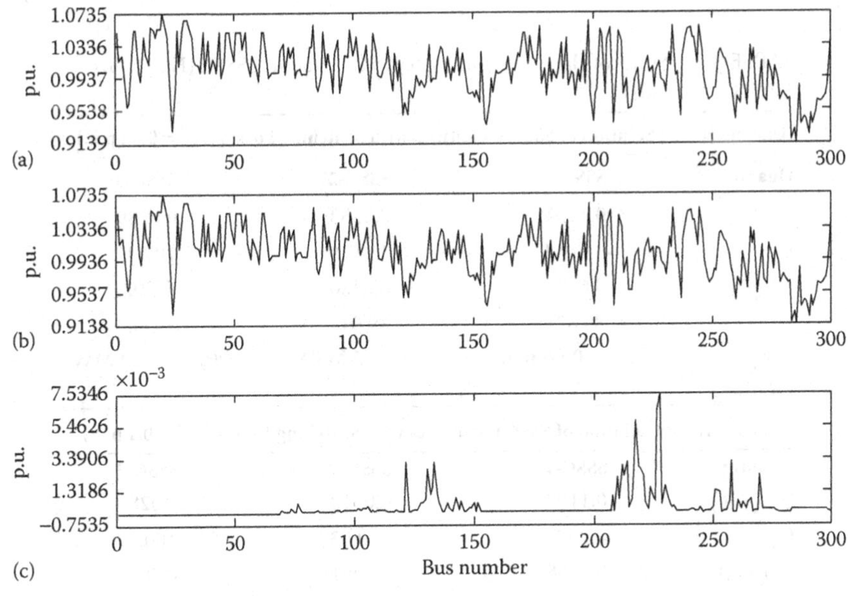

FIGURE 3.4 Bus voltage profile corresponding to the case study of Table 3.2. (a) Bus voltage magnitude without SSSC; (b) bus voltage magnitude with SSSC; (c) voltage magnitude difference.

3.6.2 Studies with Practical SSSCs without Any Device Limit Constraints

3.6.2.1 IEEE 118-Bus System

Again, in this system, three SSSCs have been considered on branches among buses 75-118 (SSSC-1), 5-11 (SSSC-2), and 95-96 (SSSC-3) for control of bus voltage, reactive power flow, and active power flow, respectively. The control reference values are kept identical to those in the corresponding case study presented in Section 3.6.1 (Table 3.1). For representing the SSSC switching losses, a value of 0.02 p.u. has been initially chosen for G_C^{eq}. Subsequently, the convergence property of the proposed technique has also been validated with different values of G_C^{eq}. The results corresponding to $G_C^{eq} = 0.02$ p.u. and $G_C^{eq} = 0.1$ p.u. are shown in Table 3.3, whereas those corresponding to $G_C^{eq} = 0.15$ p.u. and $G_C^{eq} = 0.2$ p.u. are shown in Table 3.4.

From Tables 3.3 and 3.4, it is observed that with lower values of switching losses (lower values of G_C^{eq}), the value of γ_{se} deviates a little from 90° in order to replenish the losses in the SSSC. However, with increasing values of G_C^{eq} (elevated converter switching loss), the angle γ_{se} also deviates further from 90° to account for enhanced switching losses.

The bus voltage profiles of the 118-bus system for the case studies with $G_C^{eq} = 0.1$ p.u. and $G_C^{eq} = 0.2$ p.u. are shown in Figures 3.5 and 3.6, respectively.

TABLE 3.3 Study of IEEE 118-Bus System with Practical SSSC (No Device Limit Constraints)

Unconstrained Solution of SSSC Quantities with Switching Losses ($G_C^{eq} = 0.02$ p.u.)			
Quantity	**SSSC-1**	**SSSC-2**	**SSSC-3**
V_{se} (p.u.)	0.1154	0.0553	0.0289
θ_{se}	−130.44°	−110.29°	−102.47°
I_{se} (p.u.)	0.6976	0.6726	0.202
γ_{se}	89.82°	89.91°	89.84°
CO	V_{118} = **0.96 p.u.**	$Q_{5\text{-}11}$ = **5 MVAR**	$P_{96\text{-}95}$ = **20 MW**
NI	7		
Unconstrained Solution of SSSC Quantities with Switching Losses ($G_C^{eq} = 0.1$ p.u.)			
Quantity	**SSSC-1**	**SSSC-2**	**SSSC-3**
V_{se} (p.u.)	0.1198	0.0518	0.0289
θ_{se}	−128.7°	−110.03°	−100.87°
I_{se} (p.u.)	0.7058	0.6561	0.2015
γ_{se}	89.03°	89.55°	89.18°
CO	V_{118} = **0.96 p.u.**	$Q_{5\text{-}11}$ = **5 MVAR**	$P_{96\text{-}95}$ = **20 MW**
NI	7		

TABLE 3.4 Further Study of IEEE 118-Bus System with Practical SSSC (No Device Limit Constraints)

Unconstrained Solution of SSSC Quantities with Switching Losses (G_C^{eq} = 0.15 p.u.)			
Quantity	**SSSC-1**	**SSSC-2**	**SSSC-3**
V_{se} (p.u.)	0.1232	0.05	0.0289
θ_{se}	−127.49°	−109.87°	−99.85°
I_{se} (p.u.)	0.7119	0.6471	0.2012
γ_{se}	88.51°	89.34°	88.76°
CO	V_{118} = **0.96 p.u.**	$Q_{5\text{-}11}$ = **5 MVAR**	$P_{96\text{-}95}$ = **20 MW**
NI	7		
Unconstrained Solution of SSSC Quantities with Switching Losses (G_C^{eq} = 0.2 p.u.)			
Quantity	**SSSC-1**	**SSSC-2**	**SSSC-3**
V_{se} (p.u.)	0.1273	0.0483	0.029
θ_{se}	−126.14°	−109.72°	−98.81°
I_{se} (p.u.)	0.7192	0.639	0.201
γ_{se}	87.97°	89.13°	88.35°
CO	V_{118} = **0.96 p.u.**	$Q_{5\text{-}11}$ = **5 MVAR**	$P_{96\text{-}95}$ = **20 MW**
NI	7		

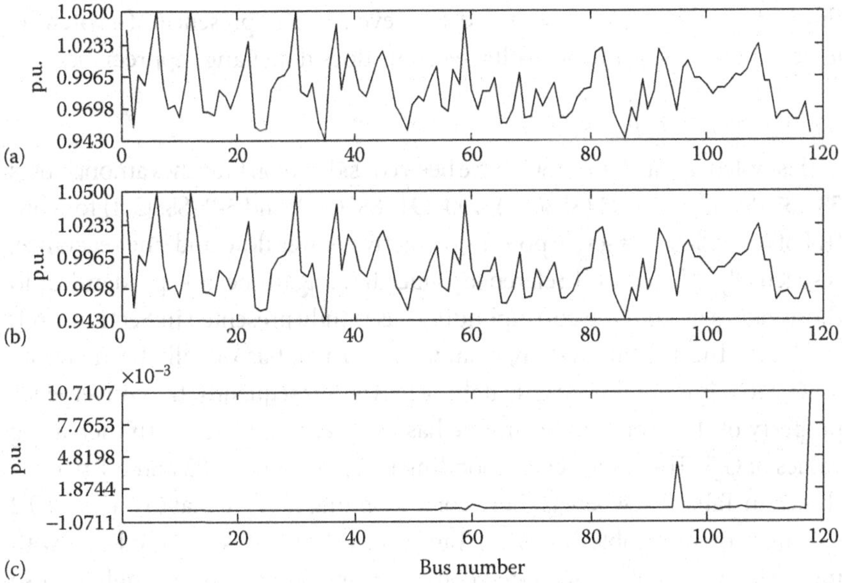

FIGURE 3.5 Bus voltage profile corresponding to the case study of Table 3.3. (a) Bus voltage magnitude without SSSC; (b) bus voltage magnitude with SSSC; (c) voltage magnitude difference.

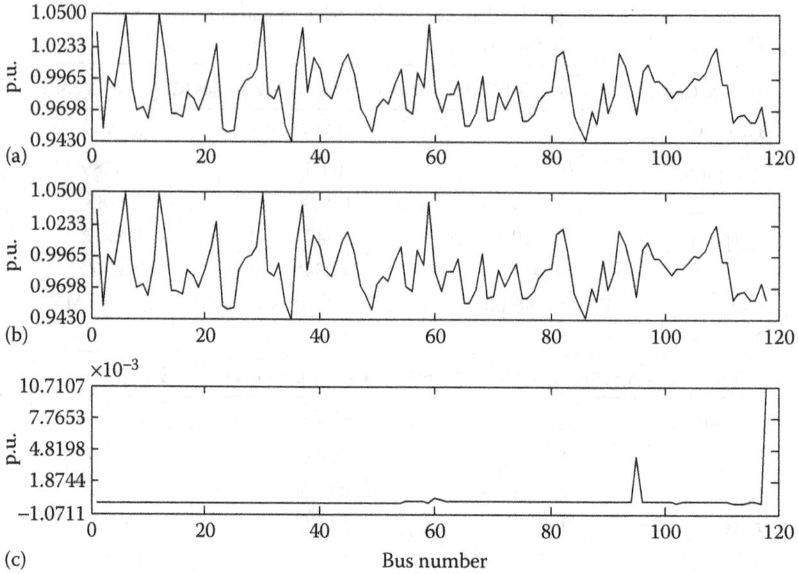

FIGURE 3.6 Bus voltage profile corresponding to the case study of Table 3.4. (a) Bus voltage magnitude without SSSC; (b) bus voltage magnitude with SSSC; (c) voltage magnitude difference.

From these figures, it is observed that even in the presence of switching losses in the SSSCs, the bus voltage profile does not change appreciably.

3.6.2.2 IEEE 300-Bus System

In this system again, four SSSCs have been considered on branches among buses 58-237 (SSSC-1), 71-83 (SSSC-2), 190-191 (SSSC-3), and 5-9 (SSSC-4) for control of bus voltage, reactive power flow, active power flow, and line reactance, respectively. The control reference values have again been kept identical to those values used in the corresponding case study presented in Section 3.6.1. Similar to the 118-bus system, a value of 0.02 p.u. has initially been chosen for G_C^{eq} to represent the SSSC switching losses. Subsequently, the convergence property of the proposed technique has also been validated with increasing values of G_C^{eq}. The results corresponding to G_C^{eq} equal to 0.02 and 0.1 p.u. are shown in Table 3.5, whereas those corresponding to G_C^{eq} equal to 0.15 and 0.2 p.u. are shown in Table 3.6. From Tables 3.5 and 3.6, it is observed that with increasing values of G_C^{eq} (elevated converter switching loss), the angle γ_{se} also deviates further from 90° to account for enhanced switching losses.

The bus voltage profiles in the 300-bus system for the case studies with $G_C^{eq} = 0.1$ p.u. and $G_C^{eq} = 0.2$ p.u. are shown in Figures 3.7 and 3.8,

TABLE 3.5 Study of IEEE 300-Bus System with Practical SSSC (No Device Limit Constraints)

Unconstrained Solution of SSSC Quantities with Switching Losses ($G_C^{eq} = 0.02$ p.u.)				
Quantity	**SSSC-1**	**SSSC-2**	**SSSC-3**	**SSSC-4**
V_{se} (p.u.)	0.0348	0.0608	0.2067	0.1217
θ_{se}	−89.57°	−73.63°	−140.71°	−98.8°
I_{se} (p.u.)	0.3668	0.4933	1.6863	1.2174
γ_{se}	89.89°	89.86°	89.86°	89.89°
CO	V_{237} = **0.97 p.u.**	$Q_{83\text{-}71}$ = **25 MVAR**	$P_{191\text{-}190}$ = **150 MW**	$X_{5\text{-}9}$ = **−0.1 p.u.**
NI	7			
Unconstrained Solution of SSSC Quantities with Switching Losses ($G_C^{eq} = 0.1$ p.u.)				
Quantity	**SSSC-1**	**SSSC-2**	**SSSC-3**	**SSSC-4**
V_{se} (p.u.)	0.0343	0.0596	0.2035	0.1215
θ_{se}	−89.19°	−72.78°	−138.97°	−97.82°
I_{se} (p.u.)	0.3668	0.4891	1.6658	1.2154
γ_{se}	89.46°	89.3°	89.3°	89.43°
CO	V_{237} = **0.97 p.u.**	$Q_{83\text{-}71}$ = **25 MVAR**	$P_{191\text{-}190}$ = **150 MW**	$X_{5\text{-}9}$ = **−0.1 p.u.**
NI	7			

TABLE 3.6 Further Study of IEEE 300-Bus System with Practical SSSC (No Device Limit Constraints)

Unconstrained Solution of SSSC Quantities with Switching Losses ($G_C^{eq} = 0.15$ p.u.)				
Quantity	**SSSC-1**	**SSSC-2**	**SSSC-3**	**SSSC-4**
V_{se} (p.u.)	0.0341	0.0589	0.2017	0.1214
θ_{se}	−88.96°	−72.26°	−37.91°	−97.21°
I_{se} (p.u.)	0.3668	0.4866	1.6538	1.2141
γ_{se}	89.2°	88.96°	88.95°	89.14°
CO	V_{237} = **0.97 p.u.**	$Q_{83\text{-}71}$ = **25 MVAR**	$P_{191\text{-}190}$ = **150 MW**	$X_{5\text{-}9}$ = **−0.1 p.u.**
NI	8			
Unconstrained Solution of SSSC Quantities with Switching Losses ($G_C^{eq} = 0.2$ p.u.)				
Quantity	**SSSC-1**	**SSSC-2**	**SSSC-3**	**SSSC-4**
V_{se} (p.u.)	0.0338	0.0582	0.1999	0.1213
θ_{se}	−88.74°	−71.75°	−136.85°	−96.6°
I_{se} (p.u.)	0.3668	0.4842	1.6425	1.2127
γ_{se}	88.94°	88.62°	88.61°	88.85°
CO	V_{237} = **0.97 p.u.**	$Q_{83\text{-}71}$ = **25 MVAR**	$P_{191\text{-}190}$ = **150 MW**	$X_{5\text{-}9}$ = **−0.1 p.u.**
NI	7			

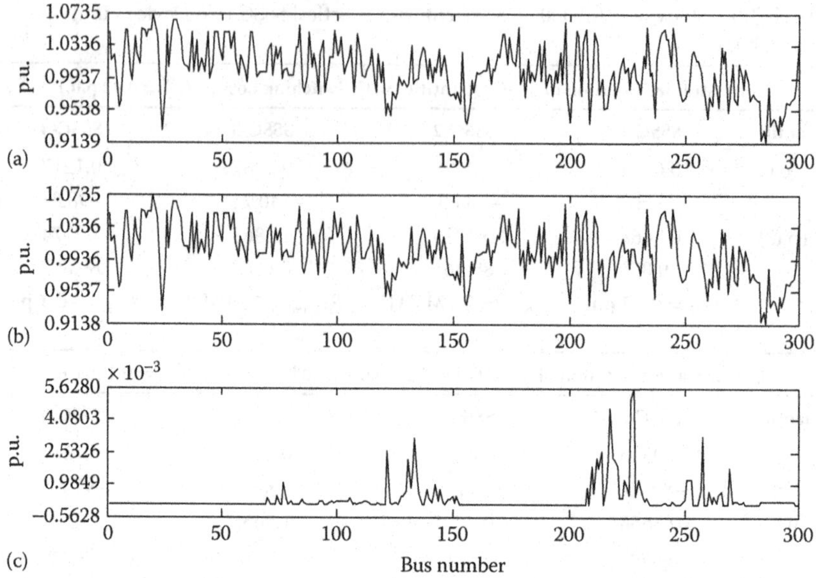

FIGURE 3.7 Bus voltage profile corresponding to the case study of Table 3.5. (a) Bus voltage magnitude without SSSC; (b) bus voltage magnitude with SSSC; (c) voltage magnitude difference.

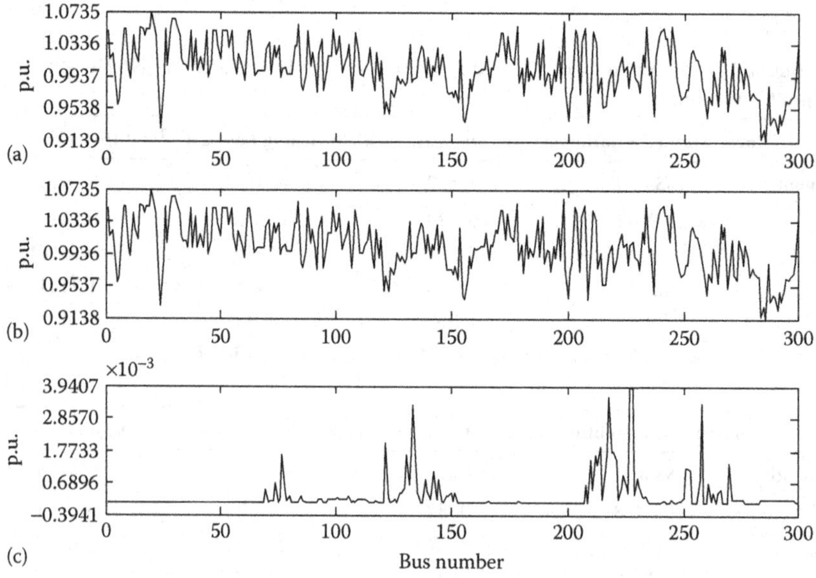

FIGURE 3.8 Bus voltage profile corresponding to the case study of Table 3.6. (a) Bus voltage magnitude without SSSC; (b) bus voltage magnitude with SSSC; (c) voltage magnitude difference.

respectively. Again from these figures, it is observed that as in the case of 118-bus system, the presence of switching losses in the SSSCs does not alter the bus voltage profile appreciably.

It is important to note that the values of G_C^{eq} chosen in Tables 3.3 through 3.6 are much higher than the practical values reported in the literature [8]. However, the only purpose of the studies reported in these tables is to test the convergence characteristic of the proposed method even with quite adverse values of switching loss. As the developed methodology is able to attain excellent convergence even with unrealistically high values of G_C^{eq}, it can be safely concluded that the proposed technique would exhibit convergence for any practical value of converter switching losses.

3.6.3 Studies with Practical SSSCs Including Device Limit Constraints

In these studies, both the switching losses and the device limit constraints have been considered. As already mentioned in Section 3.4.2, two major device limit constraints of the SSSC have been considered: the magnitudes of the SSSC injected voltage and the SSSC line current. Also, in all these cases, the SSSC control references have been kept same as those in the corresponding case studies presented in Sections 3.6.1 and 3.6.2. The procedure for enforcing the limits has already been discussed in Section 3.4.2. Although various case studies have been carried out in both the 118- and 300-bus test systems to test the convergence characteristic of the proposed method with different values of G_C^{eq}, only a few representative case studies (corresponding to the highest value of $G_C^{eq} = 0.2$ p.u.) are presented below for both the 118- and 300-bus test systems. In these case studies, the implementation of each of the two device limit constraints is realized in four different ways: (1) limit violation of a single SSSC, (2) limit violations of two SSSCs simultaneously, (3) limit violations of three SSSCs simultaneously, and (4) limit violations of all four SSSCs simultaneously (only in case of the 300-bus test system). Also, to demonstrate the robustness of the proposed technique, a wide range of values of the device limit thresholds has been chosen. From a large number of power-flow studies, it has been observed that the developed methodology is able to enforce the different device limit constraints with all values of G_C^{eq} up to 0.2 p.u. Hence, it can be safely concluded that the device limit constraints can be quite successfully enforced by the proposed technique for any practical value of converter switching losses.

Although several combinations of single and multiple device limit constraints with varying device limit thresholds were applied to both the 118- and 300-bus test systems, only some representative results are reported in this chapter. In all subsequent tables, both the COs that have been satisfied and the limit constraints that have been violated are shown in bold cases. Table 3.7 shows the results corresponding to the limit violation of a single SSSC for the IEEE 118-bus test system.

Table 3.7 shows the results of two case studies for the 118 bus system: (1) the violation of the voltage limit of SSSC-1 only and (2) the violation of the current limit of SSSC-2 only. For incorporating the voltage limit of

TABLE 3.7 First Study of IEEE 118-Bus System with Practical SSSC and Device Limits

Unconstrained Solution of SSSC Quantities with Switching Losses (G_C^{eq} = 0.2 p.u.)			
Quantity	**SSSC-1**	**SSSC-2**	**SSSC-3**
V_{se} (p.u.)	0.1273	0.0483	0.029
θ_{se}	–126.14°	–109.72°	–98.81°
I_{se} (p.u.)	0.7192	0.639	0.201
γ_{se}	87.97°	89.13°	88.35°
CO	V_{118} = **0.96 p.u.**	$Q_{5\text{-}11}$ = **5 MVAR**	$P_{96\text{-}95}$ = **20 MW**
NI	7		
Solution of SSSC Quantities with Switching Losses (G_C^{eq} = 0.2 p.u.) and Specified Voltage Limit for SSSC-1: V_{se1}^{Lim} = 0.125 p.u.			
Quantity	**SSSC-1**	**SSSC-2**	**SSSC-3**
V_{se} (p.u.)	**0.125**	0.0483	0.029
θ_{se}	–126.01°	–109.72°	–98.82°
I_{se} (p.u.)	0.7126	0.639	0.201
γ_{se}	87.99°	89.13°	88.35°
CO	V_{118} = 0.9596 p.u.	$Q_{5\text{-}11}$ = **5 MVAR**	$P_{96\text{-}95}$ = **20 MW**
NI	9		
Solution of SSSC Quantities with Switching Losses (G_C^{eq} = 0.2 p.u.) and Specified Current Limit for SSSC-2: I_{se2}^{Lim} = 0.62 p.u.			
Quantity	**SSSC-1**	**SSSC-2**	**SSSC-3**
V_{se} (p.u.)	0.1273	0.0442	0.029
θ_{se}	–126.14°	–110.25°	–98.81°
I_{se} (p.u.)	0.7192	**0.62**	0.201
γ_{se}	87.97°	89.18°	88.35°
CO	V_{118} = **0.96 p.u.**	$Q_{5\text{-}11}$ = 5.38 MVAR	$P_{96\text{-}95}$ = **20 MW**
NI	8		

SSSC-1, the CO of maintaining the voltage of bus 118 at a specified value is relaxed. Similarly, for incorporating the current limit of SSSC-2, the CO of maintaining the reactive power flow through line 5-11 at a specified value has been relaxed. The converged values of the COs met and the device limits violated are shown in bold cases. The number of iterations required to obtain convergence may increase from that obtained in the unconstrained case.

Corresponding to these studies in the 118-bus system, the bus voltage profiles with SSSC voltage and current limits are shown in Figures 3.9 and 3.10, respectively. From these figures, it is again observed that the bus voltage profiles change very little when SSSC device constraint limits are incorporated.

Table 3.8 shows the results of two case studies for the 300-bus system: (1) the violation of the voltage limit of SSSC-3 only and (2) the violation of current limit of SSSC-4 only. For incorporating the voltage limit of SSSC-3, the CO of maintaining the real power flow through line 190-191 at a specified value is relaxed. Similarly, for incorporating the current limit of SSSC-4, the CO of maintaining the reactance of line 5-9 at a specified value has

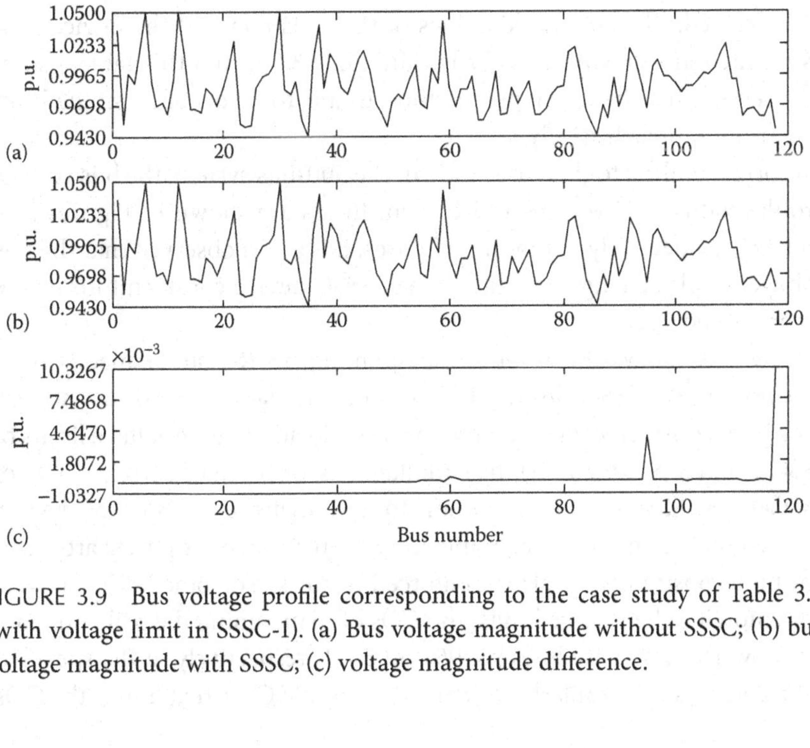

FIGURE 3.9 Bus voltage profile corresponding to the case study of Table 3.7 (with voltage limit in SSSC-1). (a) Bus voltage magnitude without SSSC; (b) bus voltage magnitude with SSSC; (c) voltage magnitude difference.

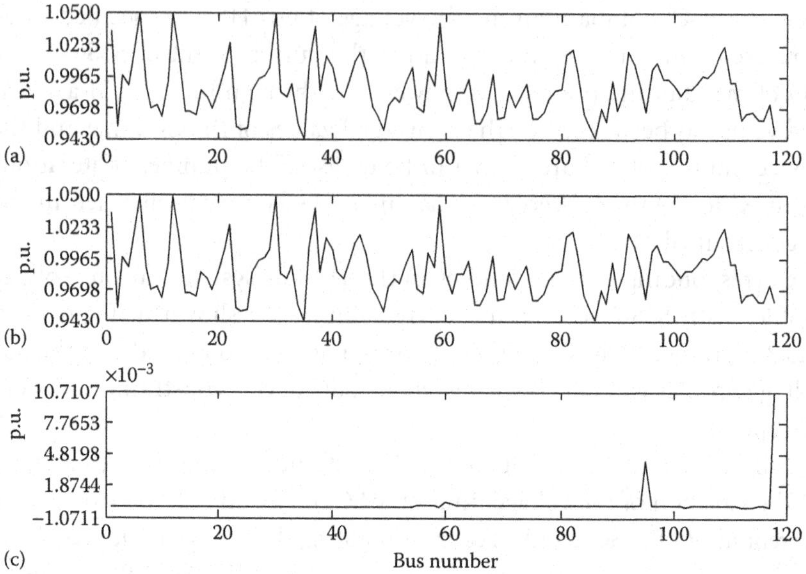

FIGURE 3.10 Bus voltage profile corresponding to the case study of Table 3.7 (with current limit on SSSC-2). (a) Bus voltage magnitude without SSSC; (b) bus voltage magnitude with SSSC; (c) voltage magnitude difference.

been relaxed. The converged values of the COs met and the device limits violated are shown in bold cases in Table 3.8. From this table also, it is observed that the developed method is able to incorporate single limit constraint quite effectively.

Corresponding to this case study in the 300-bus system, the bus voltage profiles with SSSC voltage and current limits are shown in Figures 3.11 and 3.12, respectively. From these figures, it is again observed that the bus voltage profiles change very little when SSSC device constraint limits are incorporated.

Table 3.9 shows the results corresponding to the simultaneous limit violation of two SSSCs in the IEEE 118-bus test system. Two case studies have been considered: (1) the simultaneous violation of the voltage limits of SSSC-2 and SSSC-3 and (2) the simultaneous violation of current limits of SSSC-1 and SSSC-2. It is important to note that the selection of devices (SSSCs) on which limit constraints have been enforced is purely arbitrary. The limit constraints can also be enforced very easily on other SSSCs as well. For enforcing the voltage limits, the COs of maintaining the real and reactive power flows in the lines 96–95 and 5–11, respectively, at the specified values are relaxed. Similarly, for enforcing the SSSC current limits, the COs

TABLE 3.8 First Study of IEEE 300-Bus System with Practical SSSC and Device Limits

Unconstrained Solution of SSSC Quantities with Switching Losses (G_C^{eq} = 0.2 p.u.)				
Quantity	**SSSC-1**	**SSSC-2**	**SSSC-3**	**SSSC-4**
V_{se} (p.u.)	0.0338	0.0582	0.1999	0.1213
θ_{se}	–88.74°	–71.75°	–136.85°	–96.6°
I_{se} (p.u.)	0.3668	0.4842	1.6425	1.2127
γ_{se}	88.94°	88.62°	88.61°	88.85°
CO	V_{237} = **0.97 p.u.**	$Q_{83\text{-}71}$ = **25 MVAR**	$P_{191\text{-}190}$ = **150 MW**	$X_{5\text{-}9}$ = **–0.1 p.u.**
NI	7			
Solution of SSSC Quantities with Switching Losses (G_C^{eq} = 0.2 p.u.) and Specified Voltage Limit for SSSC-3: V_{se3}^{Lim} = 0.18 p.u.				
Quantity	**SSSC-1**	**SSSC-2**	**SSSC-3**	**SSSC-4**
V_{se} (p.u.)	0.0338	0.0582	**0.18**	0.1213
θ_{se}	–88.79°	–71.74°	–136.39°	–96.6°
I_{se} (p.u.)	0.3668	0.4839	1.5587	1.2128
γ_{se}	88.94°	88.62°	88.68°	88.85°
CO	V_{237} = **0.97 p.u.**	$Q_{83\text{-}71}$ = **25 MVAR**	$P_{191\text{-}190}$ = 142.96 MW	$X_{5\text{-}9}$ = **–0.1 p.u.**
NI	9			
Solution of SSSC Quantities with Switching Losses (G_C^{eq} = 0.2 p.u.) and Specified Current Limit for SSSC-4: I_{se4}^{Lim} = 1.21 p.u.				
Quantity	**SSSC-1**	**SSSC-2**	**SSSC-3**	**SSSC-4**
V_{se} (p.u.)	0.0338	0.0582	0.1999	0.1207
θ_{se}	–88.74°	–71.75°	–136.85°	–96.63°
I_{se} (p.u.)	0.3668	0.4842	1.6425	**1.21**
γ_{se}	88.94°	88.62°	88.61°	88.86°
CO	V_{237} = **0.97 p.u.**	$Q_{83\text{-}71}$ = **25 MVAR**	$P_{191\text{-}190}$ = **150 MW**	$X_{5\text{-}9}$ = –0.0997 p.u.
NI	7			

of maintaining the voltage of bus 118 and the reactive power flow in the line 5–11 at the specified values are relaxed. The converged values of the COs met and the device limits violated are shown in bold cases. It is observed from this table that the number of iterations required to obtain convergence increases compared to that obtained in the unconstrained case.

The bus voltage profiles for the two cases depicted in Table 3.9 are shown in Figures 3.13 and 3.14, respectively. From these two figures, it is observed that the voltage profile in the 118-bus system changes very little (compared to the case with no SSSC) even with imposition of two simultaneous limit constraints.

Table 3.10 shows the results corresponding to the simultaneous limit violation of two SSSCs for the IEEE 300-bus test system. Two case studies

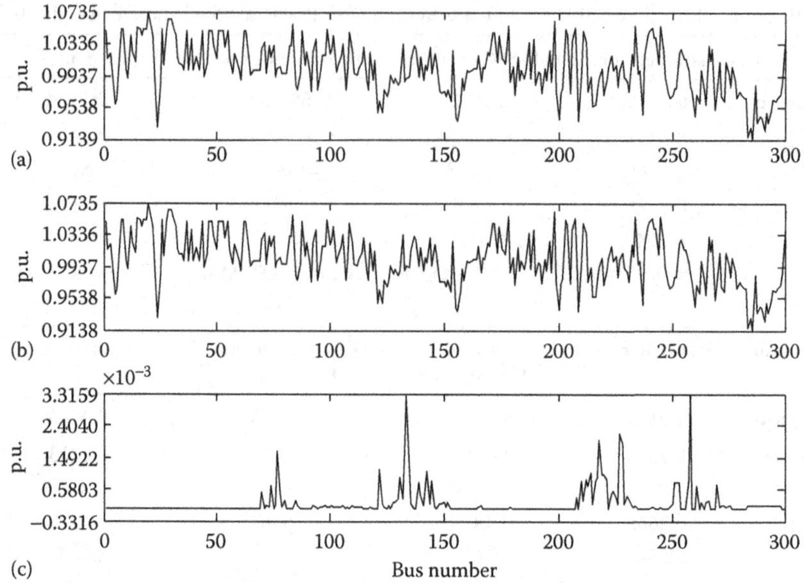

FIGURE 3.11 Bus voltage profile corresponding to the case study of Table 3.8 (with voltage limit on SSSC-3). (a) Bus voltage magnitude without SSSC; (b) bus voltage magnitude with SSSC; (c) voltage magnitude difference.

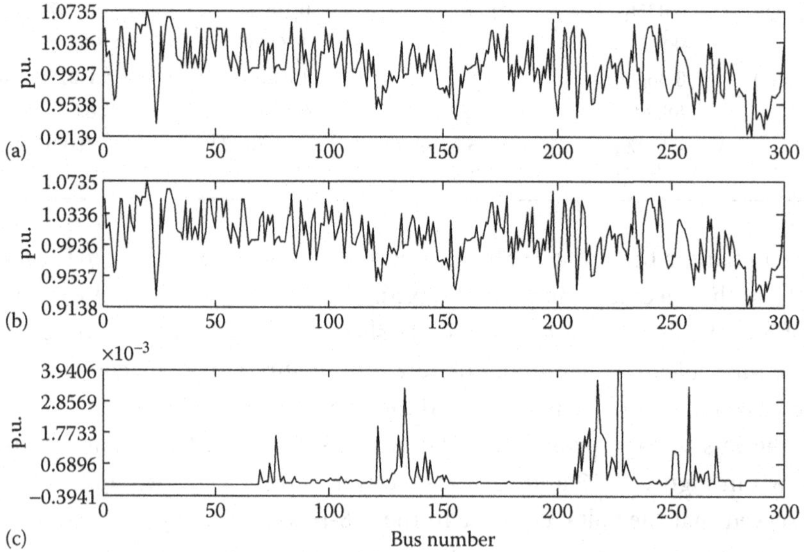

FIGURE 3.12 Bus voltage profile corresponding to the case study of Table 3.8 (with current limit on SSSC-4). (a) Bus voltage magnitude without SSSC; (b) bus voltage magnitude with SSSC; (c) voltage magnitude difference.

TABLE 3.9 Second Study of IEEE 118-Bus System with Practical SSSC and Device Limits

Solution of SSSC Quantities with Switching Losses (G_C^{eq} = 0.2 p.u.) and Specified Voltage Limits on SSSC-2 and SSSC-3: V_{se2}^{Lim} = 0.045 p.u. and V_{se3}^{Lim} = 0.028 p.u.			
Quantity	**SSSC-1**	**SSSC-2**	**SSSC-3**
V_{se} (p.u.)	0.1273	**0.045**	**0.028**
θ_{se}	–126.14°	–110.15°	–99.41°
I_{se} (p.u.)	0.7192	0.6237	0.1976
γ_{se}	87.97°	89.17°	88.38°
CO	V_{118} = **0.96 p.u.**	$Q_{5\text{-}11}$ = 5.31 MVAR	$P_{96\text{-}95}$ = 19.64 MW
NI	9		

Solution of SSSC Quantities with Switching Losses (G_C^{eq} = 0.2 p.u.) and Specified Current Limits on SSSC-1 and SSSC-2: I_{se1}^{Lim} = 0.7 p.u. and I_{se2}^{Lim} = 0.6 p.u.			
Quantity	**SSSC-1**	**SSSC-2**	**SSSC-3**
V_{se} (p.u.)	0.1207	0.04	0.029
θ_{se}	–125.78°	–110.79°	–98.84°
I_{se} (p.u.)	**0.7**	**0.6**	0.201
γ_{se}	88.23°	89.2°	88.55°
CO	V_{118} = 0.9589 p.u.	$Q_{5\text{-}11}$ = 5.73 MVAR	$P_{96\text{-}95}$ = **20 MW**
NI	8		

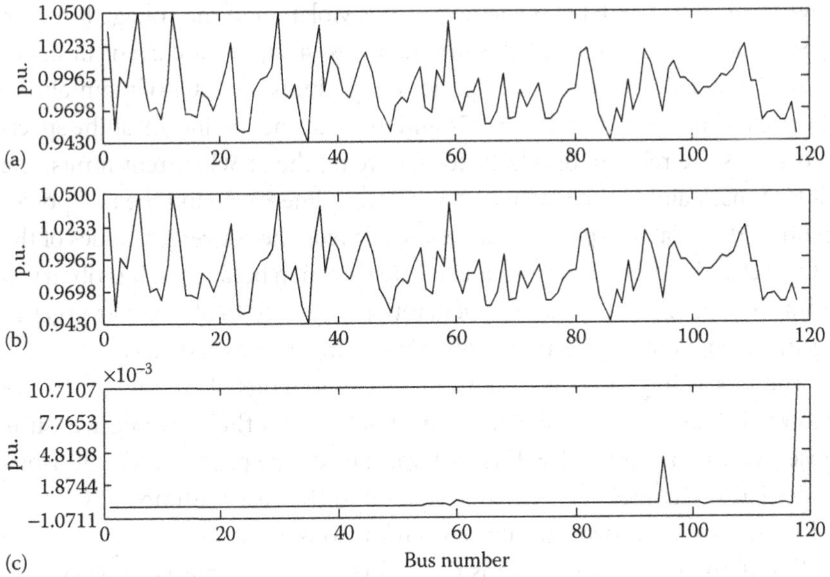

FIGURE 3.13 Bus voltage profile corresponding to the case study of Table 3.9 (with voltage limits on SSSC-2 and SSSC-3). (a) Bus voltage magnitude without SSSC; (b) bus voltage magnitude with SSSC; (c) voltage magnitude difference.

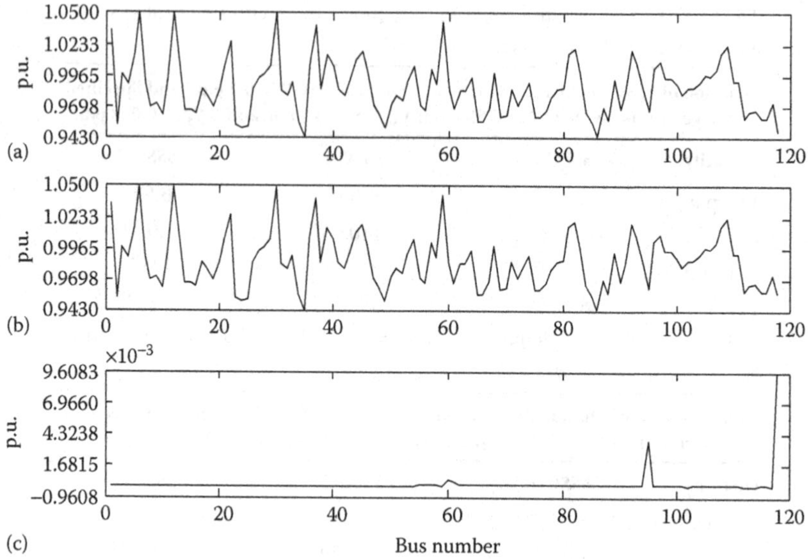

FIGURE 3.14 Bus voltage profile corresponding to the case study of Table 3.9 (with current limits on SSSC-1 and SSSC-2). (a) Bus voltage magnitude without SSSC; (b) bus voltage magnitude with SSSC; (c) voltage magnitude difference.

have been considered: (1) the simultaneous violation of the voltage limits of SSSC-2 and SSSC-4 and (2) the simultaneous violation of current limits of SSSC-2 and SSSC-3. For enforcing the voltage limits, the COs of maintaining the reactive power flow in line 83-71 and the reactance of line 5-9 at the specified values are relaxed. Similarly, for enforcing the SSSC current limits, the COs of maintaining the reactive power flow in line 83-71 and the real power of line 191-190 at the specified values are relaxed. The converged values of the COs met and the device limits violated are shown in bold cases. It is observed from Table 3.10 that the number of iterations required to obtain convergence again increases compared to that obtained in the unconstrained case.

The bus voltage profiles for the two cases depicted in Table 3.10 are shown in Figures 3.15 and 3.16, respectively. From these two figures, it is observed that as in the 118-bus system, the voltage profile in the 300-bus system also changes very little (compared to the case with no SSSC) even with imposition of two simultaneous limit constraints.

Simultaneous limit violations of all three SSSCs in the 118-bus system are considered in Table 3.11. Again, two case studies have been considered: (1) the simultaneous violation of the voltage limits of SSSC-1, SSSC-2, and SSSC-3 and (2) the simultaneous violation of current limits of SSSC-1,

TABLE 3.10 Second Study of IEEE 300-Bus System with Practical SSSC and Device Limits

Solution of SSSC Quantities with Switching Losses (G_C^{eq} = 0.2 p.u.) and Specified Voltage Limits for SSSC-2 and SSSC-4: V_{se2}^{Lim} = 0.05 p.u. and V_{se4}^{Lim} = 0.12 p.u.

Quantity	SSSC-1	SSSC-2	SSSC-3	SSSC-4
V_{se} (p.u.)	0.0338	**0.05**	0.1999	**0.12**
θ_{se}	−88.73°	−73.26°	−136.85°	−96.65°
I_{se} (p.u.)	0.3668	0.4556	1.6425	1.2068
γ_{se}	88.94°	88.74°	88.61°	88.86°
CO	V_{237} = **0.97 p.u.**	$Q_{83\text{-}71}$ = 22.48 MVAR	$P_{191\text{-}190}$ = **150 MW**	$X_{5\text{-}9}$ = −0.0994 p.u.
NI	10			

Solution of SSSC Quantities with Switching Losses (G_C^{eq} = 0.2 p.u.) and Specified Current Limits for SSSC-2 and SSSC-3: I_{se2}^{Lim} = 0.48 p.u. and I_{se3}^{Lim} = 1.64 p.u.

Quantity	SSSC-1	SSSC-2	SSSC-3	SSSC-4
V_{se} (p.u.)	0.0338	0.057	0.1993	0.1213
θ_{se}	−88.74°	−71.98°	−136.84°	−96.6°
I_{se} (p.u.)	0.3668	**0.48**	**1.64**	1.2127
γ_{se}	88.94°	88.64°	88.61°	88.85°
CO	V_{237} = **0.97 p.u.**	$Q_{83\text{-}71}$ = 24.62 MVAR	$P_{191\text{-}190}$ = 149.8 MW	$X_{5\text{-}9}$ = **−0.1 p.u.**
NI	8			

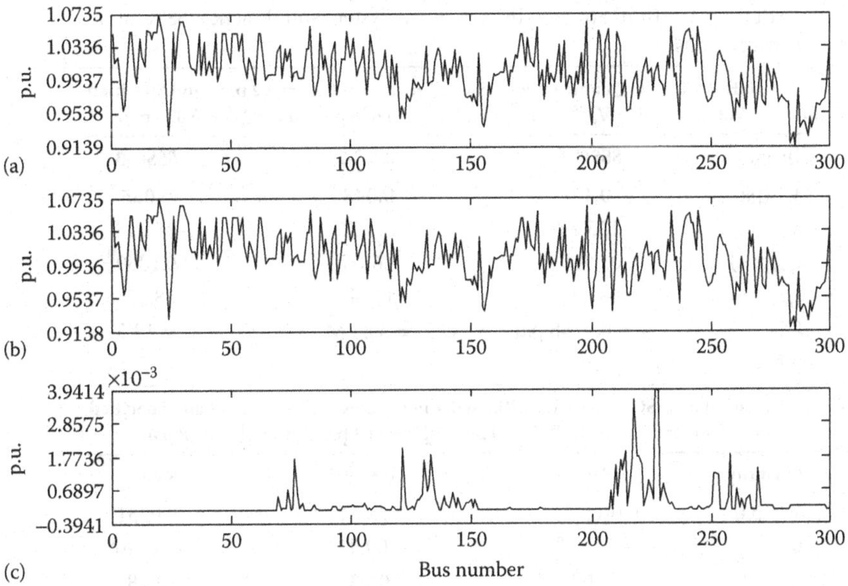

FIGURE 3.15 Bus voltage profile corresponding to the case study of Table 3.10 (with voltage limits on SSSC-2 and SSSC-4). (a) Bus voltage magnitude without SSSC; (b) bus voltage magnitude with SSSC; (c) voltage magnitude difference.

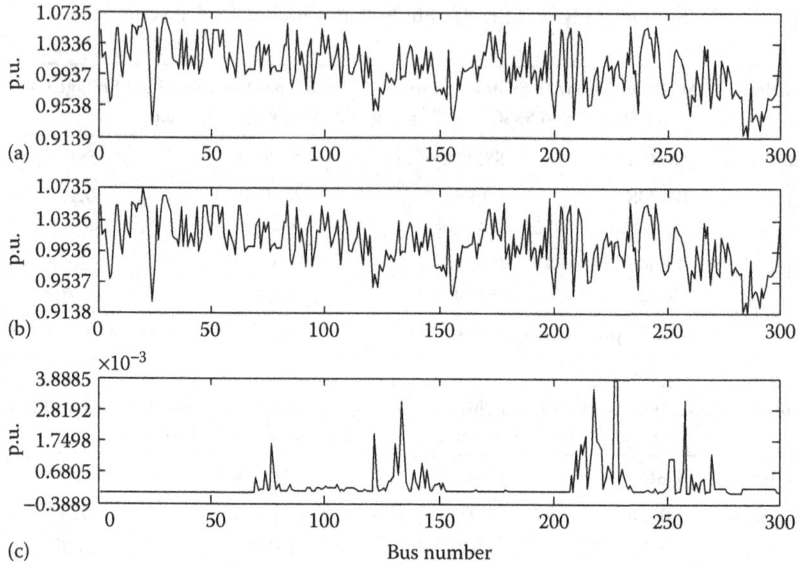

FIGURE 3.16 Bus voltage profile corresponding to the case study of Table 3.10 (with current limits on SSSC-2 and SSSC-3). (a) Bus voltage magnitude without SSSC; (b) bus voltage magnitude with SSSC; (c) voltage magnitude difference.

TABLE 3.11 Third Study of IEEE 118-Bus System with Practical SSSC and Device Limits

Solution of SSSC Quantities with Switching Losses ($G_C^{eq} = 0.2$ p.u.) and Specified Voltage Limits: $V_{se1}^{Lim} = 0.12$ p.u., $V_{se2}^{Lim} = 0.048$ p.u., and $V_{se3}^{Lim} = 0.025$ p.u.			
Quantity	**SSSC-1**	**SSSC-2**	**SSSC-3**
V_{se} (p.u.)	**0.12**	**0.048**	**0.025**
θ_{se}	−125.74°	−109.76°	−101.28°
I_{se} (p.u.)	0.6979	0.6377	0.1869
γ_{se}	88.03°	89.14°	88.47°
CO	$V_{118} = 0.9588$ p.u.	$Q_{5\text{-}11} = 5.03$ MVAR	$P_{96\text{-}95} = 18.5$ MW
NI	10		
Solution of SSSC Quantities with Switching Losses ($G_C^{eq} = 0.2$ p.u.) and Specified Current Limits: $I_{se1}^{Lim} = 0.69$ p.u., $I_{se2}^{Lim} = 0.63$ p.u., and $I_{se3}^{Lim} = 0.18$ p.u.			
Quantity	**SSSC-1**	**SSSC-2**	**SSSC-3**
V_{se} (p.u.)	0.1173	0.0464	0.0231
θ_{se}	−125.6°	−109.97°	−102.46°
I_{se} (p.u.)	**0.69**	**0.63**	**0.18**
γ_{se}	88.05°	89.16°	88.53°
CO	$V_{118} = 0.9583$ p.u.	$Q_{5\text{-}11} = 5.19$ MVAR	$P_{96\text{-}95} = 17.76$ MW
NI	9		

SSSC-2, and SSSC-3. To accommodate these three limit violations, all the three COs have been relaxed. The limit violated quantities are again shown in bold cases. From Table 3.12, the increased deviations of the converged values of the COs (from the set values) and the increase in the number of iterations compared to that in the unconstrained case can be observed.

The bus voltage profiles for the two cases depicted in Table 3.11 are shown in Figures 3.17 and 3.18, respectively. From these two figures again, it is observed that even in the presence of three simultaneous limit violations, the bus voltage profile changes very little.

Table 3.12 shows the results for simultaneous limit violations of three SSSCs in the 300-bus system for two cases: (1) the simultaneous violation of the voltage limits of SSSC-2, SSSC-3, and SSSC-4 and (2) the simultaneous violation of current limits of SSSC-2, SSSC-3, and SSSC-4. Again, it is important to note that the selection of devices (SSSCs) on which limit constraints have been enforced is purely arbitrary. The limit constraints can also be enforced very easily on other SSSCs. For enforcing these limits, the objectives of controlling the real power flow in line 191-190, the reactive power flow in line 83-71, and the reactance of line 5-9 at the specified values are relaxed. Also, the number of iterations increases compared to that in the unconstrained case.

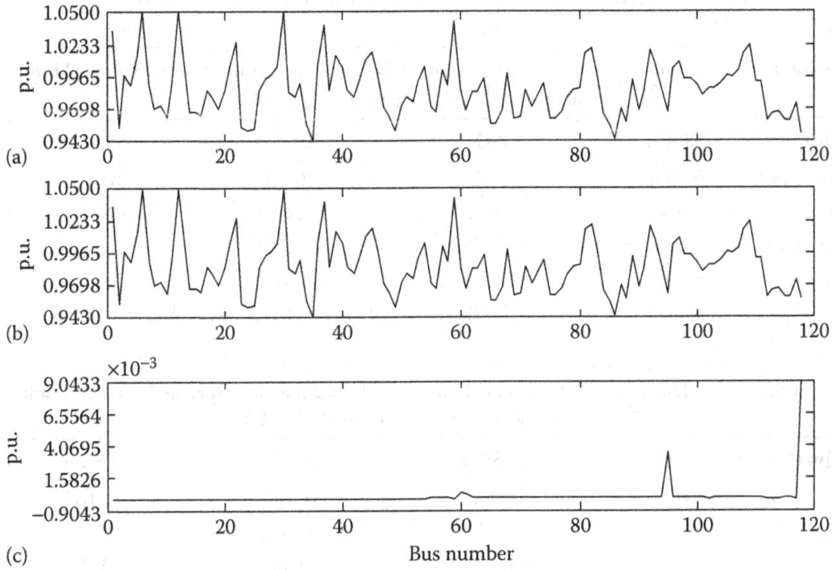

FIGURE 3.17 Bus voltage profile corresponding to the case study of Table 3.11 (with three voltage limit constraints). (a) Bus voltage magnitude without SSSC; (b) bus voltage magnitude with SSSC; (c) voltage magnitude difference.

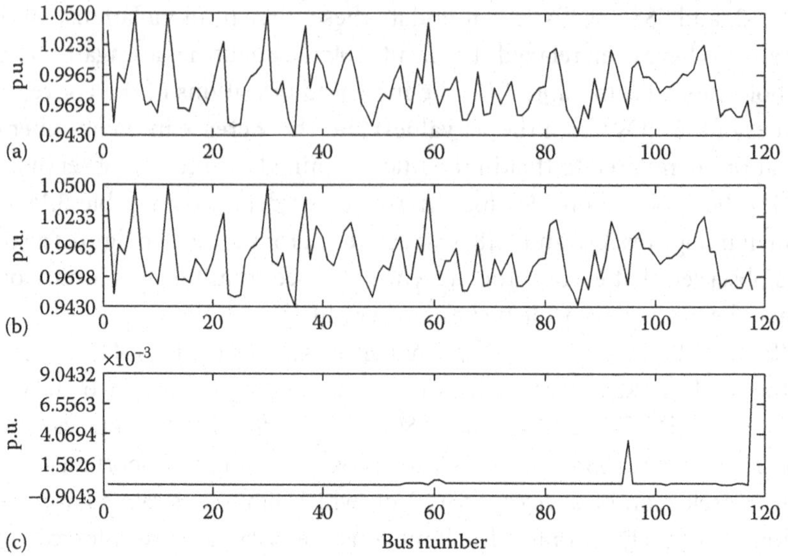

FIGURE 3.18 Bus voltage profile corresponding to the case study of Table 3.11 (with three current limit constraints). (a) Bus voltage magnitude without SSSC; (b) bus voltage magnitude with SSSC; (c) voltage magnitude difference.

TABLE 3.12 Third Study of IEEE 300-Bus System with Practical SSSC and Device Limits

Solution of SSSC Quantities with Switching Losses ($G_C^{eq} = 0.2$ p.u.) and Specified Voltage Limits: $V_{se2}^{Lim} = 0.05$ p.u., $V_{se3}^{Lim} = 0.18$ p.u. and $V_{se4}^{Lim} = 0.11$ p.u.

Quantity	SSSC-1	SSSC-2	SSSC-3	SSSC-4
V_{se} (p.u.)	0.0338	**0.05**	**0.18**	**0.11**
θ_{se}	−88.77°	−73.22°	−136.37°	−97.11°
I_{se} (p.u.)	0.3668	0.4555	1.5587	1.162
γ_{se}	88.94°	88.74°	88.68°	88.92°
CO	V_{237} = **0.97 p.u.**	$Q_{83\text{-}71}$ = 22.51 MVAR	$P_{191\text{-}190}$ = 142.9 MW	$X_{5\text{-}9}$ = −0.0946 p.u.
NI	9			

Solution of SSSC Quantities with Switching Losses ($G_C^{eq} = 0.2$ p.u.) and Specified Current Limits: $I_{se2}^{Lim} = 0.47$ p.u., $I_{se3}^{Lim} = 1.62$ p.u. and $I_{se4}^{Lim} = 1.2$ p.u.

Quantity	SSSC-1	SSSC-2	SSSC-3	SSSC-4
V_{se} (p.u.)	0.0338	0.0542	0.1946	0.1185
θ_{se}	−88.75°	−72.49°	−136.72°	−96.73°
I_{se} (p.u.)	0.3668	**0.47**	**1.62**	**1.2**
γ_{se}	88.94°	88.68°	88.62°	88.87°
CO	V_{237} = **0.97 p.u.**	$Q_{83\text{-}71}$ = 23.74 MVAR	$P_{191\text{-}190}$ = 148.1 MW	$X_{5\text{-}9}$ = −0.0987 p.u.
NI	8			

The bus voltage profiles for the two cases depicted in Table 3.12 are shown in Figures 3.19 and 3.20, respectively. From these two figures again it is observed that as in 118-bus system, in the 300-bus system also, even in the presence of three simultaneous limit violations, the bus voltage profile changes very little.

Results corresponding to the simultaneous limit violations of all four SSSCs in the 300-bus system are shown in Table 3.13. Two cases are considered: (1) the simultaneous violation of the voltage limits of all the four SSSCs and (2) the simultaneous violation of current limits of all the four SSSCs. The device limit thresholds have been set quite low than the corresponding unconstrained values. Consequently, the converged values of the COs deviate more. This demonstrates the robustness of the developed technique for stringent limit violations. Also, the number of iterations increases compared to that in the unconstrained case. It is to be noted that for incorporating these four limit constraints, all the four COs have been relaxed.

The bus voltage profiles for the two cases depicted in Table 3.13 are shown in Figures 3.21 and 3.22, respectively. From these two figures again, it is observed that in the 300-bus system, even in the presence of four simultaneous limit violations, the bus voltage profile changes very little.

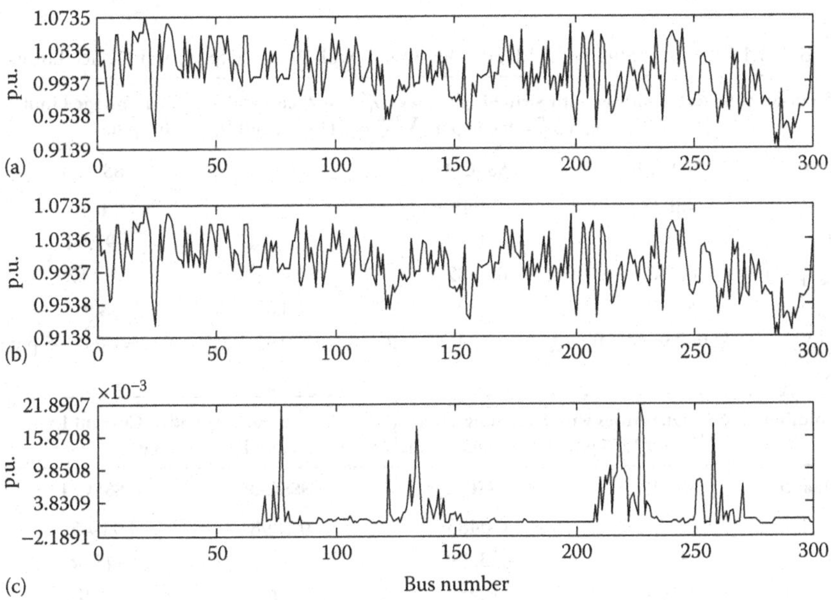

FIGURE 3.19 Bus voltage profile corresponding to the case study of Table 3.12 (with three voltage limit constraints). (a) Bus voltage magnitude without SSSC; (b) bus voltage magnitude with SSSC; (c) voltage magnitude difference.

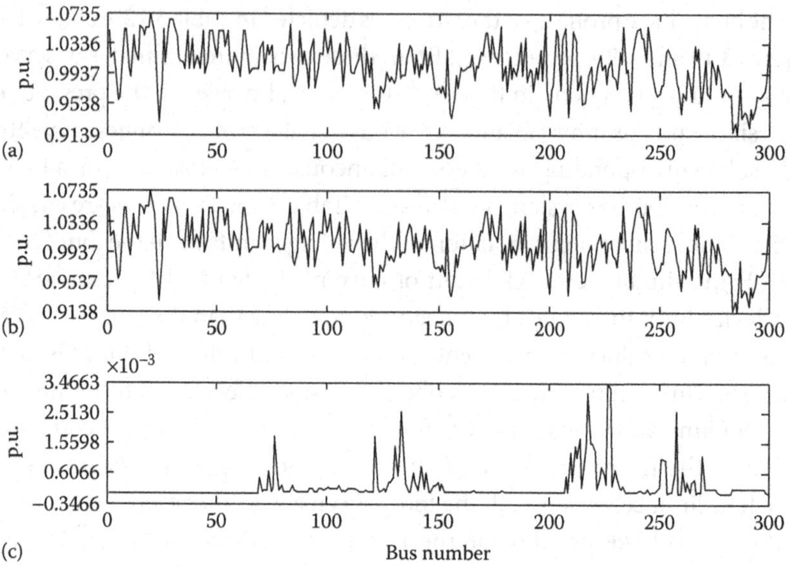

FIGURE 3.20 Bus voltage profile corresponding to the case study of Table 3.12 (with three current limit constraints). (a) Bus voltage magnitude without SSSC; (b) bus voltage magnitude with SSSC; (c) voltage magnitude difference.

TABLE 3.13 Fourth Study of IEEE 300-Bus System with Practical SSSC and Device Limits

Solution of SSSC Quantities with Switching Losses (G_C^{eq} = 0.2 p.u.) and Specified Voltage Limits: V_{se1}^{Lim} = 0.033 p.u., V_{se2}^{Lim} = 0.045 p.u., V_{se3}^{Lim} = 0.15 p.u., and V_{se4}^{Lim} = 0.1 p.u.

Quantity	SSSC-1	SSSC-2	SSSC-3	SSSC-4
V_{se} (p.u.)	**0.033**	**0.045**	**0.15**	**0.1**
θ_{se}	−88.9°	−74.06°	−135.74°	−97.54°
I_{se} (p.u.)	0.3666	0.4382	1.434	1.117
γ_{se}	88.97°	88.82°	88.8°	88.97°
CO	V_{237} = 0.9704 p.u.	$Q_{83\text{-}71}$ = 21.1 MVAR	$P_{191\text{-}190}$ = 132.3 MW	$X_{5\text{-}9}$ = −0.0895 p.u.
NI	10			

Solution of SSSC Quantities with Switching Losses (G_C^{eq} = 0.2 p.u.) and Specified Current Limits: I_{se1}^{Lim} = 0.3665 p.u., I_{se2}^{Lim} = 0.45 p.u., I_{se3}^{Lim} = 1.6 p.u., and I_{se4}^{Lim} = 1.0 p.u.

Quantity	SSSC-1	SSSC-2	SSSC-3	SSSC-4
V_{se} (p.u.)	0.0325	0.0485	0.1898	0.0741
θ_{se}	−88.82°	−73.48°	−136.57°	−98.68°
I_{se} (p.u.)	**0.3665**	**0.45**	**1.6**	**1.0**
γ_{se}	88.98°	88.77°	88.64°	89.15°
CO	V_{237} = 0.9707 p.u.	$Q_{83\text{-}71}$ = 22 MVAR	$P_{191\text{-}190}$ = 146.5 MW	$X_{5\text{-}9}$ = −0.0741 p.u.
NI	9			

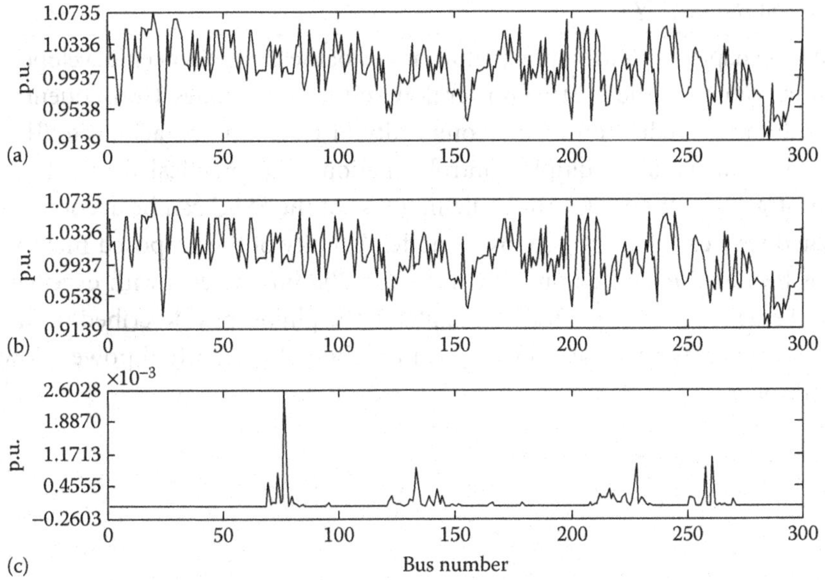

FIGURE 3.21 Bus voltage profile corresponding to the case study of Table 3.13 (with four voltage limit constraints). (a) Bus voltage magnitude without SSSC; (b) bus voltage magnitude with SSSC; (c) voltage magnitude difference.

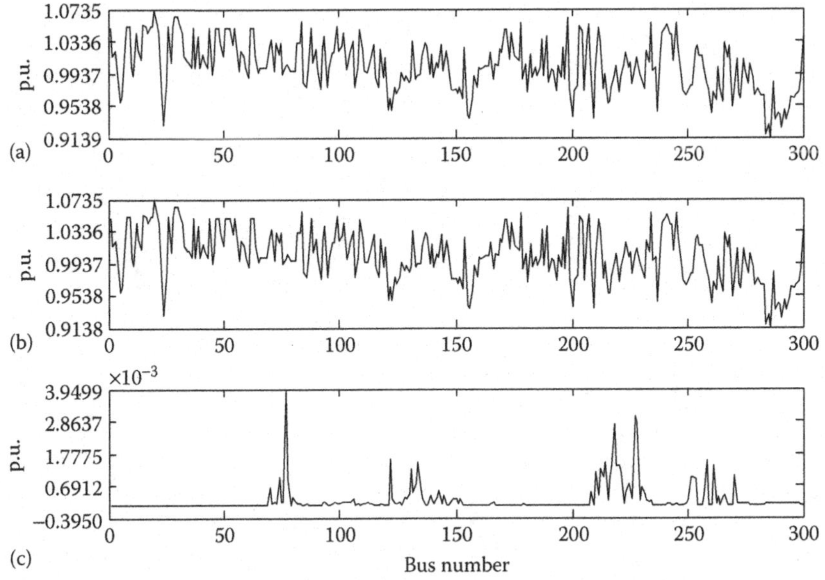

FIGURE 3.22 Bus voltage profile corresponding to the case study of Table 3.13 (with four current limit constraints). (a) Bus voltage magnitude without SSSC; (b) bus voltage magnitude with SSSC; (c) voltage magnitude difference.

3.7 SUMMARY

In this chapter, a Newton power flow model of the SSSC has been developed, which can reuse the existing power flow and Jacobian codes. Consequently, a substantial reduction in the complexity of codes can be achieved. This model can handle multiple control functions and practical device limit constraints of the SSSC. The switching loss of the SSSC can be accommodated very easily in the proposed model. Validity of the proposed method has been demonstrated on IEEE 118- and 300-bus systems with excellent convergence characteristics. In Chapter 4, the philosophy described in this chapter is extended for power flow modeling of the unified power flow controller.

CHAPTER 4

Newton Power Flow Model of the Unified Power Flow Controller

4.1 INTRODUCTION

Among the voltage-sourced converter-based flexible AC transmission system controllers, the unified power flow controller (UPFC) is one of the most comprehensive and versatile ones, which has ushered in new horizons in power transmission control [15,16]. Within its operating limits, a UPFC can independently control three power system parameters [15–21].

For proper utilization of the UPFC in power system planning, operation, and control, a power flow solution of the network incorporating UPFC(s) is a fundamental requirement. As a result, the development of a suitable power flow model for analyzing the behavior of the UPFC in large power systems has been a challenge for power system engineers worldwide. In this regard, a lot of research work has been carried out in the literature [55–76] for developing efficient power flow algorithms for the UPFC. In a way similar to the static synchronous series compensator (SSSC), it is observed from these works that the complexities of software codes are increased manifold when a UPFC is modeled in an existing Newton–Raphson power flow algorithm. In fact, compared with the SSSC, the problem is aggravated in case of the UPFC. In a UPFC, there are two representative voltage sources—one each for the shunt and series converters. Contributions from these two voltage sources necessitate modifications in the existing power flow equations for the sending end (SE) and receiving end (RE) buses of the line incorporating the UPFC. Further, an entirely new expression for the real power handled by the UPFC has to be written. Moreover, in the Jacobian matrix, multiple subblocks exclusively related to the shunt and series voltage sources of the UPFC appear, and as a result, entirely new codes need to be written for computation of each of these subblocks. Due to these factors, the complexity of software codes is greatly enhanced. As the number of UPFCs in a system increases, the problem becomes more acute.

This problem has not been directly addressed by any of the existing works on power flow modeling of the UPFC. In this regard, efforts to obtain reusability of the original Newton–Raphson power flow codes due to the incorporation of a UPFC were demonstrated in [76]. However, in this method too, new codes related to the UPFC have to be written, although separately, in a special routine (such as a class or a function), to preserve the original codes. Thus, no substantial reduction in the programming complexity is achieved. To address this issue, an indirect approach for power flow modeling of a UPFC [84,86] is proposed in this chapter. By this modeling approach, an existing power system installed with UPFCs is transformed to an equivalent augmented network without any UPFC. This results in a substantial reduction in the programming complexity because of the following reasons:

1. In the absence of any UPFC, the terms contributed by the representative shunt and series voltage sources of the UPFC cease to exist. Thus, all bus power injections can be computed in the proposed model using the existing Newton–Raphson codes.
2. The equation(s) for the UPFC real power(s) are transformed to sum(s) of real powers of additional power-flow buses, which in turn can be computed using existing codes.
3. In the proposed model, only three Jacobian subblocks need be evaluated—two of which can be evaluated using existing Jacobian codes directly, whereas the third one can be computed with very minor modifications of the existing Jacobian codes.

4.2 UPFC MODEL FOR NEWTON POWER FLOW ANALYSIS

Figure 4.1 shows an n-bus power system network in which a UPFC is connected in the branch i–j between buses i and j of the network. The UPFC is connected in series at the sending end (SE) of the transmission line.

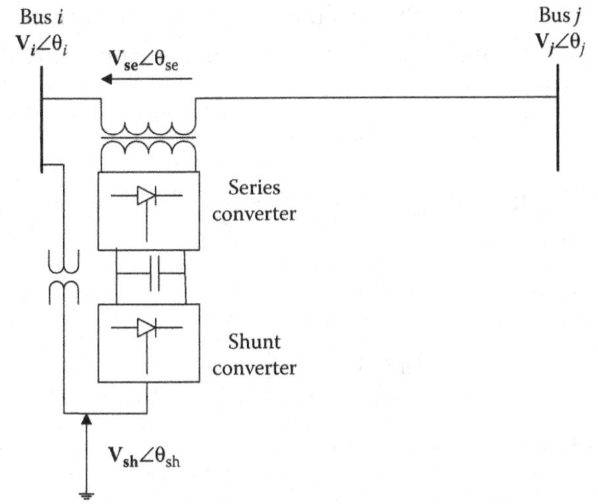

FIGURE 4.1 UPFC connected between buses i and j of an n-bus power system.

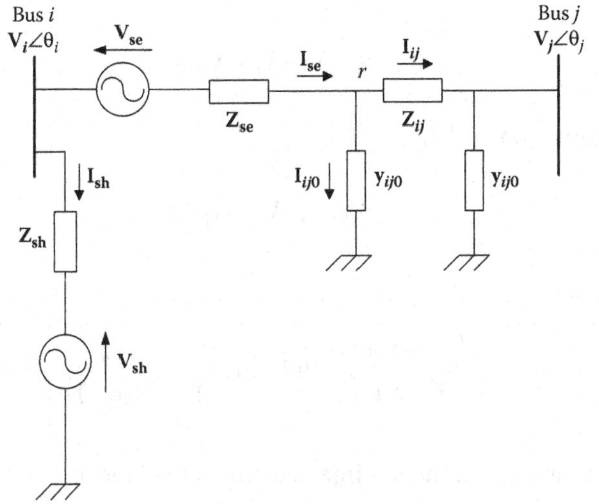

FIGURE 4.2 Equivalent circuit of the UPFC-incorporated power system network.

The equivalent circuit of the aforementioned network is shown in Figure 4.2 in which the UPFC is represented by two voltage sources. The voltage source $\mathbf{V}_{se}$ is connected in series with the transmission line, represented by its equivalent pi representation, at node r. $\mathbf{V}_r$ is the voltage of node r. As shown in Figure 4.2, the total current through the series coupling transformer is $\mathbf{I}_{se}$, which consists of two parts: (1) $\mathbf{I}_{ij}$ flowing through the line series impedance and (2) $\mathbf{I}_{ij0}$ which is the line charging current.

The current flowing through the shunt coupling transformer is $\mathbf{I}_{\mathrm{sh}}$. Also, $\mathbf{Z}_{\mathrm{se}}$ and $\mathbf{Z}_{\mathrm{sh}}$ are the impedances of the series and shunt coupling transformer, respectively.

Now,

$$\text{let}\,\mathbf{Z}_{ij} = \mathbf{R}_{ij} + j\mathbf{X}_{ij}, \mathbf{y}_{ij} = \frac{1}{\mathbf{Z}_{ij}}, \mathbf{y}_{\mathrm{se}} = \frac{1}{\mathbf{Z}_{\mathrm{se}}} \text{ and } \mathbf{y}_{\mathrm{sh}} = \frac{1}{\mathbf{Z}_{\mathrm{sh}}} \tag{4.1}$$

Also,

$$\mathbf{Y}_{ij} = \mathbf{Y}_{ji} = -\mathbf{y}_{ij} = -\mathbf{y}_{ji} \tag{4.2}$$

From Figure 4.2,

$$\mathbf{I}_{\mathrm{se}} = \mathbf{I}_{ij0} + \mathbf{I}_{ij} = \mathbf{y}_{ij0}\mathbf{V}_r + \mathbf{y}_{ij}(\mathbf{V}_r - \mathbf{V}_j) \tag{4.3}$$

Also,

$$\mathbf{I}_{\mathrm{se}} = \mathbf{y}_{\mathrm{se}}(\mathbf{V}_i - \mathbf{V}_r - \mathbf{V}_{\mathrm{se}}) \tag{4.4}$$

From Equations 4.3 and 4.4,

$$\mathbf{I}_{\mathrm{se}} = \alpha_1\mathbf{V}_i - \beta_1\mathbf{V}_j - \alpha_1\mathbf{V}_{\mathrm{se}} \tag{4.5}$$

where:

$$\alpha_1 = \frac{\mathbf{y}_{\mathrm{se}}(\mathbf{y}_{ij} + \mathbf{y}_{ij0})}{\mathbf{y}_{ij} + \mathbf{y}_{ij0} + \mathbf{y}_{\mathrm{se}}} \text{ and } \beta_1 = \frac{\mathbf{y}_{\mathrm{se}}\mathbf{y}_{ij}}{\mathbf{y}_{ij} + \mathbf{y}_{ij0} + \mathbf{y}_{\mathrm{se}}} \tag{4.6}$$

Again, from Figure 4.2, the net injected current at bus i is

$$\mathbf{I}_i = \left(\mathbf{Y}_{ii}^{\mathrm{old}} - \mathbf{y}_{ij} - \mathbf{y}_{ij0}\right)\mathbf{V}_i + \sum_{k=1,k\neq i,k\neq j}^{n} \mathbf{Y}_{ik}\mathbf{V}_k + \mathbf{I}_{\mathrm{se}} + \mathbf{I}_{\mathrm{sh}} \tag{4.7}$$

In the above equation, $\mathbf{Y}_{ii}^{\mathrm{old}} = \sum_{k=1,k\neq i}^{n} \mathbf{y}_{ik} + \mathbf{y}_{i0}^{p} + \mathbf{y}_{ij0}$ is the self-admittance of bus i for the existing n-bus system without any UPFC connected and $\mathbf{y}_{i0}^{p}$ accounts for the shunt capacitances of all transmission lines connected to bus i, except that of line i–j. Substituting $\mathbf{I}_{\mathrm{sh}} = \mathbf{y}_{\mathrm{sh}}\left(\mathbf{V}_i - \mathbf{V}_{\mathrm{sh}}\right)$ and Equation 4.5 in Equation 4.7 one gets,

$$\mathbf{I}_i = \sum_{k=1}^{n+2} \mathbf{Y}_{ik}\mathbf{V}_k \tag{4.8}$$

In the above equation, $\mathbf{Y}_{ii} = \mathbf{Y}_{ii}^{\text{old}} - \mathbf{y}_{ij} - \mathbf{y}_{ij0} + \alpha_1 + \mathbf{y}_{\text{sh}}$ is the new value of self-admittance for the ith bus with UPFC and

$$\mathbf{V}_{n+1} = \mathbf{V}_{\text{se}}, \mathbf{V}_{n+2} = \mathbf{V}_{\text{sh}}, \mathbf{Y}_{i(n+1)} = -\alpha_1, \mathbf{Y}_{i(n+2)} = -\mathbf{y}_{\text{sh}}, \text{ and } \mathbf{Y}_{ij} = -\beta_1 \tag{4.9}$$

Similarly, the net injected current at bus j can be written as

$$\mathbf{I}_j = \sum_{k=1}^{n+2} \mathbf{Y}_{jk}\mathbf{V}_k \tag{4.10}$$

provided

$$\mathbf{Y}_{j(n+1)} = \beta_1, \mathbf{Y}_{j(n+2)} = \mathbf{0}, \text{ and } \mathbf{Y}_{ji} = -\beta_1 \tag{4.11}$$

Thus, from Equations 4.8 and 4.10, it can be observed that the effect of incorporating a UPFC in a transmission line is equivalent to the addition of two more buses $(n + 1)$ and $(n + 2)$ to the existing n-bus system, provided Equations 4.9 and 4.11 are satisfied. The buses $(n + 1)$ and $(n + 2)$ are representative of the effects of the series and shunt voltage sources of the UPFC, respectively.

Now, from Figure 4.2, the net injected currents at the two fictitious power-flow buses $(n + 1)$ and $(n + 2)$ are equal to the currents flowing into the transmission system from these buses. Hence, for the equivalent $(n + 2)$ bus system,

$$\mathbf{I}_{n+1} = -\mathbf{I}_{\text{se}} = \alpha_1\mathbf{V}_{\text{se}} - \alpha_1\mathbf{V}_i + \beta_1\mathbf{V}_j = \sum_{k=1}^{n+2} \mathbf{Y}_{(n+1)k}\mathbf{V}_k \tag{4.12}$$

provided

$$\mathbf{Y}_{(n+1)(n+1)} = \alpha_1, \mathbf{Y}_{(n+1)i} = -\alpha_1, \mathbf{Y}_{(n+1)j} = \beta_1, \mathbf{Y}_{(n+1)(n+2)} = 0 \tag{4.13}$$

Also,

$$\mathbf{I}_{n+2} = -\mathbf{I}_{\text{sh}} = \mathbf{y}_{\text{sh}}(\mathbf{V}_{\text{sh}} - \mathbf{V}_i) = \sum_{k=1}^{n+2} \mathbf{Y}_{(n+2)k}\mathbf{V}_k \tag{4.14}$$

where:

$$\mathbf{Y}_{(n+2)(n+2)} = \mathbf{y}_{\text{sh}}, \mathbf{Y}_{(n+2)i} = -\mathbf{y}_{\text{sh}}, \mathbf{Y}_{(n+2)j} = 0, \mathbf{Y}_{(n+2)(n+1)} = 0 \tag{4.15}$$

4.3 POWER FLOW EQUATIONS IN THE PROPOSED UPFC MODEL

To bring out the advantages of the proposed UPFC model, the relevant equations from the existing UPFC models are first described followed by the corresponding equations obtained by the developed UPFC model. Let there be p UPFCs incorporated in an existing n-bus system. With existing UPFC models, from the equations of the net injected currents at buses i and j (e.g., Equation 4.7), the expression for the active power injection at any (SE or RE) bus a can be written as

$$P_a = \sum_{k=1}^{n} V_a V_k Y_{ak} \cos\left(\theta_a - \theta_k - \varphi_{yak}\right), \quad a \leq n \tag{4.16}$$

if no UPFC is connected to bus a;

$$\begin{aligned} P_a = & \sum_{k=1,\, k \neq b}^{n} V_a V_k Y_{ak} \cos\left(\theta_a - \theta_k - \varphi_{yak}\right) \\ & - V_a V_b \beta_c \cos\left(\theta_a - \theta_b - \varphi_{y\beta c}\right) \\ & - V_a V_{\text{se}\,c} \alpha_c \cos\left(\theta_a - \theta_{\text{se}\,c} - \varphi_{y\alpha c}\right) \\ & - V_a V_{\text{sh}c}\, y_{\text{sh}c} \cos\left(\theta_a - \theta_{\text{sh}\,c} - \varphi_{y\text{sh}c}\right) \end{aligned} \tag{4.17}$$

if UPFC c is connected in branch a–b with a as SE bus $\left(c \leq p\right)$;

$$\begin{aligned} P_a = & \sum_{k=1, k \neq b}^{n} V_a V_k Y_{ak} \cos\left(\theta_a - \theta_k - \varphi_{yak}\right) - V_a V_b \beta_c \cos\left(\theta_a - \theta_b - \varphi_{y\beta c}\right) \\ & + V_a V_{\text{se}\,c} \beta_c \cos\left(\theta_a - \theta_{\text{se}\,c} - \varphi_{y\beta c}\right) \end{aligned} \tag{4.18}$$

if UPFC c is connected in branch a–b with a as RE bus.

Thus, from the above equation, it is observed that with existing UPFC models, additional terms arising out of the contributions from the series and shunt voltage sources representing the UPFC have come into the picture.

These necessitate modifications in the existing power flow software codes. This is also true for the reactive power injection equation at any bus.

However, in the proposed model, there would be a total of $(n + 2p)$ buses, without any UPFC. The effect of the cth UPFC comes into the picture as two additional fictitious power-flow buses $(n + 2c - 1)$ and $(n + 2c)$ representative of the series and the shunt converters, respectively. The expression for the active power at any bus a (SE or RE) can be written using Equations 4.8 and 4.10 as

$$P_a = \sum_{k=1}^{n+2p} V_a V_k Y_{ak} \cos\left(\theta_a - \theta_k - \varphi_{yak}\right), \quad a \le (n+2p) \tag{4.19}$$

Similarly, the expression for the reactive power at any bus a (SE or RE) can be written as

$$Q_a = \sum_{k=1}^{n+2p} V_a V_k Y_{ak} \sin\left(\theta_a - \theta_k - \varphi_{yak}\right), \quad a \le (n+2p) \tag{4.20}$$

From the above two equations, it can be observed that both the active and reactive power injections at any bus can be computed using existing power flow codes.

Now, with existing models, the expression for the real power delivered by any UPFC c connected between buses a and b is

$$\begin{aligned} P_{\text{UPFC}c} &= \text{Re}\left[\mathbf{V}_{\text{se}c}(-\mathbf{I}^*_{\text{se}c}) + \mathbf{V}_{\text{sh}c}(-\mathbf{I}^*_{\text{sh}c})\right] \\ &= V_{\text{se}c} V_b \beta_c \cos\left(\theta_{\text{se}c} - \theta_b - \varphi_{y\beta c}\right) \\ &\quad - V_{\text{se}c} V_a \alpha_c \cos\left(\theta_{\text{se}c} - \theta_a - \varphi_{y\alpha c}\right) \\ &\quad - V_{\text{sh}c} V_a y_{\text{sh}c} \cos\left(\theta_{\text{sh}c} - \theta_a - \varphi_{y\text{sh}c}\right) \\ &\quad + V^2_{\text{sh}c} y_{\text{sh}c} \cos \varphi_{y\text{sh}c} + V^2_{\text{se}c} \alpha_c \cos \varphi_{y\alpha c} \end{aligned} \tag{4.21}$$

From the above equation, it is observed that fresh codes are required for its evaluation. In the proposed model, the expression of $P_{\text{UPFC}c}$ can be written using Equations 4.12 and 4.14 as

$$\begin{aligned} P_{\text{UPFC}c} &= \text{Re}\left[\mathbf{V}_{\mathbf{sec}}(-\mathbf{I}_{\mathbf{sec}}^{*}) + \mathbf{V}_{\mathbf{sh}c}(-\mathbf{I}_{\mathbf{sh}c}^{*})\right] \\ &= \text{Re}\left[\mathbf{V}_{n+2c-1}(\mathbf{I}_{n+2c-1}^{*}) + \mathbf{V}_{n+2c}(\mathbf{I}_{n+2c}^{*})\right] \\ &= \text{Re}\left(\mathbf{S}_{n+2c-1} + \mathbf{S}_{n+2c}\right) = P_{n+2c-1} + P_{n+2c} \end{aligned} \tag{4.22}$$

From the above equation, it can be observed that in the proposed model, the sum of active power injections at two buses $(n + 2c - 1)$ and $(n + 2c)$ (computed from Equation 4.19 with existing power flow codes) yields the UPFC real power.

Apart from the equations corresponding to real and reactive power injections, those corresponding to various control objectives can also be evaluated using existing codes in the proposed model. Generally, a UPFC can independently control three power system parameters. Normally, the SE bus voltage, the line active power flow, and the line reactive power flow comprise these three control objectives [72]. In this chapter too, these three same control objectives have been considered.

With existing UPFC models, using Equation 4.5, the expression for the line active power flow with the UPFC c connected between buses a and b is

$$P_{ab} = \text{Re}\left(\mathbf{V}_a \mathbf{I}_{ab}^{*}\right) = \text{Re}\left(\mathbf{V}_a \mathbf{I}_{\mathbf{sec}}^{*}\right)$$

or

$$\begin{aligned} P_{ab} &= V_a^2 \alpha_c \cos\varphi_{y\alpha c} - V_a V_b \beta_c \cos\left(\theta_a - \theta_b - \varphi_{y\beta c}\right) \\ &- V_a V_{\text{sec}} \alpha_c \cos\left(\theta_a - \theta_{\text{sec}} - \varphi_{y\alpha c}\right) \end{aligned} \tag{4.23}$$

Similarly, the expression for the line reactive power flow is given by

$$\begin{aligned} Q_{ab} &= -V_a^2 \alpha_c \sin\varphi_{y\alpha c} - V_a V_b \beta_c \sin\left(\theta_a - \theta_b - \varphi_{y\beta c}\right) \\ &- V_a V_{\text{sec}} \alpha_c \sin\left(\theta_a - \theta_{\text{sec}} - \varphi_{y\alpha c}\right) \end{aligned} \tag{4.24}$$

Both the above two equations have new terms involving the UPFC series voltage source and the modified admittances due to the UPFC series coupling transformers. These necessitate modifications in the existing power flow codes.

In the proposed model, using Equation 4.12, the expression for active power flow becomes

$$P_{ab} = \mathrm{Re}\left(\mathbf{V}_a \mathbf{I}_{ab}^*\right) = \mathrm{Re}\left(\mathbf{V}_a \mathbf{I}_{sec}^*\right) = \mathrm{Re}\left[\mathbf{V}_a\left(-\mathbf{I}_{n+2c-1}^*\right)\right]$$

or

$$P_{ab} = -\sum_{k=1}^{n+2p} V_a V_k Y_{(n+2c-1)k} \cos\left(\theta_a - \theta_k - \varphi_{y(n+2c-1)k}\right) \tag{4.25}$$

as the series converter of the cth UPFC is transformed to $(n + 2c - 1)$th power flow bus in the proposed model. In a similar way,

$$Q_{ab} = -\sum_{k=1}^{n+2p} V_a V_k Y_{(n+2c-1)k} \sin\left(\theta_a - \theta_k - \varphi_{y(n+2c-1)k}\right) \tag{4.26}$$

From the above two equations, it is observed that both the line active and reactive power flows can be evaluated in the proposed model using very minor modifications in the existing power flow codes.

4.4 IMPLEMENTATION IN NEWTON POWER FLOW ANALYSIS

If the number of voltage-controlled buses is $(m - 1)$, the power-flow problem for an n-bus system with p UPFCs can be formulated as follows:

Solve $\boldsymbol{\theta}$, $\mathbf{V}$, $\boldsymbol{\theta}_{\mathbf{se}}$, $\mathbf{V}_{\mathbf{se}}$, $\boldsymbol{\theta}_{\mathbf{sh}}$, and $\mathbf{V}_{\mathbf{sh}}$.

Specified $\mathbf{P}$, $\mathbf{Q}$, $\mathbf{P_U}$, $\mathbf{V_B}$, $\mathbf{P_L}$, and $\mathbf{Q_L}$

where:

$$\boldsymbol{\theta} = [\theta_2 \ldots \theta_n]^{\mathrm{T}} \quad \mathbf{V} = [V_{m+1} \ldots V_n]^{\mathrm{T}} \tag{4.27}$$

$$\boldsymbol{\theta}_{\mathbf{se}} = [\theta_{se1} \ldots \theta_{sep}]^{\mathrm{T}} \quad \mathbf{V}_{\mathbf{se}} = [V_{se1} \ldots V_{sep}]^{\mathrm{T}} \tag{4.28}$$

$$\boldsymbol{\theta}_{\mathbf{sh}} = [\theta_{sh1} \ldots \theta_{shp}]^{\mathrm{T}} \quad \mathbf{V}_{\mathbf{sh}} = [V_{sh1} \ldots V_{shp}]^{\mathrm{T}} \tag{4.29}$$

$$\mathbf{P} = [P_2 \ldots P_n]^{\mathrm{T}} \quad \mathbf{Q} = [Q_{m+1} \ldots Q_n]^{\mathrm{T}} \tag{4.30}$$

$$\mathbf{P_U} = [P_{UPFC1} \ldots P_{UPFCp}]^T \quad \mathbf{V_B} = [V_{BUS1} \ldots V_{BUSp}]^T \tag{4.31}$$

$$\mathbf{P_L} = [P_{LINE1} \ldots P_{LINEp}]^T \quad \mathbf{Q_L} = [Q_{LINE1} \ldots Q_{LINEp}]^T \tag{4.32}$$

In Equations 4.27 through 4.32 $\mathbf{P_U}$, $\mathbf{V_B}$, $\mathbf{P_L}$, and $\mathbf{Q_L}$ represent the vectors for the specified real powers (supplied by the UPFCs), SE bus voltages (at which the p UPFCs are connected), line active power flows, and line reactive power flows (in the p lines incorporating the UPFCs) of the p UPFCs, respectively. Without any loss of generality in Equations 4.27 through 4.32, it is assumed that there are m generators connected at the first m buses of the system with bus 1 being the slack bus. Thus, the basic power-flow equation for the Newton power flow solution is represented as

$$\begin{bmatrix} \frac{\partial \mathbf{P}}{\partial \theta} & \frac{\partial \mathbf{P}}{\partial \mathbf{V}} & \frac{\partial \mathbf{P}}{\partial \theta_{se}} & \frac{\partial \mathbf{P}}{\partial \mathbf{V}_{se}} & \frac{\partial \mathbf{P}}{\partial \theta_{sh}} & \frac{\partial \mathbf{P}}{\partial \mathbf{V}_{sh}} \\ \frac{\partial \mathbf{Q}}{\partial \theta} & \frac{\partial \mathbf{Q}}{\partial \mathbf{V}} & \frac{\partial \mathbf{Q}}{\partial \theta_{se}} & \frac{\partial \mathbf{Q}}{\partial \mathbf{V}_{se}} & \frac{\partial \mathbf{Q}}{\partial \theta_{sh}} & \frac{\partial \mathbf{Q}}{\partial \mathbf{V}_{sh}} \\ \frac{\partial \mathbf{P_U}}{\partial \theta} & \frac{\partial \mathbf{P_U}}{\partial \mathbf{V}} & \frac{\partial \mathbf{P_U}}{\partial \theta_{se}} & \frac{\partial \mathbf{P_U}}{\partial \mathbf{V}_{se}} & \frac{\partial \mathbf{P_U}}{\partial \theta_{sh}} & \frac{\partial \mathbf{P_U}}{\partial \mathbf{V}_{sh}} \\ \frac{\partial \mathbf{V_B}}{\partial \theta} & \frac{\partial \mathbf{V_B}}{\partial \mathbf{V}} & \frac{\partial \mathbf{V_B}}{\partial \theta_{se}} & \frac{\partial \mathbf{V_B}}{\partial \mathbf{V}_{se}} & \frac{\partial \mathbf{V_B}}{\partial \theta_{sh}} & \frac{\partial \mathbf{V_B}}{\partial \mathbf{V}_{sh}} \\ \frac{\partial \mathbf{P_L}}{\partial \theta} & \frac{\partial \mathbf{P_L}}{\partial \mathbf{V}} & \frac{\partial \mathbf{P_L}}{\partial \theta_{se}} & \frac{\partial \mathbf{P_L}}{\partial \mathbf{V}_{se}} & \frac{\partial \mathbf{P_L}}{\partial \theta_{sh}} & \frac{\partial \mathbf{P_L}}{\partial \mathbf{V}_{sh}} \\ \frac{\partial \mathbf{Q_L}}{\partial \theta} & \frac{\partial \mathbf{Q_L}}{\partial \mathbf{V}} & \frac{\partial \mathbf{Q_L}}{\partial \theta_{se}} & \frac{\partial \mathbf{Q_L}}{\partial \mathbf{V}_{se}} & \frac{\partial \mathbf{Q_L}}{\partial \theta_{sh}} & \frac{\partial \mathbf{Q_L}}{\partial \mathbf{V}_{sh}} \end{bmatrix} \begin{bmatrix} \Delta\theta \\ \Delta\mathbf{V} \\ \Delta\theta_{se} \\ \Delta\mathbf{V}_{se} \\ \Delta\theta_{sh} \\ \Delta\mathbf{V}_{sh} \end{bmatrix} = \begin{bmatrix} \Delta\mathbf{P} \\ \Delta\mathbf{Q} \\ \Delta\mathbf{P_U} \\ \Delta\mathbf{V_B} \\ \Delta\mathbf{P_L} \\ \Delta\mathbf{Q_L} \end{bmatrix} \tag{4.33}$$

or

$$\begin{bmatrix} \multicolumn{2}{c}{\mathbf{J}_{old}} & \mathbf{J}_1 & \mathbf{J}_2 & \mathbf{J}_3 & \mathbf{J}_4 \\ & & \mathbf{J}_5 & \mathbf{J}_6 & \mathbf{J}_7 & \mathbf{J}_8 \\ \mathbf{J}_9 & \mathbf{J}_{10} & \mathbf{J}_{11} & \mathbf{J}_{12} & \mathbf{J}_{13} & \mathbf{J}_{14} \\ 0 & \mathbf{J}_{15} & 0 & 0 & 0 & 0 \\ \mathbf{J}_{16} & \mathbf{J}_{17} & \mathbf{J}_{18} & \mathbf{J}_{19} & \mathbf{J}_{20} & \mathbf{J}_{21} \\ \mathbf{J}_{22} & \mathbf{J}_{23} & \mathbf{J}_{24} & \mathbf{J}_{25} & \mathbf{J}_{26} & \mathbf{J}_{27} \end{bmatrix} \begin{bmatrix} \Delta\theta \\ \Delta\mathbf{V} \\ \Delta\theta_{se} \\ \Delta\mathbf{V}_{se} \\ \Delta\theta_{sh} \\ \Delta\mathbf{V}_{sh} \end{bmatrix} = \begin{bmatrix} \Delta\mathbf{P} \\ \Delta\mathbf{Q} \\ \Delta\mathbf{P_U} \\ \Delta\mathbf{V_B} \\ \Delta\mathbf{P_L} \\ \Delta\mathbf{Q_L} \end{bmatrix} \tag{4.34}$$

In the above equation, $\mathbf{J}_{\text{old}}$ is the conventional power-flow Jacobian subblock corresponding to the angle and voltage magnitude variables of the n-buses.

Now, in the proposed model, there are $(n + 2p)$ buses, including $2p$ fictitious power flow buses. Thus, the quantities to be solved for power flow are $\boldsymbol{\theta}^{\text{new}}$ and $\mathbf{V}^{\text{new}}$.

$$\boldsymbol{\theta}^{\text{new}} = [\theta_2 \ldots \theta_{n+2p}]^{\text{T}} \text{ and } \mathbf{V}^{\text{new}} = [V_{m+1} \ldots V_{n+2p}]^{\text{T}} \tag{4.35}$$

Thus, Equation 4.34 is transformed in the proposed model as

$$\begin{bmatrix} \mathbf{JX1} \\ \hline \mathbf{JX2} \\ \hline \mathbf{0} \quad \mathbf{JX3} \\ \hline \mathbf{JX4} \end{bmatrix} \begin{bmatrix} \Delta\boldsymbol{\theta}^{\text{new}} \\ \hline \Delta\mathbf{V}^{\text{new}} \end{bmatrix} = \begin{bmatrix} \Delta\mathbf{PQ} \\ \hline \Delta\mathbf{P}_{\text{U}} \\ \hline \Delta\mathbf{V}_{\text{B}} \\ \hline \Delta\mathbf{PQ}_{\text{L}} \end{bmatrix} \tag{4.36}$$

where:

$$\Delta\mathbf{PQ} = [\Delta\mathbf{P}^{\text{T}}\ \Delta\mathbf{Q}^{\text{T}}]^{\text{T}} \quad \Delta\mathbf{PQ}_{\text{L}} = [\Delta\mathbf{P}_{\text{L}}^{\text{T}}\ \Delta\mathbf{Q}_{\text{L}}^{\text{T}}]^{\text{T}} \tag{4.37}$$

Also, **JX1**, **JX2**, **JX3**, and **JX4** are identified easily from Equation 4.36. The elements of the matrix **JX3** are either unity or zero, depending on whether an element of the vector $\mathbf{V}_B$ is also an element of the vector $\mathbf{V}^{\text{new}}$ or not. Thus, **JX3** is a constant matrix known *a priori* and does not need to be computed.

Now, it can be shown that in Equation 4.36,

1. The matrices **JX1** and **JX2** can be computed using existing Jacobian codes.
2. The matrix **JX4** can be computed using very minor modifications of the existing Jacobian codes.

The justification of the above two statements are shown as follows.

To compute the two matrices **JX1** and **JX2** in the proposed model with $(n + 2p)$ buses, let

$$\mathbf{P}^{\text{new}} = [P_2 \ldots P_{n+2p}]^{\text{T}} \tag{4.38}$$

Subsequently, a new Jacobian matrix is defined as

$$\mathbf{J}^{\text{new}} = \left[\begin{array}{cc} \dfrac{\partial \mathbf{P}^{\text{new}}}{\partial \boldsymbol{\theta}^{\text{new}}} & \dfrac{\partial \mathbf{P}^{\text{new}}}{\partial \mathbf{V}^{\text{new}}} \\ \hdashline \dfrac{\partial \mathbf{Q}}{\partial \boldsymbol{\theta}^{\text{new}}} & \dfrac{\partial \mathbf{Q}}{\partial \mathbf{V}^{\text{new}}} \end{array}\right] = \left[\begin{array}{c} \mathbf{J}^{\text{A}} \\ \hdashline \mathbf{J}^{\text{B}} \end{array}\right] \tag{4.39}$$

It can be observed (from Equations 4.19 and 4.20) that in the proposed model, the expressions for P_i $(2 \le i \le n+p)$ and Q_i $(m+1 \le i \le n)$ in $\mathbf{P}^{\text{new}}$ and $\mathbf{Q}$, respectively (of Equation 4.39), can be computed using existing power flow codes. Hence, the matrix $\mathbf{J}^{\text{new}}$ can be computed with existing codes for calculating the Jacobian matrix.

From Equations 4.35 and 4.38,

$$\frac{\partial \mathbf{P}^{\text{new}}}{\partial \boldsymbol{\theta}^{\text{new}}} = \left[\begin{array}{c} \dfrac{\partial \mathbf{P}}{\partial \boldsymbol{\theta}^{\text{new}}} \\ \hdashline \mathbf{JX5} \end{array}\right] \tag{4.40}$$

where:

$$\mathbf{JX5} = \left[\begin{array}{cc:c:cc} \dfrac{\partial P_{n+1}}{\partial \boldsymbol{\theta}^{\text{new}}} & \dfrac{\partial P_{n+2}}{\partial \boldsymbol{\theta}^{\text{new}}} & \cdots \quad \cdots & \dfrac{\partial P_{n+2p-1}}{\partial \boldsymbol{\theta}^{\text{new}}} & \dfrac{\partial P_{n+2p}}{\partial \boldsymbol{\theta}^{\text{new}}} \end{array}\right]^{\text{T}} \tag{4.41}$$

From Equation 4.40, it is observed that the subblock $\partial \mathbf{P}/\partial \boldsymbol{\theta}^{\text{new}}$ is contained within $\partial \mathbf{P}^{\text{new}}/\partial \boldsymbol{\theta}^{\text{new}}$. In a similar way, it can be shown that $\partial \mathbf{P}/\partial \mathbf{V}^{\text{new}}$ is contained within $\partial \mathbf{P}^{\text{new}}/\partial \mathbf{V}^{\text{new}}$. Hence, once the matrix $\mathbf{J}^{\text{new}}$ is computed, the matrix

$$\mathbf{JX1} = \left[\begin{array}{cc} \dfrac{\partial \mathbf{P}}{\partial \boldsymbol{\theta}^{\text{new}}} & \dfrac{\partial \mathbf{P}}{\partial \mathbf{V}^{\text{new}}} \\ \hdashline \multicolumn{2}{c}{\mathbf{J}^{\text{B}}} \end{array}\right]$$

can be very easily extracted from the matrix $\mathbf{J}^{\text{new}}$ using simple matrix extraction codes only. Hence, no fresh codes need to be written for computing $\mathbf{JX1}$.

Now, using Equation 4.22,

$$\frac{\partial}{\partial \boldsymbol{\theta}^{\text{new}}}[P_{n+1} + P_{n+2}] = \frac{\partial P_{\text{UPFC1}}}{\partial \boldsymbol{\theta}^{\text{new}}} \tag{4.42}$$

Thus, in Equation 4.41, the sum of the first two rows of **JX5** equals $\partial P_{\mathrm{UPFC}\,1}/\partial \theta^{\mathrm{new}}$ (the first element of the Jacobian subblock $\partial \mathbf{P}_{\mathrm{U}}/\partial \theta^{\mathrm{new}}$). Similarly, the sum of the last two rows yields the last element of $\partial \mathbf{P}_{\mathrm{U}}/\partial \theta^{\mathrm{new}}$. Therefore, $\partial \mathbf{P}_{\mathrm{U}}/\partial \theta^{\mathrm{new}}$ can be easily extracted from $\partial \mathbf{P}^{\mathrm{new}}/\partial \theta^{\mathrm{new}}$. In a similar way, it can be shown that $\partial \mathbf{P}_{\mathrm{U}}/\partial \mathbf{V}^{\mathrm{new}}$ can also be easily extracted from $\partial \mathbf{P}^{\mathrm{new}}/\partial \mathbf{V}^{\mathrm{new}}$. Hence, the matrix $\mathbf{JX2} = \left[\partial \mathbf{P}_{\mathrm{U}}/\partial \theta^{\mathrm{new}} \;\; \partial \mathbf{P}_{\mathrm{U}}/\partial \mathbf{V}^{\mathrm{new}}\right]$ does not need to be computed—it can be formed from the matrix $\mathbf{J}^{\mathrm{A}}$ of $\mathbf{J}^{\mathrm{new}}$ by matrix extraction codes (in conjunction with codes for simple matrix row addition). Hence, both **JX1** and **JX2** need not be computed and can be extracted from $\mathbf{J}^{\mathrm{new}}$ (subsequent to its computation using existing Jacobian codes).

Next, it is shown that unlike existing UPFC models, the matrix

$$\mathbf{JX4} = \begin{bmatrix} \partial \mathbf{P}_{\mathrm{L}}/\partial \theta^{\mathrm{new}} & \partial \mathbf{P}_{\mathrm{L}}/\partial \mathbf{V}^{\mathrm{new}} \\ \partial \mathbf{Q}_{\mathrm{L}}/\partial \theta^{\mathrm{new}} & \partial \mathbf{Q}_{\mathrm{L}}/\partial \mathbf{V}^{\mathrm{new}} \end{bmatrix}$$

can be computed using very minor modifications of the existing Jacobian codes.

In the proposed model, the expression for the active power flow in the cth UPFC connected line (between SE and RE buses a and b, respectively) can be written using Equation 4.25 as

$$P_{\mathrm{LINE}c} = -\sum_{k=1}^{n+2p} V_a V_k Y_{(n+2c-1)k} \cos\left(\theta_a - \theta_k - \varphi_{y(n+2c-1)k}\right) \tag{4.43}$$

In a similar way, using Equation 4.26,

$$Q_{\mathrm{LINE}c} = -\sum_{k=1}^{n+2p} V_a V_k Y_{(n+2c-1)k} \sin\left(\theta_a - \theta_k - \varphi_{y(n+2c-1)k}\right) \tag{4.44}$$

It can be observed that unlike Equations 4.23 and 4.24, Equations 4.43 and 4.44 can be computed using very minor modifications of the existing power flow codes. Consequently, the matrix **JX4**, which constitutes the partial derivatives of Equations 4.43 and 4.44 with respect to the relevant variables (θ_i, θ_j, θ_{se}, θ_{sh}, V_i, V_j, V_{se}, and V_{sh}), can also be computed with very minor modifications of the existing Jacobian codes, unlike with existing UPFC models. This reduces the complexity of software code substantially.

4.5 ACCOMMODATION OF UPFC DEVICE LIMIT CONSTRAINTS

The four major device limit constraints of the UPFC considered in this book are as follows [70]:

1. The UPFC series-injected voltage magnitude $V_{\text{se}}^{\text{Lim}}$
2. The line current through the series converter $I_{\text{se}}^{\text{Lim}}$
3. Real power transfer through the dc link $P_{\text{DC}}^{\text{Lim}}$
4. Shunt converter current $I_{\text{sh}}^{\text{Lim}}$

These four device limit constraints have been taken into account following the guidelines of [70] and [71]. The device limit constraints have been accommodated by the principle that whenever a particular constraint limit is violated, it is kept at its specified limit, whereas a control objective is relaxed. Mathematically, this signifies the replacement of the Jacobian elements pertaining to the control objective by those of the violated constraints.

The four device constraint limits of $V_{\text{se}}^{\text{Lim}}$, $I_{\text{se}}^{\text{Lim}}$, $P_{\text{DC}}^{\text{Lim}}$, and $I_{\text{sh}}^{\text{Lim}}$ can be incorporated as follows:

1. *Limit on UPFC series converter voltage* (V_{se}^{Lim})

   For considering $V_{\text{se}}^{\text{Lim}}$ for the *m*th UPFC (connected in the line between buses *i* [SE] and *j* [RE]), V_{n+2m-1} is preset at $V_{\text{se}}^{\text{Lim}}$ and either the line active ($P_{\text{LINE}m}$) or reactive power flow ($Q_{\text{LINE}m}$) control objective is relaxed. This relaxed active or reactive power mismatch is replaced by $\Delta V_{n+2m-1} = V_{\text{se}m}^{\text{Lim}} - V_{n+2m-1}$ and the corresponding row of the matrix **JX4** is rendered constant with all elements known *a priori*—all of them are equal to zero except the entry pertaining to V_{n+2m-1}, which is unity. All other elements of the matrix **JX4** can be computed using very minor modifications of the existing Jacobian codes, as demonstrated in the previous Section 4.4.

2. *Limit on line current through the UPFC series converter* (I_{se}^{Lim})

   In this case also, I_{n+2m-1} is preset at $I_{\text{se}}^{\text{Lim}}$ and either P_{LINE} or Q_{LINE} control objective is relaxed. Thus, the corresponding mismatch is replaced by $\Delta I_{n+2m-1} = I_{\text{se}m}^{\text{Lim}} - I_{n+2m-1}$, where (from Equations 4.12 and 4.13)

$$I_{n+2m-1} = [a_1 + a_2 + a_3 + a_4]^{1/2} \tag{4.45}$$

where:

$$a_1 = Y^2_{(n+2m-1)i}\, V_i^2 + Y^2_{(n+2m-1)j}\, V_j^2 + Y^2_{(n+2m-1)(n+2m-1)}\, V^2_{n+2m-1}$$
$$a_2 = 2Y_{(n+2m-1)i}\, Y_{(n+2m-1)j}\, V_i\, V_j \cos\left(\theta_i - \theta_j + \varphi_{y(n+2m-1)i} - \varphi_{y(n+2m-1)j}\right)$$
$$a_3 = 2Y_{(n+2m-1)i} Y_{(n+2m-1)(n+2m-1)} V_i V_{n+2m-1} \cos\left(\theta_i - \theta_{n+2m-1} + \varphi_{y(n+2m-1)i} - \varphi_{y(n+2m-1)(n+2m-1)}\right)$$
$$a_4 = 2Y_{(n+2m-1)j} Y_{(n+2m-1)(n+2m-1)} V_j V_{n+2m-1} \cos\left(\theta_j - \theta_{n+2m-1} + \varphi_{y(n+2m-1)j} - \varphi_{y(n+2m-1)(n+2m-1)}\right)$$

The Jacobian elements are changed accordingly.

If, however, both V_{se}^{Lim} and I_{se}^{Lim} are violated, leniency is exercised on both P_{LINE} and Q_{LINE} control objectives, which are replaced by ΔV_{n+2m-1} and ΔI_{n+2m-1}, respectively. The corresponding Jacobian elements are also replaced. The derivations of the expressions of a_1, a_2, a_3, and a_4 are given in the Appendix.

3. *Limit on real power transfer through the dc link* (P_{DC}^{Lim})

In the proposed model, the series and shunt converters of the *m*th UPFC are transformed to fictitious power-flow buses (*n* + 2*m* – 1) and (*n* + 2*m*), respectively. For the *m*th UPFC (connected in the line between buses say *i* [SE] and *j* [RE]), the real power exchange through the DC link P_{DCm} is defined (using Equations 4.5, 4.9, 4.12, 4.13, and 4.45) as

$$\begin{aligned} P_{DCm} &= \text{Re}\left[\mathbf{V}_{sem}\left(-\mathbf{I}^*_{sem}\right)\right] = \text{Re}\left(\mathbf{V}_{n+2m-1}\mathbf{I}^*_{n+2m-1}\right) \\ &= V_{n+2m-1}\, V_i\, Y_{(n+2m-1)i} \cos\left(\theta_{n+2m-1} - \theta_i - \varphi_{y(n+2m-1)i}\right) \\ &\quad + V_{n+2m-1}\, V_j\, Y_{(n+2m-1)j} \cos\left(\theta_{n+2m-1} - \theta_j - \varphi_{y(n+2m-1)j}\right) \\ &\quad + V^2_{n+2m-1}\, Y_{(n+2m-1)(n+2m-1)} \cos\varphi_{y(n+2m-1)(n+2m-1)} \end{aligned} \tag{4.46}$$

If P_{DC}^{Lim} is violated, then P_{DC} is preset at P_{DC}^{Lim} and either the P_{LINE} or Q_{LINE} control objective is relaxed, with the corresponding mismatch replaced by $\Delta P_{DCm} = P_{DCm}^{Lim} - P_{DCm}$. It is important to note that a negative value of P_{DCm} signifies that the series converter is absorbing real power from the line (and supplying it to the SE bus through the shunt converter). The Jacobian elements are changed accordingly.

4. *Limit on shunt converter current* (I_{sh}^{Lim})

   As in the previous cases, for the mth UPFC, I_{n+2m} is preset at $I_{\text{sh}}^{\text{Lim}}$ and the UPFC SE bus voltage control objective is relaxed, with the voltage mismatch replaced by $\Delta I_{n+2m} = I_{\text{sh}m}^{\text{Lim}} - I_{n+2m}$, where (from Equations 4.14 and 4.15)

$$I_{n+2m} = [b_1 + b_2]^{1/2} \tag{4.47}$$

   where:

$$b_1 = Y_{(n+2m)i}^2 V_i^2 + Y_{(n+2m)(n+2m)}^2 V_{n+2m}^2$$

$$b_2 = 2Y_{(n+2m)i} Y_{(n+2m)(n+2m)} V_i V_{n+2m} \cos\left(\theta_i - \theta_{n+2m} + \varphi_{y(n+2m)i} - \varphi_{y(n+2m)(n+2m)}\right)$$

   The Jacobian elements are changed accordingly.

   If, however, all of $P_{\text{DC}m}^{\text{Lim}}$, $I_{\text{se}m}^{\text{Lim}}$, and $I_{\text{sh}m}^{\text{Lim}}$ are violated, all the control objectives of line active and reactive power flow along with the SE bus voltage are relaxed, with the corresponding mismatches replaced by $\Delta P_{\text{DC}m}$, ΔI_{n+2m-1}, and ΔI_{n+2m}, respectively, as given before, along with modifications of the corresponding Jacobian elements. The derivations of the expressions of b_1 and b_2 are given in the Appendix.

4.6 SELECTION OF INITIAL CONDITIONS

The initial conditions for the shunt voltage source were chosen as $V_{\text{sh}}^0 \angle \theta_{\text{sh}}^0 = 1.0\angle 0^0$ following [60] whereas those for the series voltage source were chosen as $V_{\text{se}}^0 \angle \theta_{\text{se}}^0 = 0.1\angle -(\pi/2)$, as suggested in [49]. During the case studies, it was observed that the shunt converter current magnitude I_{n+2} becomes zero by adoption of the shunt voltage source initial condition. As a consequence, the Jacobians of I_{n+2} (obtained using Equation 4.47), which possess I_{n+2} terms in the denominator, were rendered indeterminate. This is shown in the Appendix. Modifying the shunt source initial condition to $V_{\text{sh}}^0 \angle \theta_{\text{sh}}^0 = 1.0\angle -(\pi/9)$ solves this problem, without any observed detrimental effect on the convergence.

4.7 CASE STUDIES AND RESULTS

The proposed method was applied to IEEE 118- and 300-bus systems to validate its feasibility. In each test system, multiple UPFCs were included and studies have been carried out for (1) UPFCs without any device limit constraints and (2) UPFCs with device limit constraints. As each UPFC can independently control up to three power system parameters, in this chapter, each UPFC is used for the control of the SE bus voltage and the line active and reactive power flows. In all the case studies, a convergence tolerance of 10^{-12} p.u. has been chosen. Although a large number of case studies confirmed the validity of the model, a few sets of representative results are presented below. As in Chapter 3, in all the subsequent tables, the symbol NI denotes the number of iterations taken by the algorithm.

4.7.1 Studies of UPFCs without Any Device Limit Constraints

4.7.1.1 Case I: IEEE 118-Bus System

In this system, three UPFCs have been considered on the lines among buses 5-11 (UPFC-1), 37-39 (UPFC-2), and 96-95 (UPFC-3). The control references chosen are $V_5^{SP} = 1.0$ p.u., $Q_{5\text{-}11}^{SP} = 5$ MVAR, and $P_{5\text{-}11}^{SP} = 100$ MW (UPFC-1); $V_{37}^{SP} = 1.0$ p.u., $Q_{37\text{-}39}^{SP} = 10$ MVAR, and $P_{37\text{-}39}^{SP} = 100$ MW (UPFC-2); and $V_{96}^{SP} = 1.0$ p.u., $Q_{96\text{-}95}^{SP} = 5$ MVAR, and $P_{96\text{-}95}^{SP} = 50$ MW (UPFC-3). The results are shown in Table 4.1, in which the final values of the controlled variables are shown in boldfaced letters. It is also to be noted that in this table as well as in all subsequent tables, a base of 100 MVA has been used for both 118- and 300-bus systems to represent P_{DC} in p.u.

From Table 4.1, it is also observed that in the presence of UPFCs, the line active and reactive power flow levels are enhanced compared to their corresponding values in the base case. Furthermore, the voltages at the three buses (at which the UPFCs are connected) are maintained precisely at 1.0 p.u. The bus voltage profiles without and with UPFCs are shown in Figure 4.3a and b, respectively. Further, the difference between the bus voltage magnitudes with and without UPFC is shown in Figure 4.3c. From this figure, it is observed that in the presence of UPFC, there is very little change in the bus voltage profile.

4.7.1.2 Case II: IEEE 300-Bus System

In this system too, three UPFCs are incorporated in branches among buses 2-3 (UPFC-1), 86-102 (UPFC-2), and 212-211 (UPFC-3). The control references chosen are $V_2^{SP} = 1.04$ p.u., $Q_{2\text{-}3}^{SP} = 50$ MVAR,

TABLE 4.1 Study of IEEE 118-Bus System with UPFC (No Device Limit Constraints)

Solution of Base Case Power Flow (without Any UPFC)			
Parameter	**Line 1**	**Line 2**	**Line 3**
P_{LINE} (MW)	78.17	75.2	14.67
Q_{LINE} (MVAR)	3.44	7.94	4.03
V_{BUS} (p.u.)	1.0018	0.9908	1.0014
NI	6		
Unconstrained Solution of UPFC Quantities			
Quantity	**UPFC-1 (Line 1)**	**UPFC-2 (Line 2)**	**UPFC-3 (Line 3)**
V_{se} (p.u.)	0.1233	0.1773	0.1072
θ_{se}	−114.32°	−120.3°	−107.42°
V_{sh} (p.u.)	0.9733	1.1429	1.0057
θ_{sh}	−16.71°	−21.85°	−4.66°
I_{se} (p.u.)	1.0012	1.005	0.5025
I_{sh} (p.u.)	0.2669	1.4292	0.0574
P_{DC} (p.u.)	0.0103	0.0089	0.0066
P_{LINE} (MW)	**100**	**100**	**50**
Q_{LINE} (MVAR)	**5**	**10**	**5**
V_{BUS} (p.u.)	**1.0**	**1.0**	**1.0**
NI	6		

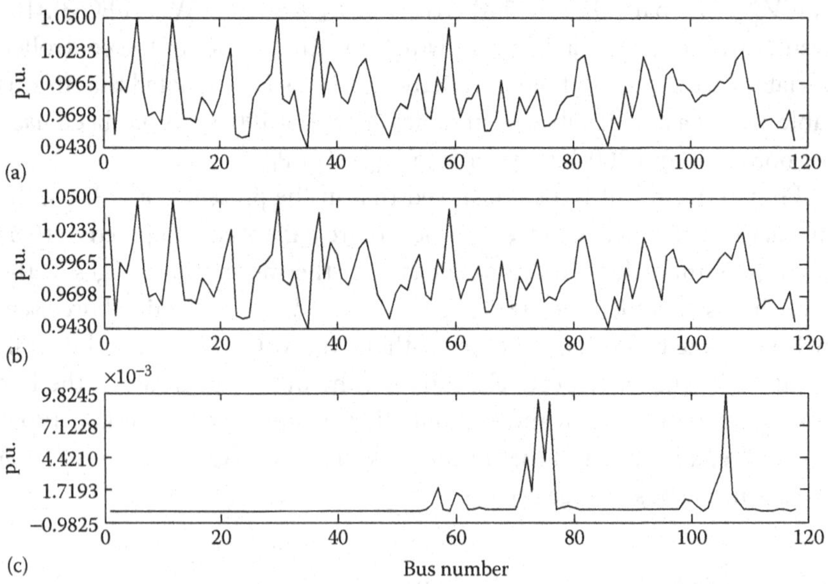

FIGURE 4.3 Bus voltage profile corresponding to the case study of Table 4.1. (a) Bus voltage magnitude without UPFC; (b) bus voltage magnitude with UPFC; (c) voltage magnitude difference.

TABLE 4.2 Study of IEEE 300-Bus System with UPFC (No Device Limit Constraints)

Solution of Base Case Power Flow (without Any UPFC)			
Parameter	**Line 1**	**Line 2**	**Line 3**
P_{LINE} (MW)	36.03	37.95	127.25
Q_{LINE} (MVAR)	73.23	19.93	63.45
V_{BUS} (p.u.)	1.0352	0.991	1.0129
NI	7		
Unconstrained Solution of UPFC Quantities			
Quantity	**UPFC-1 (Line 1)**	**UPFC-2 (Line 2)**	**UPFC-3 (Line 3)**
V_{se} (p.u.)	0.0668	0.0744	0.1851
θ_{se}	–112.42°	–98.52°	–125.71°
V_{sh} (p.u.)	0.9836	1.024	1.0277
θ_{sh}	7.55°	–14.76°	–22.73°
I_{se} (p.u.)	0.6799	0.5099	1.5501
I_{sh} (p.u.)	0.5644	0.2402	0.0812
P_{DC} (p.u.)	–0.0119	0.0034	–0.028
P_{LINE} (MW)	**50**	**50**	**150**
Q_{LINE} (MVAR)	**50**	**10**	**50**
V_{BUS} (p.u.)	**1.04**	**1.0**	**1.02**
NI	7		

and $P_{2\text{-}3}^{SP}$ = 50 MW (UPFC-1); V_{86}^{SP} = 1.0 p.u., $Q_{102\text{-}86}^{SP}$ = 10 MVAR, and $P_{86\text{-}102}^{SP}$ = 50 MW (UPFC-2); and V_{212}^{SP} = 1.02 p.u., $Q_{212\text{-}211}^{SP}$ = 50 MVAR, and $P_{212\text{-}211}^{SP}$ = 150 MW (UPFC-3). The results are shown in Table 4.2. Again in this table also, the final values of the controlled variables are shown in boldfaced letters.

From Table 4.2, it is also observed that in the presence of UPFCs, the line active power flow levels are enhanced compared to their corresponding values in the base case. The higher active power flow levels in the lines are maintained with the values of line reactive power flow levels, which are lesser than their corresponding values in the base case, in the presence of UPFCs. In this context, it is important to note that the values of line reactive power flows chosen are entirely arbitrary. These values could be chosen to be higher than the base case values as well. Furthermore, the voltages at the three buses (at which the UPFCs are connected) are maintained at values higher than those in the base case. The bus voltage profiles in this case are shown in Figure 4.4. From this figure, it is observed that in the presence of UPFC, there is very little change in the bus voltage profile.

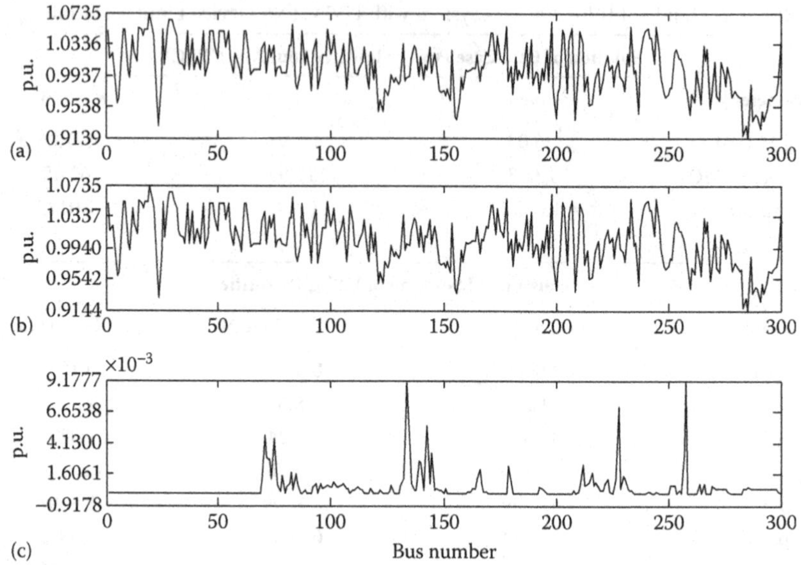

FIGURE 4.4 Bus voltage profile corresponding to the case study of Table 4.2. (a) Bus voltage magnitude without UPFC; (b) bus voltage magnitude with UPFC; (c) voltage magnitude difference.

4.7.2 Studies of UPFCs with Device Limit Constraints

In these case studies, various device limit constraints are considered for both the 118- and 300-bus test systems. As already mentioned in Section 4.5, four major device constraint limits have been considered. In all these case studies, the UPFC control references (prior to enforcement of the specified limit constraints) are kept identical to those in the corresponding case studies presented in Section 4.7.1. Although a large number of case studies were carried out for implementation of these limit constraints, only a few sets of representative results are presented below.

Case 1: In this case, the simultaneous limit violations of only the series injected voltage V_{se}^{Lim} have been studied. To accommodate this violation, the line active power flow control objective has been relaxed. In this context, it is important to note that exercising leniency on the line active power flow control objective is purely arbitrary. The line reactive power flow control objective can be relaxed as well. The results for the 118- and 300-bus systems are given in Tables 4.3 and 4.4, respectively.

The converged final values of the constrained variables, that is, the UPFC series injected voltages are shown in bold cases, along with the values of the line reactive power and SE bus voltage control objectives.

TABLE 4.3 Study of IEEE 118-Bus System with UPFC and V_{se}^{Lim} Constraint

Solution of UPFC Quantities with Limits on Series Injected Voltage Only			
Specified voltage limits: V_{se1}^{Lim}= 0.12 p.u., V_{se2}^{Lim}= 0.17 p.u., and V_{se3}^{Lim}= 0.1 p.u.			
Quantity	**UPFC-1 (Line 1)**	**UPFC-2 (Line 2)**	**UPFC-3 (Line 3)**
V_{se} (p.u.)	**0.12**	**0.17**	**0.1**
θ_{se}	–114.28°	–120.24°	–107.66°
V_{sh} (p.u.)	0.973	1.1424	1.005
θ_{sh}	–16.67°	–21.81°	–4.63°
I_{se} (p.u.)	0.9854	0.9873	0.4751
I_{sh} (p.u.)	0.2697	1.4239	0.0508
P_{DC} (p.u.)	0.0098	0.008	0.0058
P_{LINE} (MW)	98.42	98.22	47.24
Q_{LINE} (MVAR)	**5**	**10**	**5**
V_{BUS} (p.u.)	**1.0**	**1.0**	**1.0**
NI	6		

TABLE 4.4 Study of IEEE 300-Bus System with UPFC and V_{se}^{Lim} Constraint

Solution of UPFC Quantities with Limits on Series Injected Voltage Only			
Specified voltage limits: V_{se1}^{Lim} = 0.065 p.u., V_{se2}^{Lim} = 0.07 p.u., and V_{se3}^{Lim} = 0.18 p.u.			
Quantity	**UPFC-1 (Line 1)**	**UPFC-2 (Line 2)**	**UPFC-3 (Line 3)**
V_{se} (p.u.)	**0.065**	**0.07**	**0.18**
θ_{se}	–113.29°	–97.84°	–125.93°
V_{sh} (p.u.)	0.9835	1.0238	1.0274
θ_{sh}	7.57°	–14.69°	–22.68°
I_{se} (p.u.)	0.6719	0.4954	1.5308
I_{sh} (p.u.)	0.5649	0.2378	0.0783
P_{DC} (p.u.)	–0.0112	0.0029	–0.0267
P_{LINE} MW	48.82	48.52	147.92
Q_{LINE} (MVAR)	**50**	**10**	**50**
V_{BUS} (p.u.)	**1.04**	**1.0**	**1.02**
NI	7		

Corresponding to these studies in the 118- and 300-bus systems, the bus voltage profiles with UPFC series injected voltage limits are shown in Figures 4.5 and 4.6, respectively. From these figures, it is again observed that the bus voltage profiles change very little even with the simultaneous imposition of the limit constraint of series injected voltage on the three UPFCs.

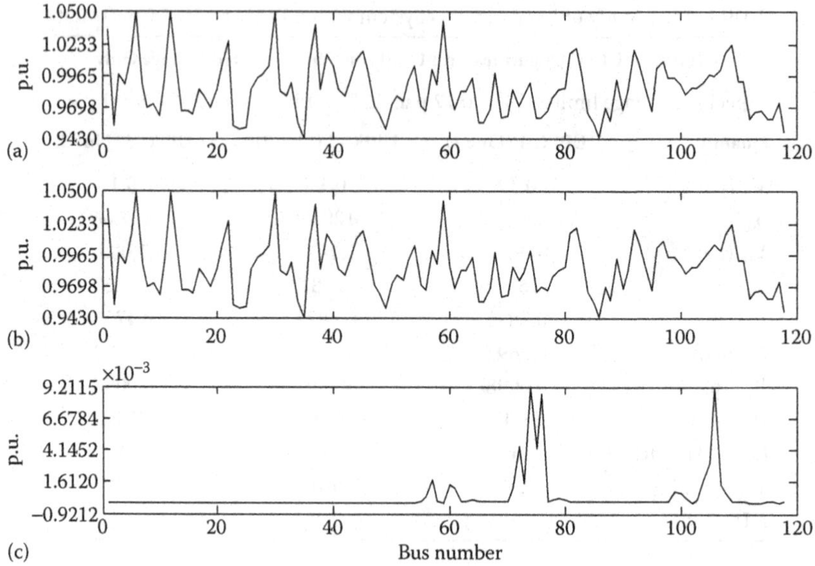

FIGURE 4.5 Bus voltage profile corresponding to the case study of Table 4.3. (a) Bus voltage magnitude without UPFC; (b) bus voltage magnitude with UPFC; (c) voltage magnitude difference.

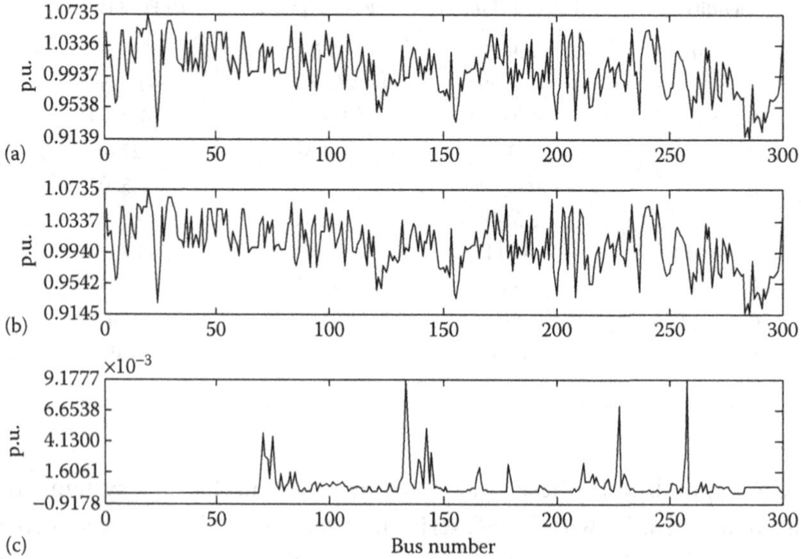

FIGURE 4.6 Bus voltage profile corresponding to the case study of Table 4.4. (a) Bus voltage magnitude without UPFC; (b) bus voltage magnitude with UPFC; (c) voltage magnitude difference.

Case 2: In this case, the violations of only the dc link power transfer limits P_{DC}^{Lim} have been studied. To accommodate this violation, the line reactive power flow control objective has been relaxed. The results for the 118-bus system are given in Table 4.5. It can be observed that the number of iterations needed to obtain convergence increases over that in the unconstrained case.

From Table 4.5, it can be observed that the power flow solution obtained is not a realistic one (unrealistic values of complex bus voltages, converter currents, and power flows). For UPFC-3, corresponding to a very small reduction in the value of P_{DC} from 0.0066 p.u. (unconstrained case) to 0.0062 p.u., the value of Q_{Line} undergoes a wide variation (5 to −10.78 MVAR). This is very improbable. The same is true for the value of V_{se}, I_{se}, and V_{sh}. Also, the convergence is observed to be very slow. This is due to the adoption of the initial condition of $V_{se}^0 \angle \theta_{se}^0 = 0.1\angle -(\pi/2)$. These unrealistic power flow solutions (corresponding to UPFC-3) are highlighted in bold cases in Table 4.5.

To investigate the convergence pattern and the final power flow solution *vis-à-vis* different values of V_{se}^0, further case studies were again carried out. Some representative results (for UPFC-3) are presented in Table 4.6. Those with double asterisks (**) indicate unrealistic power flow solutions.

TABLE 4.5 Initial Study of IEEE 118-Bus System with UPFC and P_{DC}^{Lim} Constraint

Solution of UPFC Quantities with Limit on DC Link Power Transfer Only

Specified dc link power transfer limits: $P_{DC1}^{Lim} = 0.0099$ p.u., $P_{DC2}^{Lim} = 0.0085$ p.u., and $P_{DC3}^{Lim} = 0.0062$ p.u.

Quantity	UPFC-1 (Line 1)	UPFC-2 (Line 2)	UPFC-3 (Line 3)
V_{se} (p.u.)	0.1233	0.1773	**0.1109**
θ_{se}	−113.75°	−119.92°	−88.73°
V_{sh} (p.u.)	0.9727	1.1425	**0.9943**
θ_{sh}	−16.71°	−21.84°	−4.66°
I_{se} (p.u.)	1.0009	1.0046	**0.5115**
I_{sh} (p.u.)	0.2735	1.4247	0.057
P_{DC} (p.u.)	0.0099	0.0085	0.0062
P_{LINE} (MW)	100	100	50
Q_{LINE} (MVAR)	4.34	9.56	**−10.78**
V_{BUS} (p.u.)	1.0	1.0	1.0
NI	21		

TABLE 4.6 Effect of Variation of V_{se}^{0} on the Power Flow Solution for the Study of Table 4.5

	Initial Condition for V_{se}								
V_{se}^{0} (p.u.)	0.09	0.08	0.07	0.06	0.05	0.04	0.03	0.02	0.01
	Final Power Flow Solution for UPFC-3								
V_{se3} (p.u.)	0.1109	0.111	–	0.111	0.1075	0.1075	0.111	0.1111	–
I_{se3} (p.u.)	0.5115	0.5117	–	0.5117	0.5007	0.5007	0.5117	0.5125	–
Q_{LINE3} (MVAR)	−10.78	−10.88	–	−10.88	−2.6	−2.6	−10.88	−11.27	–
NI	27**	47**	div	54**	14	15	45**	119**	div

TABLE 4.7 Final Study of IEEE 118-Bus System with UPFC and P_{DC}^{Lim} Constraint

Solution of UPFC Quantities with Limit on DC Link Power Transfer Only

Specified dc link power transfer limits: $P_{DC1}^{Lim} = 0.0099$ p.u., $P_{DC2}^{Lim} = 0.0085$ p.u., and $P_{DC3}^{Lim} = 0.0062$ p.u.

Quantity	UPFC-1 (Line 1)	UPFC-2 (Line 2)	UPFC-3 (Line 3)
V_{se} (p.u.)	0.1233	0.1773	0.1075
θ_{se}	−113.75°	−119.92°	−98.26°
V_{sh} (p.u.)	0.9727	1.1425	1.0002
θ_{sh}	−16.71°	−21.84°	−4.66°
I_{se} (p.u.)	1.0009	1.0046	0.5007
I_{sh} (p.u.)	0.2735	1.4247	0.0066
P_{DC} (p.u.)	**0.0099**	**0.0085**	**0.0062**
P_{LINE} (MW)	**100**	**100**	**50**
Q_{LINE} (MVAR)	4.34	9.56	−2.6
V_{BUS} (p.u.)	**1.0**	**1.0**	**1.0**
NI	14 (with $V_{se}^{0} = 0.05$)		

The realistic final power flow solutions (corresponding to $V_{se}^{0} = 0.05$ p.u) for the 118-bus system are shown in Table 4.7. The converged final values of the constrained variable along with the values of the control objectives are shown in bold cases.

The bus voltage profile for the case depicted in Table 4.7 is shown in Figure 4.7. From this figure, it is observed that the voltage profile in the 118-bus system changes very little (compared to the case with no UPFC) even with simultaneous imposition of the limit constraint of dc link power transfer on the three UPFCs.

Similar to the 118-bus system, in the 300-bus system also, adoption of the initial condition of $V_{se}^{0}\angle\theta_{se}^{0} = 0.1\angle-(\pi/2)$ yields an unrealistic power

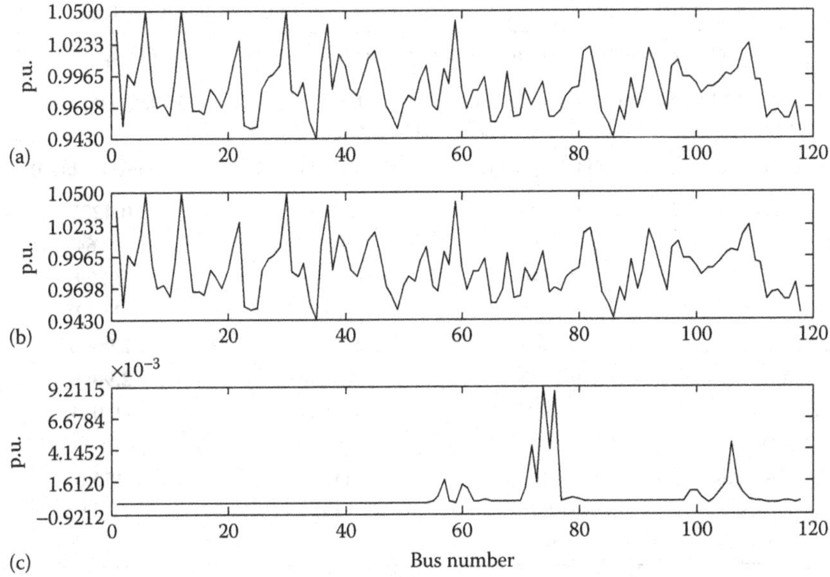

FIGURE 4.7 Bus voltage profile corresponding to the case study of Table 4.7. (a) Bus voltage magnitude without UPFC; (b) bus voltage magnitude with UPFC; (c) voltage magnitude difference.

flow solution (unrealistic values of complex bus voltages, converter currents, and power flows) for UPFC-3. The convergence was also observed to be very slow. The results are given in Table 4.8. From this table, the unrealistic power flow solutions (corresponding to UPFC-3) can be observed, which are highlighted in bold cases.

To investigate the convergence pattern and the final power flow solution *vis-à-vis* different values of V_{se}^{0}, further case studies were again carried out. Some representative results (for UPFC-3) are presented in Table 4.9. Those with double asterisks (**) indicate unrealistic power flow solutions.

The realistic final power flow solutions (corresponding to $V_{se}^{0} = 0.05$ p.u) for the 300-bus systems are shown in Table 4.10. The converged final values of the constrained variable along with the values of the control objectives are shown in bold cases.

The bus voltage profile for the case depicted in Table 4.10 is shown in Figure 4.8. From this figure, it is observed that even with the simultaneous imposition of device limit constraints, the voltage profile in the 300-bus system changes very little (compared to the case with no UPFC).

Case 3: In this case, the violations of only the shunt converter current limits I_{sh}^{Lim} have been considered. To accommodate this, the UPFC SE bus

TABLE 4.8 Initial Study of IEEE 300-Bus System with UPFC and P_{DC}^{Lim} Constraint

Solution of UPFC Quantities with Limit on DC Link Power Transfer Only

Specified dc link power transfer limits: $P_{DC1}^{Lim} = -0.0117$ p.u., $P_{DC2}^{Lim} = 0.0032$ p.u., and $P_{DC3}^{Lim} = -0.026$ p.u.

Quantity	UPFC-1 (Line 1)	UPFC-2 (Line 2)	UPFC-3 (Line 3)
V_{se} (p.u.)	0.0675	0.0751	**0.4225**
θ_{se}	−113.89°	−97.15°	−52.64°
V_{sh} (p.u.)	0.9846	1.023	**0.729**
θ_{sh}	7.16°	−15.2°	−23.12°
I_{se} (p.u.)	0.6868	0.5127	**2.9095**
I_{sh} (p.u.)	0.5543	0.2298	**2.9099**
P_{DC} (p.u.)	−0.0117	0.0032	−0.026
P_{LINE} (MW)	50	50	150
Q_{LINE} (MVAR)	51.01	11.34	**−256.07**
V_{BUS} (p.u.)	1.04	1.0	1.02
NI	46		

TABLE 4.9 Effect of Variation of V_{se}^0 on the Power Flow Solution for the Study of Table 4.8

	Initial Condition for V_{se}								
V_{se}^0 (p.u.)	0.09	0.08	0.07	0.06	0.05	0.04	0.03	0.02	0.01
	Final Power Flow Solution for UPFC-3								
V_{se3} (p.u.)	0.4225	0.4225		0.4225	0.1873	0.187		0.4266	
V_{sh3} (p.u.)	0.729	0.729		0.729	1.0359	1.0388		0.7257	
I_{se3} (p.u.)	2.9095	2.9095		2.9095	1.5781	1.5786		2.9389	
I_{sh3} (p.u.)	2.9099	2.9099	div	2.9099	0.1608	0.1898	div	2.9433	div
Q_{LINE3} (MVAR)	−256.07	−256.07		−256.07	58.41	58.55		−259.54	
NI	47**	48**		52**	30	20**		44**	

voltage control objective has been relaxed. The results for the 118- and 300-bus systems are given in Tables 4.11 and 4.12, respectively.

Corresponding to these studies in the 118- and 300-bus systems, the bus voltage profiles with UPFC shunt converter current limits are shown in Figures 4.9 and 4.10, respectively. From these figures, it is again observed that the bus voltage profiles change very little even with the simultaneous imposition of the shunt converter current limit constraint on the three UPFCs.

Case 4: In this case, violations of the limits of both the dc link power transfers and the series injected voltages V_{se}^{Lim} have been considered, with

TABLE 4.10 Final Study of IEEE 300-Bus System with UPFC and P_{DC}^{Lim} Constraint

Solution of UPFC Quantities with Limit on DC Link Power Transfer Only

Specified dc link power transfer limits: $P_{DC1}^{Lim} = -0.0117$ p.u., $P_{DC2}^{Lim} = 0.0032$ p.u., and $P_{DC3}^{Lim} = -0.026$ p.u.

Quantity	UPFC-1 (Line 1)	UPFC-2 (Line 2)	UPFC-3 (Line 3)
V_{se} (p.u.)	0.0675	0.0749	0.1873
θ_{se}	−113.42°	−96.86°	−129.12°
V_{sh} (p.u.)	0.9844	1.0228	1.0359
θ_{sh}	7.55°	−14.76°	−22.76°
I_{se} (p.u.)	0.6861	0.5124	1.5781
I_{sh} (p.u.)	0.5562	0.2281	0.1608
P_{DC} (p.u.)	**−0.0117**	**0.0032**	**−0.026**
P_{LINE} (MW)	**50**	**50**	**150**
Q_{LINE} (MVAR)	50.91	11.22	58.41
V_{BUS} (p.u.)	**1.04**	**1.0**	**1.02**
NI	30 (with $V_{se}^0 = 0.05$ p.u)		

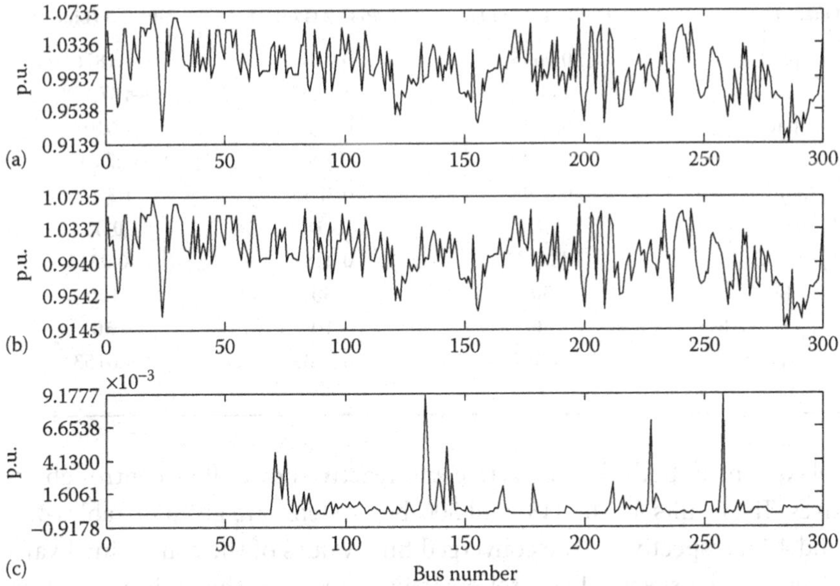

FIGURE 4.8 Bus voltage profile corresponding to the case study of Table 4.10. (a) Bus voltage magnitude without UPFC; (b) bus voltage magnitude with UPFC; (c) voltage magnitude difference.

TABLE 4.11 Study of IEEE 118-Bus System with UPFC and I_{sh}^{Lim} Constraint

Solution of UPFC Quantities with Limit on Shunt Converter Current Limits

Specified shunt converter current limits: $I_{sh1}^{Lim} = 0.2$ p.u., $I_{sh2}^{Lim} = 1.38$ p.u., and $I_{sh3}^{Lim} = 0.045$ p.u.

Quantity	UPFC-1 (Line 1)	UPFC-2 (Line 2)	UPFC-3 (Line 3)
V_{se} (p.u.)	0.123	0.1836	0.1076
θ_{se}	−114.23°	−126.77°	−108.71°
V_{sh} (p.u.)	0.9804	0.8422	0.9922
θ_{sh}	−16.79°	−21.89°	−4.61°
I_{se} (p.u.)	1.0009	1.0253	0.5042
I_{sh} (p.u.)	**0.2**	**1.38**	**0.045**
P_{DC} (p.u.)	0.0099	0.0304	0.0079
P_{LINE} (MW)	**100**	**100**	**50**
Q_{LINE} (MVAR)	**5**	**10**	**5**
V_{BUS} (p.u.)	1.0004	0.9802	0.9966
NI	9		

TABLE 4.12 Study of IEEE 300-Bus System with UPFC and I_{sh}^{Lim} Constraint

Solution of UPFC Quantities with Limit on Shunt Converter Current Limits

Specified shunt converter current limits: $I_{sh1}^{Lim} = 0.48$ p.u., $I_{sh2}^{Lim} = 0.18$ p.u., and $I_{sh3}^{Lim} = 0.05$ p.u.

Quantity	UPFC-1 (Line 1)	UPFC-2 (Line 2)	UPFC-3 (Line 3)
V_{se} (p.u.)	0.0665	0.0785	0.1876
θ_{se}	−111.68°	−120.33°	−127.17°
V_{sh} (p.u.)	0.9927	0.9523	1.0108
θ_{sh}	7.56°	−14.86°	−22.81°
I_{se} (p.u.)	0.6795	0.5255	1.5573
I_{sh} (p.u.)	**0.48**	**0.18**	**0.05**
P_{DC} (p.u.)	−0.0123	0.0187	−0.0213
P_{LINE} (MW)	**50**	**50**	**150**
Q_{LINE} (MVAR)	**50**	**10**	**50**
V_{BUS} (p.u.)	1.0407	0.9702	1.0153
NI	7		

relaxations of both the line active and reactive power flow control objectives. The results for the 118- and 300-bus systems are given in Tables 4.13 and 4.14, respectively. The converged final values of the constrained variables and the specified control objectives are again shown in bold cases. The number of iterations needed to obtain convergence also shows an increase. The bus voltage profiles for the studies with the 118- and 300-bus systems are shown in Figures 4.11 and 4.12, respectively. From these

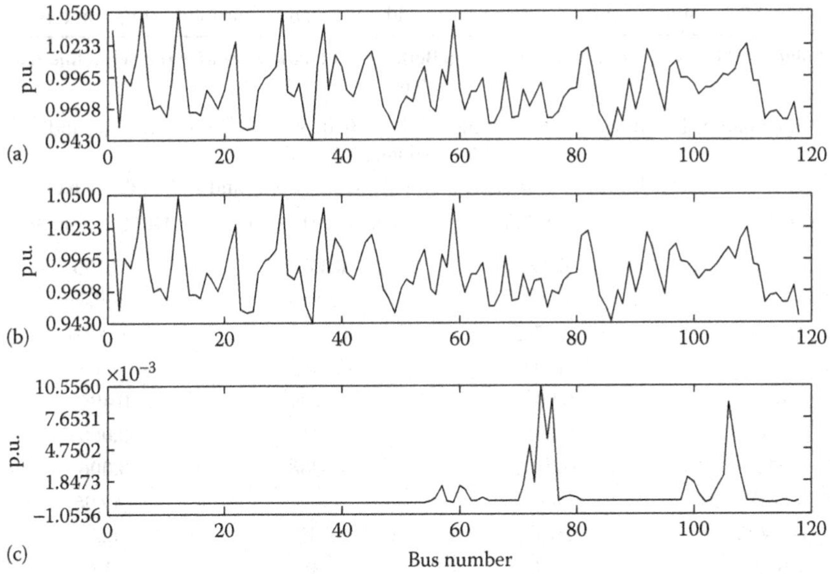

FIGURE 4.9 Bus voltage profile corresponding to the case study of Table 4.11. (a) Bus voltage magnitude without UPFC; (b) bus voltage magnitude with UPFC; (c) voltage magnitude difference.

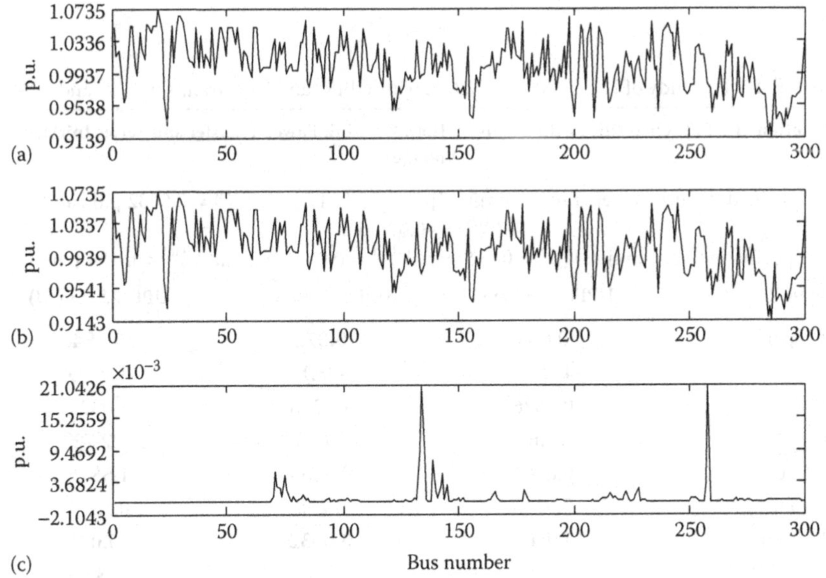

FIGURE 4.10 Bus voltage profile corresponding to the case study of Table 4.12. (a) Bus voltage magnitude without UPFC; (b) bus voltage magnitude with UPFC; (c) voltage magnitude difference.

TABLE 4.13 Study of IEEE 118-Bus System with UPFC and Constraints of V_{se}^{Lim} and P_{DC}^{Lim}

Solution of UPFC Quantities with Limits on Both DC Link Power Transfer and Series Injected Voltage

Specified dc link power transfer limits: $P_{DC1}^{Lim} = 0.0102$ p.u., $P_{DC2}^{Lim} = 0.0088$ p.u., and $P_{DC3}^{Lim} = 0.0065$ p.u.
Specified voltage limits: $V_{se1}^{Lim} = 0.121$ p.u., $V_{se2}^{Lim} = 0.175$ p.u., and $V_{se3}^{Lim} = 0.105$ p.u.

Quantity	UPFC-1 (Line 1)	UPFC-2 (Line 2)	UPFC-3 (Line 3)
V_{se} (p.u.)	**0.121**	**0.175**	**0.105**
θ_{se}	–114.67°	–120.43°	–109.41°
V_{sh} (p.u.)	0.9735	1.1429	1.0066
θ_{sh}	–16.69°	–21.84°	–4.65°
I_{se} (p.u.)	0.9904	0.9995	0.4949
I_{sh} (p.u.)	0.2646	1.4292	0.0664
P_{DC} (p.u.)	**0.0102**	**0.0088**	**0.0065**
P_{LINE} (MW)	98.9	99.43	49.06
Q_{LINE} (MVAR)	5.42	10.17	6.56
V_{BUS} (p.u.)	**1.0**	**1.0**	**1.0**
NI	11		

TABLE 4.14 Study of IEEE 300-Bus System with UPFC and Constraints of V_{se}^{Lim} and P_{DC}^{Lim}

Solution of UPFC Quantities with Limits on Both DC Link Power Transfer and Series Injected Voltage

Specified dc link power transfer limits: $P_{DC1}^{Lim} = -0.0117$ p.u., $P_{DC2}^{Lim} = 0.0032$ p.u., and $P_{DC3}^{Lim} = -0.026$ p.u.
Specified voltage limits: $V_{se1}^{Lim} = 0.065$ p.u., $V_{se2}^{Lim} = 0.073$ p.u., and $V_{se3}^{Lim} = 0.184$ p.u.

Quantity	UPFC-1 (Line 1)	UPFC-2 (Line 2)	UPFC-3 (Line 3)
V_{se} (p.u.)	**0.065**	**0.073**	**0.184**
θ_{se}	–111.65°	–98.1°	–128.09°
V_{sh} (p.u.)	0.9826	1.0238	1.0329
θ_{sh}	7.56°	–14.74°	–22.72°
I_{se} (p.u.)	0.6681	0.5053	1.5586
I_{sh} (p.u.)	0.5745	0.238	0.1315
P_{DC} (p.u.)	**–0.0117**	**0.0032**	**–0.026**
P_{LINE} (MW)	49.35	49.51	148.98
Q_{LINE} (MVAR)	48.91	10.15	55.49
V_{BUS} (p.u.)	**1.0**	**1.0**	**1.0**
NI	12		

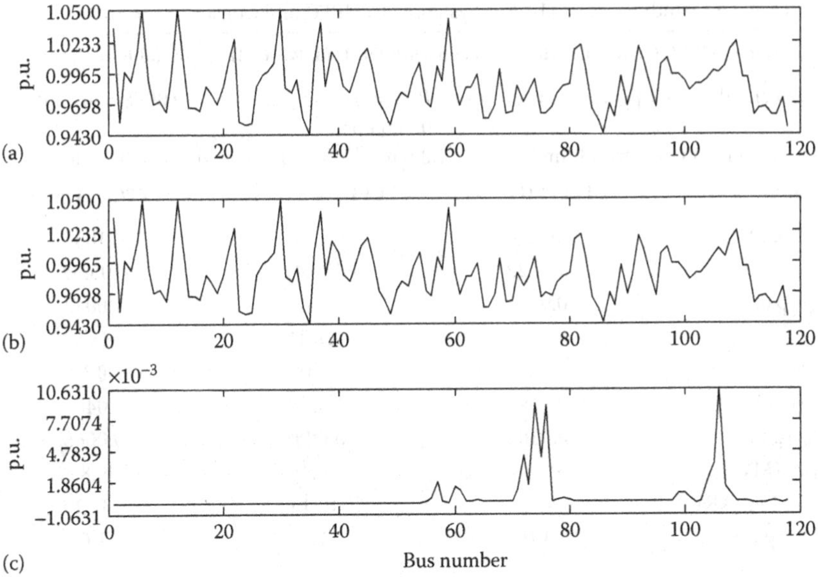

FIGURE 4.11 Bus voltage profile corresponding to the case study of Table 4.13. (a) Bus voltage magnitude without UPFC; (b) bus voltage magnitude with UPFC; (c) voltage magnitude difference.

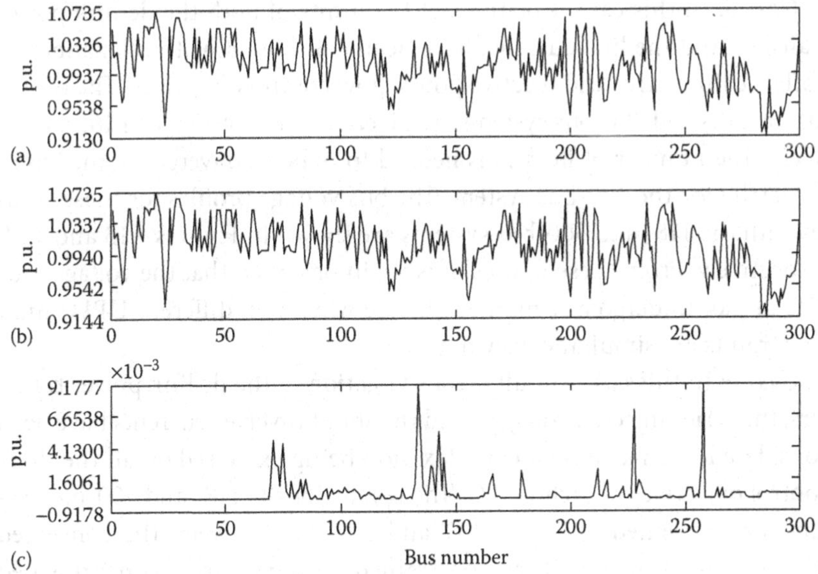

FIGURE 4.12 Bus voltage profile corresponding to the case study of Table 4.14. (a) Bus voltage magnitude without UPFC; (b) bus voltage magnitude with UPFC; (c) voltage magnitude difference.

TABLE 4.15 Study of IEEE 118-Bus System with UPFC and Constraints of I_{se}^{Lim} and P_{DC}^{Lim}

Solution of UPFC Quantities with Limits on Both DC Link Power Transfer and Line Current

Specified dc link power transfer limits: $P_{DC1}^{Lim} = 0.0102$ p.u., $P_{DC2}^{Lim} = 0.0088$ p.u., and $P_{DC3}^{Lim} = 0.0065$ p.u.

Specified line current limits: $I_{se1}^{Lim} = 1.0$ p.u., $I_{se2}^{Lim} = 1.0$ p.u., and $I_{se3}^{Lim} = 0.5$ p.u.

Quantity	UPFC-1 (Line 1)	UPFC-2 (Line 2)	UPFC-3 (Line 3)
V_{se} (p.u.)	0.123	0.1752	0.1067
θ_{se}	–114.22°	–120.41°	–105.99°
V_{sh} (p.u.)	0.9732	1.1429	1.0048
θ_{sh}	–16.7°	–21.84°	–4.66°
I_{se} (p.u.)	**1.0**	**1.0**	**0.5**
I_{sh} (p.u.)	0.2682	1.4291	0.0486
P_{DC} (p.u.)	**0.0102**	**0.0088**	**0.0065**
P_{LINE} (MW)	99.88	99.48	49.85
Q_{LINE} (MVAR)	4.89	10.14	3.82
V_{BUS} (p.u.)	**1.0**	**1.0**	**1.0**
NI	13		

figures, it is again observed that simultaneous imposition of two different UPFC limit constraints hardly changes the voltage profiles.

Case 5: In this case, violations of the limits of both the dc link power transfers and the line currents have been considered, with relaxations of both the line active and reactive power flow control objectives. The results for the 118- and 300-bus systems are given in Tables 4.15 and 4.16, respectively. The number of iterations needed to obtain convergence increases, especially for the 300-bus system. The bus voltage profiles for these studies with the 118- and 300-bus systems are shown in Figures 4.13 and 4.14, respectively. From these figures, it is again observed that the voltage profiles do not undergo any marked change when two different UPFC limit constraints are simultaneously imposed.

Case 6: In this case, simultaneous violation of the dc link power transfers, the series injected voltages, and the shunt inverter currents have been considered, with a consequent relaxation being exercised on all the three control objectives of each UPFC. The results for the 118- and 300-bus systems are presented in Tables 4.17 and 4.18, respectively. The converged final values of all the three constrained variables are again presented in bold cases. In both the systems, the number of iterations needed to obtain convergence shows a marked increase. The bus voltage profiles for these studies are shown in Figures 4.15 and 4.16, respectively. From these

TABLE 4.16 Study of IEEE 300-Bus System with UPFC and Constraints of I_{se}^{Lim} and P_{DC}^{Lim}

Solution of UPFC Quantities with Limits on Both DC Link Power Transfer and Line Current

Specified dc link power transfer limits: $P_{DC1}^{Lim} = -0.0118$ p.u., $P_{DC2}^{Lim} = 0.0033$ p.u., and $P_{DC3}^{Lim} = -0.027$ p.u.

Specified line current limits: $I_{se1}^{Lim} = 0.678$ p.u., $I_{se2}^{Lim} = 0.508$ p.u., and $I_{se3}^{Lim} = 1.55$ p.u.

Quantity	UPFC-1 (Line 1)	UPFC-2 (Line 2)	UPFC-3 (Line 3)
V_{se} (p.u.)	0.0665	0.0738	0.1837
θ_{se}	–112.37°	–98.32°	–126.71°
V_{sh} (p.u.)	0.9835	1.0239	1.0297
θ_{sh}	7.56°	–14.75°	–22.72°
I_{se} (p.u.)	**0.678**	**0.508**	**1.55**
I_{sh} (p.u.)	0.5657	0.2391	0.1006
P_{DC} (p.u.)	**–0.0118**	**0.0033**	**–0.027**
P_{LINE} (MW)	49.85	49.79	149.23
Q_{LINE} (MVAR)	49.87	10.08	52.21
V_{BUS} (p.u.)	**1.0**	**1.0**	**1.0**
NI	42		

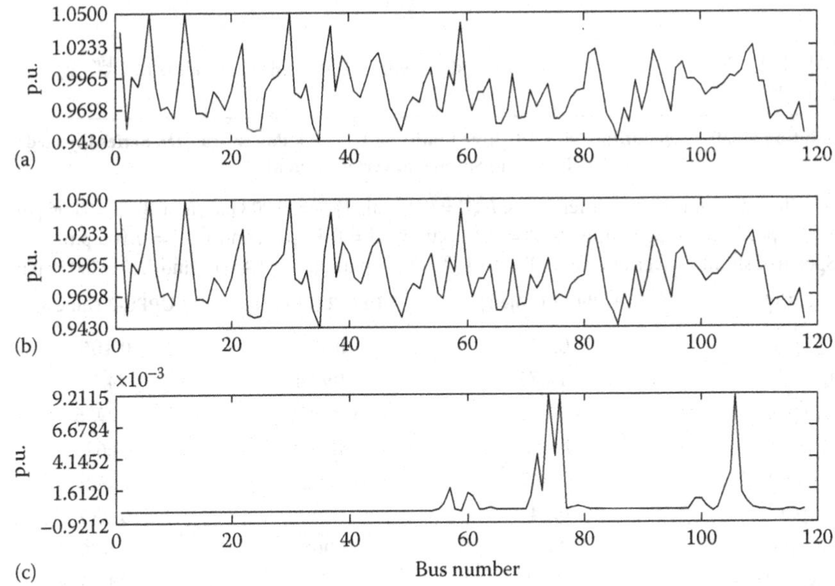

FIGURE 4.13 Bus voltage profile corresponding to the case study of Table 4.15. (a) Bus voltage magnitude without UPFC; (b) bus voltage magnitude with UPFC; (c) voltage magnitude difference.

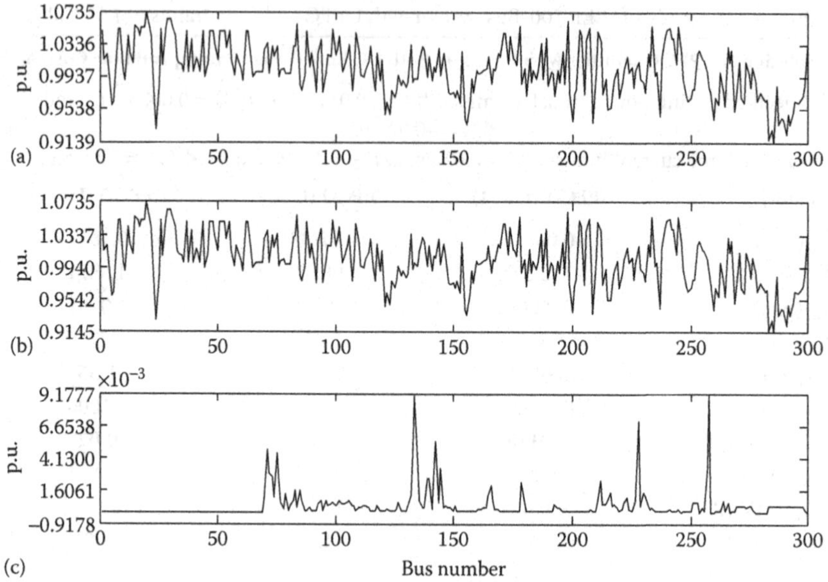

FIGURE 4.14 Bus voltage profile corresponding to the case study of Table 4.16. (a) Bus voltage magnitude without UPFC; (b) bus voltage magnitude with UPFC; (c) voltage magnitude difference.

TABLE 4.17 Study of IEEE 118-Bus System with UPFC and Constraints of P_{DC}^{Lim}, V_{se}^{Lim}, and I_{se}^{Lim}

Solution of UPFC Quantities with All Three Limits on DC Link Power Transfer, Series Injected Voltage, and Shunt Converter Current

Specified dc link power transfer limits: $P_{DC1}^{Lim} = 0.01$ p.u., $P_{DC2}^{Lim} = 0.008$ p.u., and $P_{DC3}^{Lim} = 0.006$ p.u.
Specified voltage limits: $V_{se1}^{Lim} = 0.12$ p.u., $V_{se2}^{Lim} = 0.17$ p.u., and $V_{se3}^{Lim} = 0.105$ p.u.
Specified shunt converter current limits: $I_{sh1}^{Lim} = 0.24$ p.u., $I_{sh2}^{Lim} = 1.4$ p.u., and $I_{sh3}^{Lim} = 0.055$ p.u.

Quantity	UPFC-1 (Line 1)	UPFC-2 (Line 2)	UPFC-3 (Line 3)
V_{se} (p.u.)	**0.12**	**0.17**	**0.105**
θ_{se}	−114.72°	−109.54°	−94.17°
V_{sh} (p.u.)	0.9761	0.8415	0.9936
θ_{sh}	−16.71°	−21.66°	−4.62°
I_{se} (p.u.)	0.9867	0.9883	0.4905
I_{sh} (p.u.)	**0.24**	**1.4**	**0.055**
P_{DC} (p.u.)	**0.01**	**0.008**	**0.006**
P_{LINE} (MW)	98.53	96.66	48.63
Q_{LINE} (MVAR)	5.52	−8.25	−6.06
V_{BUS} (p.u.)	1.0001	0.9815	0.999
NI	9		

TABLE 4.18 Study of IEEE 300-Bus System with UPFC and Constraints of P_{DC}^{Lim}, V_{se}^{Lim}, and I_{se}^{Lim}

Solution of UPFC Quantities with All Three Limits on DC Link Power Transfer, Series Converter Voltage, and Shunt Converter Current

Specified dc link power transfer limits: $P_{DC1}^{Lim} = -0.0118$ p.u., $P_{DC2}^{Lim} = 0.0033$ p.u., and $P_{DC3}^{Lim} = -0.027$ p.u.

Specified voltage limits: $V_{se1}^{Lim} = 0.066$ p.u., $V_{se2}^{Lim} = 0.074$ p.u., and $V_{se3}^{Lim} = 0.185$ p.u.

Specified shunt converter current limits: $I_{sh1}^{Lim} = 0.564$ p.u., $I_{sh2}^{Lim} = 0.238$ p.u., and $I_{sh3}^{Lim} = 0.078$ p.u.

Quantity	UPFC-1 (Line 1)	UPFC-2 (Line 2)	UPFC-3 (Line 3)
V_{se} (p.u.)	**0.066**	**0.074**	**0.185**
θ_{se}	−112.22°	−70.11°	−122.61°
V_{sh} (p.u.)	0.9836	0.9613	1.0104
θ_{sh}	7.59°	−14.49°	−22.74°
I_{se} (p.u.)	0.6753	0.5797	1.5295
I_{sh} (p.u.)	**0.564**	**0.238**	**0.078**
P_{DC} (p.u.)	**−0.0118**	**0.0033**	**−0.027**
P_{LINE} (MW)	49.61	44.51	150.22
Q_{LINE} (MVAR)	49.72	35.77	40.78
V_{BUS} (p.u.)	1.04	0.9851	1.0177
NI	9		

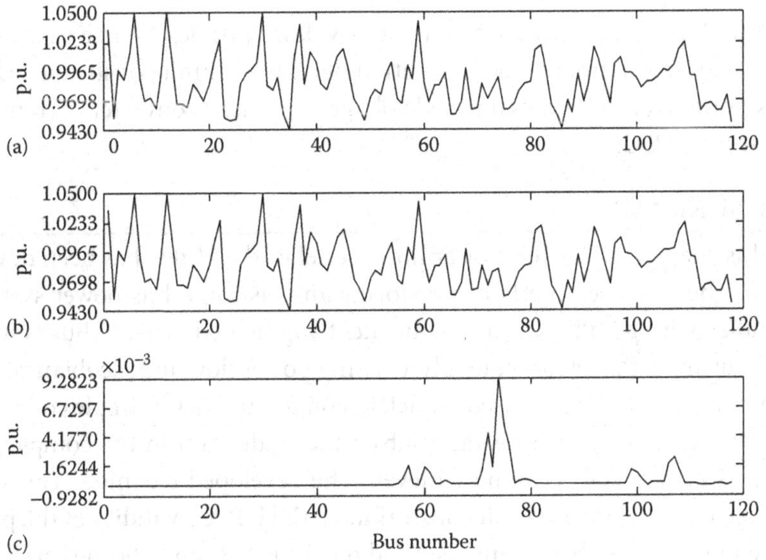

FIGURE 4.15 Bus voltage profile corresponding to the case study of Table 4.17. (a) Bus voltage magnitude without UPFC; (b) bus voltage magnitude with UPFC; (c) voltage magnitude difference.

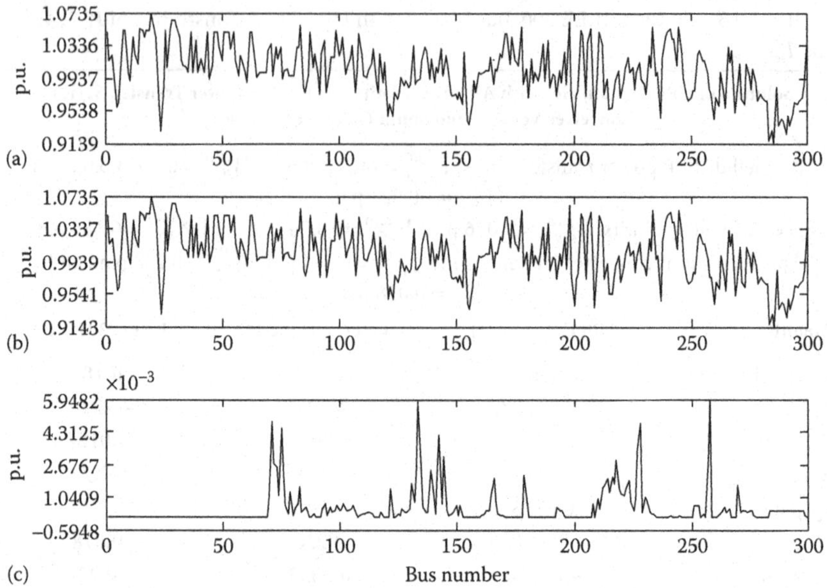

FIGURE 4.16 Bus voltage profile corresponding to the case study of Table 4.18. (a) Bus voltage magnitude without UPFC; (b) bus voltage magnitude with UPFC; (c) voltage magnitude difference.

figures, it is again observed that the bus voltage profiles change very little even with the simultaneous imposition of all three limit constraints of dc link power transfer, series injected voltage, and shunt converter current on the three UPFCs.

4.8 SUMMARY

In this chapter, a Newton power flow model of the UPFC has been developed. The proposed method transforms an existing n-bus power system installed with p UPFCs into an equivalent augmented $(n + 2p)$-bus system without any UPFC. Consequently, existing power flow and Jacobian codes can be reused in the proposed model, in conjunction with simple codes for matrix extraction. As a result, a substantial reduction in the complexity of the software codes can be achieved. The developed technique can also handle practical device limit constraints of the UPFC. Validity of the proposed method has been demonstrated on IEEE 118- and 300-bus systems with excellent convergence characteristics.

CHAPTER 5

Newton Power Flow Model of the Interline Power Flow Controller

5.1 INTRODUCTION

Unlike a static synchronous series compensator (SSSC) or a unified power flow controller (UPFC), which can control the power flow in a single transmission line, the interline power flow controller (IPFC) can address the problem of compensating multiple transmission lines simultaneously. The IPFC employs a number of dc-to-ac converters linked together at their dc terminals, each controlling the power flow of a different line, in conjunction with its series coupling transformer [22]. An IPFC with p series converters will have $(2p - 1)$ degrees of freedom—2 degrees of freedom for each of the $(p - 1)$ converters and 1 degree of freedom for the remaining one (because the real power exchange among the p series converters is balanced at all times). Usually, these 2 degrees of freedom correspond to the active and reactive power flow in a line.

For studying the behavior of the IPFC, power flow solutions of the networks incorporating IPFC(s) are required. Some excellent research works carried out in the literature are present in [77–78] for developing efficient power-flow algorithms for the IPFC. In a way similar to the SSSC and the UPFC, it is observed from these works that the complexity of software codes is increased manifold when an IPFC is modeled in an existing Newton–Raphson power flow algorithm. In fact, unlike the SSSC and the UPFC, the IPFC possesses multiple series converters. Contributions from these converters necessitate modifications in the existing power injection equations of the concerned buses (sending end [SE] and receiving end [RE]) in all the transmission lines incorporating an IPFC. Further, an entirely new expression has to be written for the real power exchanged among the series inverters of an IPFC. Moreover, the Jacobian matrix comprises multiple subblocks exclusively related to the IPFC converters, and for computation of each of these subblocks, new codes have to be written. Due to these above factors, the complexity of software codes is greatly enhanced. The problem aggravates as the number of IPFCs in a system increases.

None of the existing works on power flow modeling of the IPFC directly addresses this problem. To reduce the complexities of the software codes for implementing Newton power flow algorithm, in this chapter, a novel approach [84,87] for an IPFC power flow model is proposed. By this modeling approach, an existing power system installed with IPFCs is transformed to an equivalent augmented network without any IPFC. This results in a substantial reduction in the programming complexity because of the following reasons:

1. The power injections for the buses concerned can be computed in the proposed model using existing power flow codes, as it is devoid of contributions from any IPFC series converter.
2. In the proposed model, existing power flow codes can be used to compute the exchange of active power flow among the series inverters of the IPFC itself, which equals the sum of bus active power injections of additional power-flow buses.
3. Only three Jacobian subblocks need to be evaluated in the proposed model. Two of these subblocks can be evaluated using existing Jacobian codes directly, whereas the third can be computed with very minor modifications of the existing Jacobian codes.

5.2 IPFC MODEL FOR NEWTON POWER FLOW ANALYSIS

Figure 5.1 shows an n-bus power system network in which p series converters of a single IPFC are connected. Without loss of generality, it is assumed that the converters are connected among buses i–j, $(i+1)$–$(j+1)$, and so on, up to buses $(i+p-1)$–$(j+p-1)$. It is also further assumed that the gth converter $(1 \leq g \leq p)$ is connected at the SE, that is, at $(i+g-1)$th bus of the corresponding transmission line. The equivalent circuit of Figure 5.1 is shown in Figure 5.2.

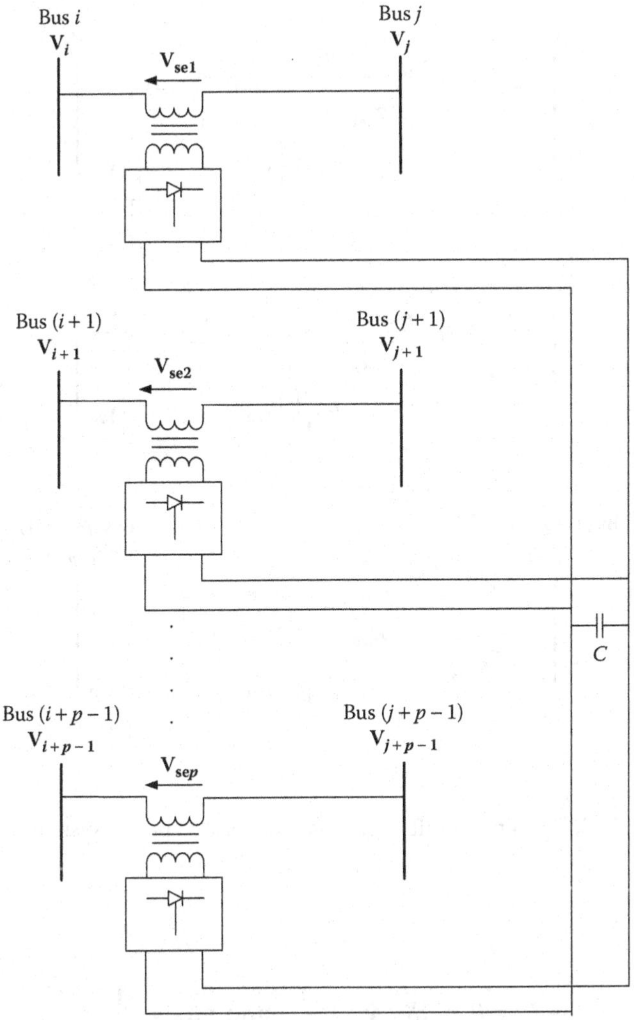

FIGURE 5.1 IPFC with p series converters connected to a power system.

In Figure 5.2, the IPFC is represented by p (series) voltage sources. The voltage source $\mathbf{V}_{se\,g}$ (not shown) (representing the gth converter) is in series with the impedance $\mathbf{Z}_{se\,g}$ (representing the impedance of the coupling transformer of the gth converter) and is connected in series with the gth transmission line (which is represented by its equivalent pi circuit). The total current through the gth $(1 \leq g \leq p)$ series coupling transformer is $\mathbf{I}_{se\,g}$, which consists of two parts: (1) $\mathbf{I}_g$ flowing into the transmission line and (2) $\mathbf{I}_{g0}$, the line charging current.

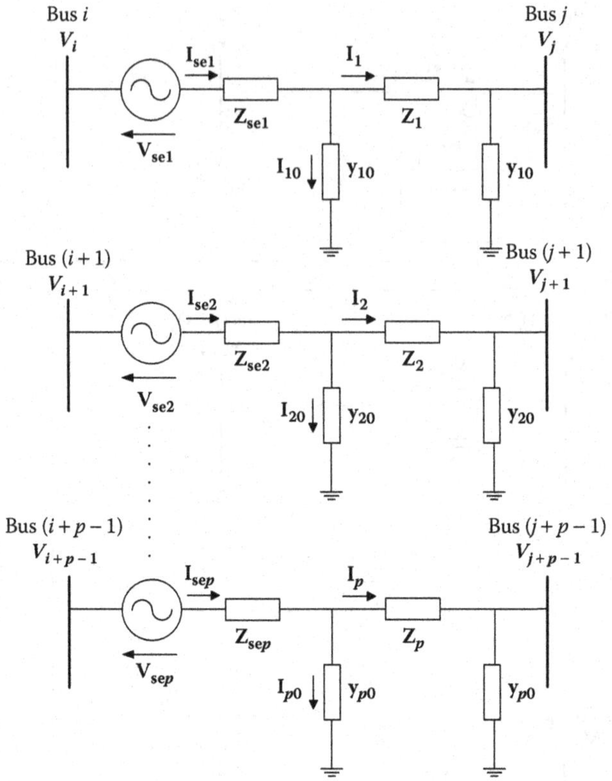

FIGURE 5.2 Equivalent circuit of IPFC-incorporated power system network.

Now, let

$$\mathbf{Z}_g = R_g + jX_g, \mathbf{y}_g = \frac{1}{\mathbf{Z}_g}, \text{ and } \mathbf{y}_{seg} = \frac{1}{\mathbf{Z}_{seg}} \tag{5.1}$$

Also,

$$\mathbf{Y}_{ij} = \mathbf{Y}_{ji} = -\mathbf{y}_{ij} = -\mathbf{y}_{ji} \tag{5.2}$$

From Figure 5.2, it can be shown that for the first converter

$$\mathbf{I}_{se1} = \alpha_1 \mathbf{V}_i - \beta_1 \mathbf{V}_j - \alpha_1 \mathbf{V}_{se1} \tag{5.3}$$

where:

$$\alpha_1 = \frac{\mathbf{y}_{se1}(\mathbf{y}_1 + \mathbf{y}_{10})}{\mathbf{y}_1 + \mathbf{y}_{10} + \mathbf{y}_{se1}} \text{ and } \beta_1 = \frac{\mathbf{y}_{se1}\mathbf{y}_1}{\mathbf{y}_1 + \mathbf{y}_{10} + \mathbf{y}_{se1}} \tag{5.4}$$

Again, from Figure 5.2, the net injected current at bus i is

$$\mathbf{I}_i = [\mathbf{Y}_{ii}^{\text{old}} - (\mathbf{y}_1 + \mathbf{y}_{10})]\mathbf{V}_i + \sum_{q=1, q\neq i, q\neq j}^{n} \mathbf{Y}_{iq}\mathbf{V}_q + \mathbf{I}_{\text{se}1} \tag{5.5}$$

In the above equation, $\mathbf{Y}_{ii}^{\text{old}} = \sum_{q=1,q\neq i,q\neq j}^{n} \mathbf{y}_{iq} + \mathbf{y}_{i0}^{p} + \mathbf{y}_1 + \mathbf{y}_{10}$ is the self-admittance of bus i for the existing n-bus system without any IPFC connected and $\mathbf{y}_{i0}^{p}$ accounts for the shunt capacitances of all transmission lines connected to bus i, except the line in branch (i–j). From Equations 5.3 and 5.5, the net injected current at bus i with IPFC becomes

$$\mathbf{I}_i = \sum_{q=1}^{n+1} \mathbf{Y}_{iq}\mathbf{V}_q \tag{5.6}$$

where $\mathbf{Y}_{ii} = \mathbf{Y}_{ii}^{\text{old}} - (\mathbf{y}_1 + \mathbf{y}_{10}) + \boldsymbol{\alpha}_1$ is the new value of self-admittance for the ith bus with IPFC and

$$\mathbf{V}_{n+1} = \mathbf{V}_{\text{se}1}, \mathbf{Y}_{i(n+1)} = -\boldsymbol{\alpha}_1, \text{ and } \mathbf{Y}_{ij} = -\boldsymbol{\beta}_1 \tag{5.7}$$

Similarly, the net injected current at bus j can be written as

$$\mathbf{I}_j = \sum_{q=1}^{n+1} \mathbf{Y}_{jq}\mathbf{V}_q \tag{5.8}$$

where $\mathbf{Y}_{jj} = \mathbf{Y}_{jj}^{\text{old}} - \mathbf{y}_1 + \boldsymbol{\gamma}_1$ is the new value of self-admittance for bus j with IPFC and

$$\boldsymbol{\gamma}_1 = \frac{\mathbf{y}_1(\mathbf{y}_{\text{se}_1} + \mathbf{y}_{10})}{\mathbf{y}_1 + \mathbf{y}_{10} + \mathbf{y}_{\text{se}1}}, \mathbf{Y}_{j(n+1)} = \boldsymbol{\beta}_1, \text{ and } \mathbf{Y}_{ji} = -\boldsymbol{\beta}_1 \tag{5.9}$$

Thus, the effect of the first series converter of the IPFC is equivalent to an additional (n + 1)th bus without any IPFC. Now, from Figure 5.2, the net injected current at this fictitious (n + 1)th bus equals the current flowing into the transmission system from this bus, that is,

$$\mathbf{I}_{n+1} = -\mathbf{I}_{\text{se}1} = \boldsymbol{\alpha}_1\mathbf{V}_{\text{se}1} - \boldsymbol{\alpha}_1\mathbf{V}_i + \boldsymbol{\beta}_1\mathbf{V}_j = \sum_{q=1}^{n+1} \mathbf{Y}_{(n+1)q}\mathbf{V}_q \tag{5.10}$$

with

$$\mathbf{V}_{n+1} = \mathbf{V}_{\text{se}1}, \mathbf{Y}_{(n+1)(n+1)} = \boldsymbol{\alpha}_1, \mathbf{Y}_{(n+1)i} = -\boldsymbol{\alpha}_1, \text{ and } \mathbf{Y}_{(n+1)j} = \boldsymbol{\beta}_1 \tag{5.11}$$

Proceeding in a similar way, it can be shown that the effect of incorporation of all the p series converters can be treated as equivalent to addition of p more buses $(n + 1)$ up to $(n + p)$ to the existing n-bus system, provided

$$\mathbf{V}_{n+1} = \mathbf{V}_{se1}, \mathbf{V}_{n+2} = \mathbf{V}_{se2}, \ldots, \mathbf{V}_{n+p} = \mathbf{V}_{sep} \tag{5.12}$$

with

$$\mathbf{Y}_{i(n+1)} = -\alpha_1, \mathbf{Y}_{(i+1)(n+2)} = -\alpha_2, \ldots, \mathbf{Y}_{(i+p-1)(n+p)} = -\alpha_p \tag{5.13}$$

$$\mathbf{Y}_{ij} = -\beta_1, \mathbf{Y}_{(i+1)(j+1)} = -\beta_2, \ldots, \mathbf{Y}_{(i+p-1)(j+p-1)} = -\beta_p \tag{5.14}$$

$$\mathbf{Y}_{j(n+1)} = \beta_1, \mathbf{Y}_{(j+1)(n+2)} = \beta_2, \ldots, \mathbf{Y}_{(j+p-1)(n+p)} = \beta_p \tag{5.15}$$

where, for the transmission line incorporating the pth converter,

$$\alpha_p = \frac{y_{sep}(y_p + y_{p0})}{y_p + y_{p0} + y_{sep}} \text{ and } \beta_p = \frac{y_{sep} y_p}{y_p + y_{p0} + y_{sep}} \tag{5.16}$$

Thus, the original n-bus system incorporating an IPFC with p converters is transformed to an equivalent $(n + p)$ bus system without any IPFC. Also, the net current injections at the p additional power-flow buses $(n + 1)$ up to $(n + p)$ representing the p series converters of the IPFC can be written as

$$\mathbf{I}_{n+1} = \sum_{q=1}^{n+p} \mathbf{Y}_{(n+1)q} \mathbf{V}_q, \text{and so on, up to } \mathbf{I}_{n+p} = \sum_{q=1}^{n+p} \mathbf{Y}_{(n+p)q} \mathbf{V}_q \tag{5.17}$$

where, corresponding to the pth converter,

$$\mathbf{Y}_{(n+p)(n+p)} = \alpha_p, \mathbf{Y}_{(n+p)(i+p-1)} = -\alpha_p, \mathbf{Y}_{(n+p)(j+p-1)} = \beta_p, \text{as } \mathbf{V}_{n+p} = \mathbf{V}_{sep} \tag{5.18}$$

Thus, in the proposed model, if the gth $(1 \le g \le p)$ voltage source of the IPFC is connected between, say, bus a (SE) and bus b (RE) with $a, b \le n$, then,

$$\mathbf{Y}_{a(n+g)} = -\alpha_g,\ \mathbf{Y}_{ab} = -\beta_g,\ \mathbf{Y}_{b(n+g)} = \beta_g,\ \mathbf{Y}_{(n+g)(n+g)} = \alpha_g, \text{ with } \mathbf{V}_{n+g} = \mathbf{V}_{seg}$$

and the net current injection at the equivalent $(n + g)$th bus is

$$\mathbf{I}_{n+g} = -\mathbf{I}_{seg} = \sum_{q=1}^{n+p} \mathbf{Y}_{(n+g)q} \mathbf{V}_q \tag{5.19}$$

5.3 POWER FLOW EQUATIONS IN THE PROPOSED IPFC MODEL

With existing IPFC models, the expression of the active power injection at any SE or RE bus can be written (using Equation 5.5) as

$$P_a = \sum_{q=1}^{n} V_a V_q Y_{aq} \cos(\theta_a - \theta_q - \varphi_{yaq}), \quad a \leq n \tag{5.20}$$

if no IPFC series converter is connected to bus a;

$$P_a = \sum_{q=1, q \neq b}^{n} V_a V_q Y_{aq} \cos(\theta_a - \theta_q - \varphi_{yaq}) - V_a V_b \beta_c \cos(\theta_a - \theta_c - \varphi_{y\beta c}) \\ - V_a V_{se\,c} \alpha_c \cos(\theta_a - \theta_{se\,c} - \varphi_{y\alpha c}) \tag{5.21}$$

if IPFC converter c is connected in branch a–b with a as SE bus;

$$P_a = \sum_{q=1, q \neq b}^{n} V_a V_q Y_{aq} \cos(\theta_a - \theta_q - \varphi_{yaq}) - V_a V_b \beta_c \cos(\theta_a - \theta_b - \varphi_{y\beta c}) \\ + V_a V_{se\,c} \beta_c \cos(\theta_a - \theta_{se\,c} - \varphi_{y\beta c}) \tag{5.22}$$

if IPFC converter c is connected in branch a–b with a as RE bus.

Thus, from the above equation, it is observed that with existing IPFC models, additional terms due to contributions from the voltage sources representing the series converters of the IPFC are present in the expression of the bus active power injection(s). This causes the existing power flow codes to be modified. This is also true for the bus reactive power injection(s).

Now, in the proposed model, it is observed from Equations 5.6, 5.8, and 5.17 that the net injected current at any of the p SE buses [buses i, $(i + 1)$, and so on, up to $(i + p - 1)$] or RE buses (buses j, $(j + 1)$, and so on, up to $(j + p - 1)$) can be written as

$$\mathbf{I}_h = \sum_{q=1}^{n+p} \mathbf{Y}_{hq} \mathbf{V}_q \quad i \leq h \leq (i + p - 1) \text{or } j \leq h \leq (j + p - 1) \tag{5.23}$$

Thus, in the proposed model, the active power injection equation at any SE or RE bus a ($a \leq n + p$) can be written using Equation 5.23 as

$$\mathrm{P}_a = \sum_{q=1}^{n+p} V_a V_q Y_{aq} \cos(\theta_a - \theta_q - \phi_{yaq}) \quad a \leq (n + p) \tag{5.24}$$

Similarly, the expression for the reactive power at any bus *a* (SE or RE) can be written as

$$Q_a = \sum_{q=1}^{n+p} V_a V_q Y_{aq} \sin(\theta_a - \theta_q - \varphi_{yaq}) \quad a \leq (n+p) \tag{5.25}$$

From the above two equations, it can be observed that both the active and reactive power injections at any bus can be computed using existing power flow codes.

Now, with existing models, the expression for the real power delivered by an IPFC with two converters *e* and *f* in series with branches *a–b* and *c–d*, respectively (for simplicity, the simplest IPFC configuration with only two converters is assumed) can be written using Equation 5.3 as

$$P_{IPFC} = \mathrm{Re}[\mathbf{V}_{see}(-\mathbf{I}^*_{see}) + \mathbf{V}_{sef}(-\mathbf{I}^*_{sef})] \tag{5.26}$$

$$\begin{aligned} P_{IPFC} = {} & V_{see}^2 \alpha_e \cos\varphi_{y\alpha e} + V_{see} V_b \beta_e \cos(\theta_{see} - \theta_b - \varphi_{y\beta e}) \\ & - V_{see} V_a \alpha_e \cos(\theta_{see} - \theta_a - \varphi_{y\alpha e}) + V_{sef} V_d \beta_f \cos(\theta_{sef} - \theta_d - \varphi_{y\beta f}) \\ & - V_{sef} V_c \alpha_f \cos(\theta_{sef} - \theta_c - \varphi_{y\alpha f}) + V_{sef}^2 \alpha_f \cos\varphi_{y\alpha f} \end{aligned} \tag{5.27}$$

From the above equation, it is observed that the IPFC real power expression cannot be computed using existing power flow codes. It will require fresh codes for its implementation.

However, in the proposed model, converters *e* and *f* of the double-converter IPFC considered are transformed to power-flow buses ($n + 1$) and ($n + 2$), and hence, using Equation 5.26,

$$P_{IPFC} = \mathrm{Re}[\mathbf{V}_{see}(-\mathbf{I}^*_{see}) + \mathbf{V}_{sef}(-\mathbf{I}^*_{sef})] = \mathrm{Re}[\mathbf{V}_{n+1}(\mathbf{I}^*_{n+1}) + \mathbf{V}_{n+2}(\mathbf{I}^*_{n+2})]$$

or

$$P_{IPFC} = \mathrm{Re}(\mathbf{S}_{n+1} + \mathbf{S}_{n+2}) = P_{n+1} + P_{n+2} \tag{5.28}$$

Hence, the sum of the active power injections of two power-flow buses (computed from Equation 5.24 using existing power flow codes) yields the IPFC real power. It can be shown that if the IPFC considered has *p* series converters, the real power delivered by it can be computed from the sum of the active power injections of *p* power-flow buses using existing codes. In such a case, Equation 5.28 becomes

$$
\begin{aligned}
P_{\text{IPFC}} &= \text{Re}[\mathbf{V}_{\text{se}1}(-\mathbf{I}^*_{\text{se}1}) + \mathbf{V}_{\text{se}2}(-\mathbf{I}^*_{\text{se}2}) + \cdots + \mathbf{V}_{\text{se}p}(-\mathbf{I}^*_{\text{se}p})] \\
&= P_{n+1} + P_{n+2} + \cdots + P_{n+p}
\end{aligned} \tag{5.29}
$$

Now, with existing IPFC models, the expression for the active power flow in the line incorporating the cth ($1 \leq c \leq p$) converter and which is connected between SE and RE buses a and b, respectively, can be written using Equation 5.3 as

$$
P_{\text{LINE}\,c} = P_{ab} = \text{Re}(\mathbf{V}_a \mathbf{I}^*_{\text{se}c})
$$

or

$$
\begin{aligned}
P_{ab} = {} & V_a^2 \alpha_c \cos\varphi_{y\alpha c} - V_a V_b \beta_c \cos(\theta_a - \theta_b - \varphi_{y\beta c}) \\
& - V_a V_{\text{se}\,c} \alpha_c \cos(\theta_a - \theta_{\text{se}\,c} - \varphi_{y\alpha c})
\end{aligned} \tag{5.30}
$$

Similarly, the expression for the line reactive power flow can be written as

$$
\begin{aligned}
Q_{ab} = {} & -V_a^2 \alpha_c \sin\varphi_{y\alpha c} - V_a V_b \beta_c \sin(\theta_a - \theta_b - \varphi_{y\beta c}) \\
& - V_a V_{\text{se}\,c} \alpha_c \sin(\theta_a - \theta_{\text{se}\,c} - \varphi_{y\alpha c})
\end{aligned} \tag{5.31}
$$

Both the above two equations have new terms involving the IPFC series converter and the modified admittances due to the series coupling transformer of the converter. These necessitate modifications in the existing power flow codes.

In the proposed model, using Equation 5.19, these expressions become

$$
P_{\text{LINE}\,c} = P_{ab} = \text{Re}(\mathbf{V}_a \mathbf{I}^*_{\text{se}c}) = \text{Re}[\mathbf{V}_a(-\mathbf{I}^*_{n+c})]
$$

or

$$
P_{ab} = -\sum_{q=1}^{n+p} V_a V_q Y_{(n+c)q} \cos(\theta_a - \theta_q - \varphi_{y(n+c)q}) \tag{5.32}
$$

as the cth converter of the IPFC is transformed to $(n + c)$th power-flow bus in the proposed model. In a similar way,

$$
Q_{ab} = -\sum_{q=1}^{n+p} V_a V_q Y_{(n+c)q} \sin(\theta_a - \theta_q - \varphi_{y(n+c)q}) \tag{5.33}
$$

From the above two equations (5.32 and 5.33), it can be observed that in the proposed model, both the line active and reactive power flows can be computed using very minor modifications of the existing power flow codes.

5.4 IMPLEMENTATION IN NEWTON POWER FLOW ANALYSIS

If the number of voltage-controlled buses is $(m - 1)$, the power-flow problem for an n-bus system incorporated with x IPFCs each having p series converters can be formulated as follows:

Solve θ, $\mathbf{V}$, θ_{se}, and $\mathbf{V}_{se}$.

Specified $\mathbf{P}$, $\mathbf{Q}$, $\mathbf{P}_{IP}$, $\mathbf{P}_L$, and $\mathbf{Q}_L$

where:

$$\theta = [\theta_2 \ldots \theta_n]^T \quad \mathbf{V} = [V_{m+1} \ldots V_n]^T \tag{5.34}$$

$$\theta_{se} = [\theta_{se1} \ldots \theta_{se\,z}]^T \quad \mathbf{V}_{se} = [V_{se1} \ldots V_{se\,z}]^T \tag{5.35}$$

$$\mathbf{P} = [P_2 \ldots P_n]^T \quad \mathbf{Q} = [Q_{m+1} \ldots Q_n]^T \tag{5.36}$$

$$\mathbf{P}_{IP} = [P_{IPFC1} \ldots P_{IPFC\,x}]^T \tag{5.37}$$

$$\mathbf{P}_L = [P_{LINE1} \ldots P_{LINE\,z}]^T \quad \mathbf{Q}_L = [Q_{LINE1} \ldots Q_{LINE\,w}]^T \tag{5.38}$$

$$w = z - x \tag{5.39}$$

$$z = \text{total number of series converters} = px \tag{5.40}$$

In the above equations, $\mathbf{P}_{IP}$, $\mathbf{P}_L$, and $\mathbf{Q}_L$ represent the vectors for the specified real powers of the x IPFCs, and the active and reactive power flows of the z and $(z\text{–}x)$ transmission lines, respectively. In these equations, it is assumed that without any loss of generality, there are m generators connected at the first m buses of the system with bus 1 being the slack bus. Thus, the basic power-flow equation for the Newton power flow solution is represented as

$$\left[\begin{array}{cc|c|c} \frac{\partial \mathbf{P}}{\partial \theta} & \frac{\partial \mathbf{P}}{\partial \mathbf{V}} & \frac{\partial \mathbf{P}}{\partial \theta_{se}} & \frac{\partial \mathbf{P}}{\partial \mathbf{V}_{se}} \\ \frac{\partial \mathbf{Q}}{\partial \theta} & \frac{\partial \mathbf{Q}}{\partial \mathbf{V}} & \frac{\partial \mathbf{Q}}{\partial \theta_{se}} & \frac{\partial \mathbf{Q}}{\partial \mathbf{V}_{se}} \\ \hline \frac{\partial \mathbf{P}_{IP}}{\partial \theta} & \frac{\partial \mathbf{P}_{IP}}{\partial \mathbf{V}} & \frac{\partial \mathbf{P}_{IP}}{\partial \theta_{se}} & \frac{\partial \mathbf{P}_{IP}}{\partial \mathbf{V}_{se}} \\ \hline \frac{\partial \mathbf{P}_L}{\partial \theta} & \frac{\partial \mathbf{P}_L}{\partial \mathbf{V}} & \frac{\partial \mathbf{P}_L}{\partial \theta_{se}} & \frac{\partial \mathbf{P}_L}{\partial \mathbf{V}_{se}} \\ \hline \frac{\partial \mathbf{Q}_L}{\partial \theta} & \frac{\partial \mathbf{Q}_L}{\partial \mathbf{V}} & \frac{\partial \mathbf{Q}_L}{\partial \theta_{se}} & \frac{\partial \mathbf{Q}_L}{\partial \mathbf{V}_{se}} \end{array}\right] \left[\begin{array}{c} \Delta\theta \\ \hline \Delta\mathbf{V} \\ \hline \Delta\theta_{se} \\ \hline \Delta\mathbf{V}_{se} \end{array}\right] = \left[\begin{array}{c} \Delta\mathbf{P} \\ \hline \Delta\mathbf{Q} \\ \hline \Delta\mathbf{P}_{IP} \\ \hline \Delta\mathbf{P}_L \\ \hline \Delta\mathbf{Q}_L \end{array}\right] \tag{5.41}$$

or

$$\left[\begin{array}{cc|c|c} \multicolumn{2}{c|}{\mathbf{J}_{\text{old}}} & \begin{array}{c}\mathbf{J}_1\\ \hline \mathbf{J}_3\end{array} & \begin{array}{c}\mathbf{J}_2\\ \hline \mathbf{J}_4\end{array} \\ \hline \mathbf{J}_5 & \mathbf{J}_6 & \mathbf{J}_7 & \mathbf{J}_8 \\ \hline \mathbf{J}_9 & \mathbf{J}_{10} & \mathbf{J}_{11} & \mathbf{J}_{12} \\ \hline \mathbf{J}_{13} & \mathbf{J}_{14} & \mathbf{J}_{15} & \mathbf{J}_{16} \end{array}\right] \left[\begin{array}{c} \Delta\boldsymbol{\theta} \\ \hline \Delta\mathbf{V} \\ \hline \Delta\boldsymbol{\theta}_{\text{se}} \\ \hline \Delta\mathbf{V}_{\text{se}} \end{array}\right] = \left[\begin{array}{c} \Delta\mathbf{P} \\ \hline \Delta\mathbf{Q} \\ \hline \Delta\mathbf{P}_{\text{IP}} \\ \hline \Delta\mathbf{P}_{\text{L}} \\ \hline \Delta\mathbf{Q}_{\text{L}} \end{array}\right] \quad (5.42)$$

In the above equation, $\mathbf{J}_{\text{old}}$ is the conventional power-flow Jacobian sub-block corresponding to the angle and voltage magnitude variables of the n-buses. The other Jacobian submatrices can be identified easily from Equation 5.42.

Now, in the proposed model, there are $(n + z)$ buses. Thus, the quantities to be solved for power-flow are $\boldsymbol{\theta}^{\text{new}}$ and $\mathbf{V}^{\text{new}}$, where

$$\boldsymbol{\theta}^{\text{new}} = [\theta_2 \ldots \theta_{n+z}]^{\text{T}} \text{ and } \mathbf{V}^{\text{new}} = [V_{m+1} \ldots V_{n+z}]^{\text{T}} \quad (5.43)$$

Thus, Equation 5.42 is transformed in the proposed model as

$$\left[\begin{array}{c} \mathbf{JX1} \\ \hline \mathbf{JX2} \\ \hline \mathbf{JX3} \end{array}\right] \left[\begin{array}{c} \Delta\boldsymbol{\theta}^{\text{new}} \\ \hline \Delta\mathbf{V}^{\text{new}} \end{array}\right] = \left[\begin{array}{c} \Delta\mathbf{PQ} \\ \hline \Delta\mathbf{P}_{\text{IP}} \\ \hline \Delta\mathbf{PQ}_{\text{L}} \end{array}\right] \quad (5.44)$$

where:

$$\Delta\mathbf{PQ} = [\Delta\mathbf{P}^{\text{T}}\ \Delta\mathbf{Q}^{\text{T}}]^{\text{T}}, \Delta\mathbf{PQ}_{\text{L}} = [\Delta\mathbf{P}_{\text{L}}^{\text{T}}\ \Delta\mathbf{Q}_{\text{L}}^{\text{T}}]^{\text{T}} \quad (5.45)$$

The matrices **JX1**, **JX2**, and **JX3** are identified easily from Equation 5.44. Now, it can be shown that in Equation 5.44,

1. The matrices **JX1** and **JX2** can be computed using existing Jacobian codes.

2. The matrix **JX3** can be computed using very minor modifications of the existing Jacobian codes.

The justification of the above two statements are shown as follows:

Let us first define for the proposed $(n + z)$ bus system,

$$\mathbf{P}^{\text{new}} = [P_2 \ldots P_{n+z}]^{\text{T}} \quad (5.46)$$

Subsequently, a new Jacobian matrix is computed as

$$\mathbf{J}^{\text{new}} = \begin{bmatrix} \dfrac{\partial \mathbf{P}^{\text{new}}}{\partial \boldsymbol{\theta}^{\text{new}}} & \dfrac{\partial \mathbf{P}^{\text{new}}}{\partial \mathbf{V}^{\text{new}}} \\ \hdashline \dfrac{\partial \mathbf{Q}}{\partial \boldsymbol{\theta}^{\text{new}}} & \dfrac{\partial \mathbf{Q}}{\partial \mathbf{V}^{\text{new}}} \end{bmatrix} = \begin{bmatrix} \mathbf{J}^{\text{A}} \\ \hdashline \mathbf{J}^{\text{B}} \end{bmatrix} \tag{5.47}$$

From Equations 5.24 and 5.25 of the proposed model, it can be observed that the expressions for P_i $(2 \le i \le n+z)$ and Q_i $(m+1 \le i \le n)$ in $\mathbf{P}^{\text{new}}$ and $\mathbf{Q}$, respectively (of Equation 5.47) can be computed using existing power flow codes. Hence, the matrix $\mathbf{J}^{\text{new}}$ can be computed with existing codes for calculating the Jacobian matrix.

From Equations 5.36, 5.43, and 5.46,

$$\frac{\partial \mathbf{P}^{\text{new}}}{\partial \boldsymbol{\theta}^{\text{new}}} = \begin{bmatrix} \dfrac{\partial \mathbf{P}}{\partial \boldsymbol{\theta}^{\text{new}}} \\ \hdashline \mathbf{JX4} \end{bmatrix} \tag{5.48}$$

where:

$$\mathbf{JX4} = \left[\frac{\partial P_{n+1}}{\partial \boldsymbol{\theta}^{\text{new}}} \quad \cdots \quad \frac{\partial P_{n+p}}{\partial \boldsymbol{\theta}^{\text{new}}} \;\vdots\; \cdots \quad \cdots \;\vdots\; \frac{\partial P_{n+z-p+1}}{\partial \boldsymbol{\theta}^{\text{new}}} \quad \cdots \quad \frac{\partial P_{n+z}}{\partial \boldsymbol{\theta}^{\text{new}}} \right]^{\text{T}} \tag{5.49}$$

From Equation 5.48, it is observed that the subblock $\partial \mathbf{P}/\partial \boldsymbol{\theta}^{\text{new}}$ is contained within the matrix $\partial \mathbf{P}^{\text{new}}/\partial \boldsymbol{\theta}^{\text{new}}$. In a similar way, it can be shown that the subblock $\partial \mathbf{P}/\partial \mathbf{V}^{\text{new}}$ is contained within the matrix $\partial \mathbf{P}^{\text{new}}/\partial \mathbf{V}^{\text{new}}$. Hence, once the matrix $\mathbf{J}^{\text{new}}$ is computed (using existing codes), the matrix

$$\mathbf{JX1} = \begin{bmatrix} \dfrac{\partial \mathbf{P}}{\partial \boldsymbol{\theta}^{\text{new}}} & \dfrac{\partial \mathbf{P}}{\partial \mathbf{V}^{\text{new}}} \\ \multicolumn{2}{c}{\mathbf{J}^{\text{B}}} \end{bmatrix}$$

can be very easily extracted from the matrix $\mathbf{J}^{\text{new}}$ using simple matrix extraction codes only. Hence, no fresh codes need to be written for computing **JX1**.

Now, from Equation 5.29,

$$\frac{\partial}{\partial \boldsymbol{\theta}^{\text{new}}}\left(P_{n+1} + P_{n+2} + \cdots + P_{n+p}\right) = \frac{\partial P_{\text{IPFC1}}}{\partial \boldsymbol{\theta}^{\text{new}}} \tag{5.50}$$

Thus, in Equation 5.49, the sum of the first p rows of **JX4** equals $\partial P_{\text{IPFC1}}/\partial \boldsymbol{\theta}^{\text{new}}$ (the first element of the Jacobian subblock $\partial \mathbf{P}_{\text{IP}}/\partial \boldsymbol{\theta}^{\text{new}}$). Similarly, the sum of the last p rows yields the last element of $\partial \mathbf{P}_{\text{IP}}/\partial \boldsymbol{\theta}^{\text{new}}$. Therefore, $\partial \mathbf{P}_{\text{IP}}/\partial \boldsymbol{\theta}^{\text{new}}$

can be easily extracted from $\partial \mathbf{P}^{\text{new}}/\partial \boldsymbol{\theta}^{\text{new}}$. In a similar way, it can be shown that $\partial \mathbf{P}_{\text{IP}}/\partial \mathbf{V}^{\text{new}}$ can also be easily extracted from $\partial \mathbf{P}^{\text{new}}/\partial \mathbf{V}^{\text{new}}$. Hence, the matrix $\mathbf{JX2} = \begin{bmatrix} \partial \mathbf{P}_{\text{IP}}/\partial \boldsymbol{\theta}^{\text{new}} & \partial \mathbf{P}_{\text{IP}}/\partial \mathbf{V}^{\text{new}} \end{bmatrix}$ need not be computed—it can be formed from the matrix $\mathbf{J}^{\text{A}}$ of $\mathbf{J}^{\text{new}}$ by matrix extraction codes (in conjunction with codes for simple matrix row addition). Hence, both **JX1** and **JX2** need not be computed and can be extracted from $\mathbf{J}^{\text{new}}$ (subsequent to its computation using existing Jacobian codes).

Now, it is shown that in the proposed model, the matrix

$$\mathbf{JX3} = \begin{bmatrix} \partial \mathbf{P}_{\text{L}}/\partial \boldsymbol{\theta}^{\text{new}} & \partial \mathbf{P}_{\text{L}}/\partial \mathbf{V}^{\text{new}} \\ \partial \mathbf{Q}_{\text{L}}/\partial \boldsymbol{\theta}^{\text{new}} & \partial \mathbf{Q}_{\text{L}}/\partial \mathbf{V}^{\text{new}} \end{bmatrix}$$

can be computed using very minor modifications of the existing Jacobian codes.

Let c $(1 \le c \le z)$ denote the IPFC series converter number. Then, for the ath $(1 \le a \le p)$ converter of the bth $(1 \le b \le x)$ IPFC, $c = (b-1)p + a$. In the proposed model, the expression for the active power flow in the line connected between the gth bus (SE) and the hth bus (RE) incorporating the cth converter can be written (using Equation 5.32) as

$$P_{\text{LINE}\,c} = P_{gh} = -\sum_{q=1}^{n+z} V_g V_q Y_{(n+c)q} \cos\left(\theta_g - \theta_q - \varphi_{y(n+c)q}\right) \tag{5.51}$$

In a similar way, the reactive power flow can be written as

$$Q_{\text{LINE}\,c} = Q_{gh} = -\sum_{q=1}^{n+z} V_g V_q Y_{(n+c)q} \sin\left(\theta_g - \theta_q - \varphi_{(n+c)q}\right) \tag{5.52}$$

It is important to note that in Equation 5.52, the variable c assumes a value of $c = [(b-1)(p-1) + a]$ corresponding to the ath $(1 \le a \le p-1)$ converter of the bth $(1 \le b \le x)$ IPFC. This is because of the fact that the reactive power flows are specified only in $(p-1)$ lines out of the p lines each having a converter (corresponding to each IPFC).

From Equations 5.51 and 5.52, it is observed that in the proposed model, both the line active power flows $P_{\text{LINE}\,i}$ $(1 \le i \le z)$ and reactive power flows $Q_{\text{LINE}\,i}$ $(1 \le i \le w)$ can be computed using very minor modifications of the existing power flow codes. Consequently, it can be shown that all the subblocks of **JX3** can be computed using very minor modifications of the existing Jacobian codes, unlike with existing IPFC models. This reduces the complexity of software codes substantially.

5.5 ACCOMMODATION OF IPFC DEVICE LIMIT CONSTRAINTS

The three major device limit constraints of the IPFC, which have been considered in this chapter, are as follows [70,71,78]:

1. Injected (series) converter voltage magnitude V_{se}^{Lim}
2. Line current through the converter I_{se}^{Lim}
3. Real power exchange through the dc link P_{DC}^{Lim}

The device limit constraints have been accommodated by the principle that whenever a particular constraint limit is violated, it is kept at its specified limit, whereas a control objective is relaxed. Mathematically, this signifies the replacement of the Jacobian elements pertaining to the control objective by those of the constraints violated. The control strategies to incorporate the above three limits are detailed below. For simplicity, we consider a single IPFC with two series converters installed in two different transmission lines. The original power flow control specifications considered are $P_{LINE\,1}$ and $Q_{LINE\,1}$ for line 1 and $P_{LINE\,2}$ (or $Q_{LINE\,2}$) for line 2.

1. In this case,

$$V_{se} = V_{se}^{Lim} \tag{5.53}$$

   If V_{se1}^{Lim} is violated, V_{n+1} is fixed at the limit and either the line active (P_{LINE1}) or the reactive power flow (Q_{LINE1}) control objective is relaxed for line 1. The corresponding relaxed active or reactive power mismatch is replaced by $\Delta V_{n+1} = V_{se1}^{Lim} - V_{n+1}$. The Jacobian elements are changed accordingly. If V_{se2}^{Lim} is violated, V_{n+2} is fixed at the limit and the control objective in vogue (line active [P_{LINE2}] or reactive power flow [Q_{LINE2}]) is relaxed for line 2 with its mismatch replaced by $\Delta V_{n+2} = V_{se2}^{Lim} - V_{n+2}$. Again, the Jacobian elements are changed accordingly. For each violation of V_{se}^{Lim}, the corresponding row of the matrix **JX3** is rendered constant with all elements known *a priori*—all of them equal zero except the entry pertaining to V_{se}, which is unity. All other elements of the matrix **JX3** can be computed using very minor modifications of the existing Jacobian codes, as demonstrated in the previous Section 5.4.

2. If I_{se1}^{Lim} is violated, I_{n+1} is fixed at the limit and either the line active (P_{LINE1}) or the reactive power flow (Q_{LINE1}) control objective is

relaxed for line 1. Thus, the corresponding mismatch is replaced by $\Delta I_{n+1} = I_{\text{se1}}^{\text{Lim}} - I_{n+1}$, where (from Equations 5.10 and 5.11)

$$I_{n+1} = \left|\mathbf{I}_{n+1}\right| = \left[d_1 + d_2 + d_3 + d_4\right]^{1/2} \tag{5.54}$$

where:

$$d_1 = Y_{(n+1)i}^2 V_i^2 + Y_{(n+1)j}^2 V_j^2 + Y_{(n+1)(n+1)}^2 V_{n+1}^2$$

$$d_2 = 2Y_{(n+1)i} Y_{(n+1)j} V_i V_j \cos\left(\theta_i - \theta_j + \varphi_{y(n+1)i} - \varphi_{y(n+1)j}\right)$$

$$d_3 = 2Y_{(n+1)i} Y_{(n+1)(n+1)} V_i V_{n+1} \cos\left(\theta_i - \theta_{n+1} + \varphi_{y(n+1)i} - \varphi_{y(n+1)(n+1)}\right)$$

$$d_4 = 2Y_{(n+1)j} Y_{(n+1)(n+1)} V_j V_{n+1} \cos\left(\theta_j - \theta_{n+1} + \varphi_{y(n+1)j} - \varphi_{y(n+1)(n+1)}\right)$$

The Jacobian elements are changed accordingly. If $I_{\text{se2}}^{\text{Lim}}$ is violated, I_{n+2} is fixed at the limit and the line active (P_{LINE2}) or reactive power flow (Q_{LINE2}) control objective is relaxed for line 2, with its mismatch and Jacobian elements changed accordingly. If, however, both $V_{\text{se1}}^{\text{Lim}}$ and $I_{\text{se1}}^{\text{Lim}}$ are violated, leniency is exercised on both P_{LINE1} and Q_{LINE1} control objectives, which are replaced by $\Delta V_{n+1} = V_{\text{se1}}^{\text{Lim}} - V_{n+1}$ and $\Delta I_{n+1} = I_{\text{se1}}^{\text{Lim}} - I_{n+1}$. The corresponding Jacobian elements are also replaced. The derivations of the expressions of d_1, d_2, d_3, and d_4 are given in the Appendix.

3. If $P_{\text{DC}}^{\text{Lim}}$ is violated, then P_{DC} is fixed at the limit and either the P_{LINE} or the Q_{LINE} control objective is relaxed. Thus, the corresponding line power flow (active or reactive) mismatch replaced by $\Delta P_{\text{DC}} = P_{\text{DC}}^{\text{Lim}} - P_{\text{DC}}$, where (using Equations 5.10 and 5.11)

$$\begin{aligned} P_{\text{DC}} &= \text{Re}\left[\mathbf{V}_{\text{se1}}(-\mathbf{I}_{\text{se1}}^*)\right] = \text{Re}(\mathbf{V}_{n+1}\mathbf{I}_{n+1}^*) \\ &= V_{n+1} V_i Y_{(n+1)i} \cos(\theta_{n+1} - \theta_i - \varphi_{y(n+1)i}) \\ &\quad + V_{n+1} V_j Y_{(n+1)j} \cos(\theta_{n+1} - \theta_j - \varphi_{y(n+1)j}) \\ &\quad + V_{n+1}^2 Y_{(n+1)(n+1)} \cos\varphi_{y(n+1)(n+1)} \end{aligned} \tag{5.55}$$

The Jacobian elements are changed accordingly. If, however, all three quantities $P_{\text{DC}}^{\text{Lim}}$, $I_{\text{se1}}^{\text{Lim}}$, and $I_{\text{se2}}^{\text{Lim}}$ (or say, $P_{\text{DC}}^{\text{Lim}}$, $V_{\text{se}}^{\text{Lim}}$, and $I_{\text{se}}^{\text{Lim}}$) are violated, all the control objectives, that is, the line active and reactive power flows of the first line, along with the line active (reactive) power flow of the second line, are relaxed. The corresponding mismatches, along with the modification of the corresponding Jacobian elements, are replaced.

5.6 SELECTION OF INITIAL CONDITIONS

In this chapter, the initial conditions for the series voltage source(s) were chosen as $V_{se}^{0}\angle\theta_{se}^{0} = 0.1\angle-(\pi/2)$ p.u. following suggestions of [49].

5.7 CASE STUDIES AND RESULTS

The proposed method was applied to IEEE 118- and IEEE 300-bus systems to validate its feasibility. In each test system, multiple IPFCs were included and studies have been carried out for (1) IPFCs without any device limit constraints and (2) IPFCs with device limit constraints. In all the case studies, a convergence tolerance of 10^{-12} p.u. has been chosen. Although a large number of case studies confirmed the validity of the model, a few sets of representative results are presented in Sections 5.7.1 and 5.7.2. As in Chapters 3 and 4, in all subsequent tables, the symbol NI denotes the number of iterations taken by the algorithm.

5.7.1 Studies of IPFCs without Any Device Limit Constraints

5.7.1.1 IEEE 118-Bus System

In this system, an IPFC with three series converters has been considered on the transmission line branches among buses 80-99 (converter 1), 80-97 (converter 2), and 80-98 (converter 3). The control references chosen are as follows: $P_{80\text{-}99}^{SP} = 40$ MW and $Q_{80\text{-}99}^{SP} = 10$ MVAR (line 1); $P_{80\text{-}97}^{SP} = 40$ MW and $Q_{80\text{-}97}^{SP} = 25$ MVAR (line 2); and $P_{80\text{-}98}^{SP} = 45$ MW (line 3). The results are shown in Table 5.1.

From the table, it is also observed that in the presence of the IPFC, the line active and reactive power flow levels are enhanced compared to their corresponding values in the base case. Furthermore, the number of iterations needed to obtain convergence marginally increases in the presence of IPFC. The bus voltage profiles without and with IPFCs are shown in Figure 5.3a and b, respectively. Further, the difference between the bus voltage magnitudes with and without IPFC is shown in Figure 5.3c. From this figure, it is observed that in the presence of IPFC, there is very little change in the bus voltage profile.

5.7.1.2 IEEE 300-Bus System

In this system too, an IPFC with three series converters has been considered on the transmission line branches among buses 3-7 (converter 1), 3-19 (converter 2), and 3-150 (converter 3). The control references chosen are as follows: $P_{3\text{-}7}^{SP} = 280$ MW and $Q_{3\text{-}7}^{SP} = 100$ MVAR (line 1); $P_{3\text{-}19}^{SP} = 150$ MW

TABLE 5.1 Study of IEEE 118-Bus System with Three-Converter IPFC (No Device Limit Constraints)

Solution of Base Case Power Flow (without Any IPFC)			
Parameter	**Line 1**	**Line 2**	**Line 3**
P_{LINE} (MW)	33.48	35.41	42.96
Q_{LINE} (MVAR)	8.74	20.55	7.6
NI	6		
Unconstrained Solution of IPFC Quantities			
Quantity	**Converter 1**	**Converter 2**	**Converter 3**
V_{se} (p.u.)	0.0759	0.0676	0.0739
θ_{se}	−113.9°	−130.79°	−72.61°
I_{se} (p.u.)	0.3965	0.4536	0.4344
P_{DC} (p.u.)	0.0042	0.0037	−0.0078
P_{Line} (MW)	**40**	**40**	**45**
Q_{Line} (MVAR)	**10**	**25**	−4.05
NI	7		

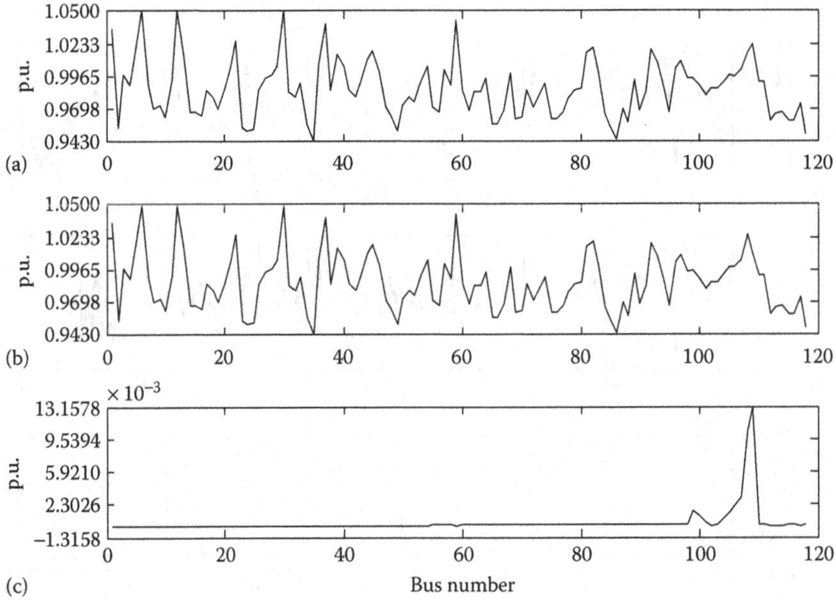

FIGURE 5.3 Bus voltage profile corresponding to the case study of Table 5.1. (a) Bus voltage magnitude without IPFC; (b) bus voltage magnitude with IPFC; (c) voltage magnitude difference.

TABLE 5.2 Study of IEEE 300-Bus System with Three-Converter IPFC (No Device Limit Constraints)

Solution of Base Case Power Flow (without Any IPFC)			
Parameter	**Line 1**	**Line 2**	**Line 3**
P_{LINE} (MW)	260	138.75	95.06
Q_{LINE} (MVAR)	120.55	11.19	138.26
NI	6		

Unconstrained Solution of IPFC Quantities			
Quantity	**Converter 1**	**Converter 2**	**Converter 3**
V_{se} (p.u.)	0.3094	0.1727	0.1877
θ_{se}	–102.76°	–92.12°	–137.22°
I_{se} (p.u.)	2.9823	1.5079	1.7862
P_{DC} (p.u.)	–0.0082	0.0213	–0.0131
P_{Line} (MW)	**280**	**150**	**100**
Q_{Line} (MVAR)	**100**	**10**	147.34
NI	8		

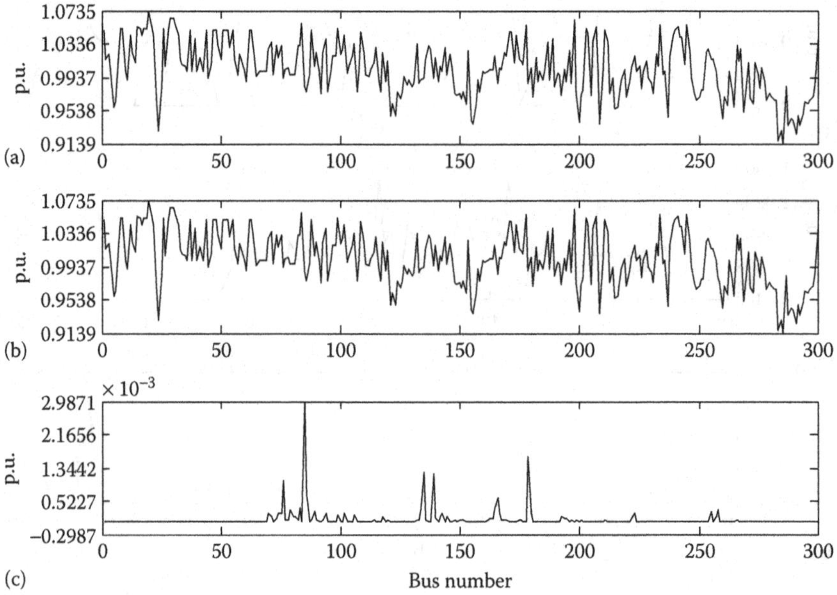

FIGURE 5.4 Bus voltage profile corresponding to the case study of Table 5.2. (a) Bus voltage magnitude without IPFC; (b) bus voltage magnitude with IPFC; (c) voltage magnitude difference.

and $Q_{3\text{-}19}^{SP} = 10$ MVAR (line 2); and $P_{3\text{-}150}^{SP} = 100$ MW (line 3). The results are shown in Table 5.2.

From the table, it is again observed that in the presence of the IPFC, the line active and reactive power flow levels are modified compared to their corresponding values in the base case. Furthermore, the number of iterations needed to obtain convergence marginally increases in the presence of IPFC.

The bus voltage profiles for this study in the 300-bus system without and with IPFCs are shown in Figure 5.4. From this figure, it is observed that in the presence of IPFC, there is very little change in the bus voltage profile.

5.7.2 Studies of IPFCs with Device Limit Constraints

In these case studies, various device limit constraints have been considered for both the 118- and 300-bus test systems. As already mentioned in Section 5.5, three major device constraint limits have been considered in this chapter. In each of the 118- and 300-bus test systems, case studies were carried out for implementation of these limit constraints in three different ways: (1) limit violation of a single constraint, (2) limit violations of two separate constraints simultaneously, and (3) limit violations of all three separate constraints simultaneously. For simplicity, in each test system, an IPFC with two series converters (the minimum possible configuration) has been considered. At first, the power flow solutions for each of these double-converter IPFCs without any device limit constraint are obtained in both the 118- and 300-bus test systems. The details of the IPFC converters considered in each of these two test systems are given in Sections 5.7.2.1 and 5.7.2.2.

5.7.2.1 IEEE 118-Bus System

In this system, a double-converter IPFC has been considered on the transmission line branches between buses 11-13 (converter 1) and 37-39 (converter 2). The control references chosen are as follows: $P_{11\text{-}13}^{SP} = 50$ MW and $Q_{11\text{-}13}^{SP} = 10$ MVAR (line 1) and $P_{37\text{-}39}^{SP} = 80$ MW (line 2). The results are shown in Table 5.3. From this table, it is again observed that in the presence of the IPFC, the line active and reactive power flow levels are modified from their corresponding values in the base case.

The bus voltage profiles for this study in the 118-bus system without and with IPFCs are shown in Figure 5.5. From this figure, it is observed that in the presence of IPFC, the bus voltage profiles are not disturbed much.

TABLE 5.3 Study of IEEE 118-Bus System with Two-Converter IPFC (No Device Limit Constraints)

Solution of Base Case Power Flow (without Any IPFC)		
Parameter	**Line 1**	**Line 2**
P_{LINE} (MW)	36.11	75.20
Q_{LINE} (MVAR)	12.05	7.94
NI	6	
Unconstrained Solution of IPFC Quantities		
Quantity	**Converter 1**	**Converter 2**
V_{se} (p.u.)	0.1111	0.0974
θ_{se}	−125.2°	−108.49°
I_{se} (p.u.)	0.5179	0.8072
P_{DC} (p.u.)	0.0041	−0.0041
P_{Line} (MW)	**50**	**80**
Q_{Line} (MVAR)	**10**	0.12
NI	6	

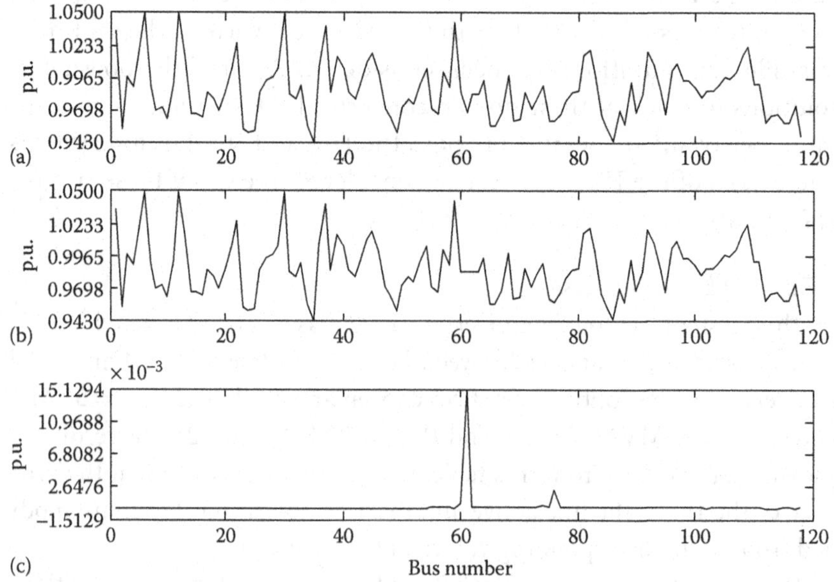

FIGURE 5.5 Bus voltage profile corresponding to the case study of Table 5.3. (a) Bus voltage magnitude without IPFC; (b) bus voltage magnitude with IPFC; (c) voltage magnitude difference.

5.7.2.2 IEEE 300-Bus System

In this system too, a double-converter IPFC has been considered on the transmission line branches between buses 2-3 (converter 1) and 86-102 (converter 2). The control references chosen are as follows: $P_{2\text{-}3}^{SP} = 50$ MW and $Q_{2\text{-}3}^{SP} = 50$ MVAR (line 1) and $P_{86\text{-}102}^{SP} = 50$ MW (line 2). The results shown in Table 5.4 confirm that by the action of the IPFCs, the line active and reactive power flow levels are modified from their corresponding values in the base case.

The bus voltage profiles for this study in the 300-bus system without and with IPFCs are shown in Figure 5.6, which again shows that the bus voltage profile is not disturbed much in the presence of IPFC.

Subsequent to the power flow solutions obtained for the 118- and 300-bus test systems without any device limit constraints (shown in Tables 5.3 and 5.4, respectively), the various constraint limits of the IPFC are imposed in both the 118- and 300-bus test systems. For this, the same line converter configurations as shown in Tables 5.3 and 5.4 have been maintained. The control objectives (as applicable while imposing single, double, or multiple device limit constraints) are also maintained identical to those adopted for Tables 5.3 and 5.4. Although a large number of case studies have been carried out for implementation of these limit constraints, only a few sets of representative results are presented in the text that follows.

TABLE 5.4 Study of IEEE 300-Bus System with Two-Converter IPFC (No Device Limit Constraints)

Solution of Base Case Power Flow (without Any IPFC)		
Parameter	**Line 1**	**Line 2**
P_{LINE} (MW)	36.03	37.95
Q_{LINE} (MVAR)	73.23	19.93
NI	6	
Unconstrained Solution of IPFC Quantities		
Quantity	**Converter 1**	**Converter 2**
V_{se} (p.u.)	0.0651	0.0763
θ_{se}	−108.42°	−120.43°
I_{se} (p.u.)	0.6769	0.5131
P_{DC} (p.u.)	−0.0145	0.0145
P_{Line} (MW)	**50**	**50**
Q_{Line} (MVAR)	**50**	5.18
NI	7	

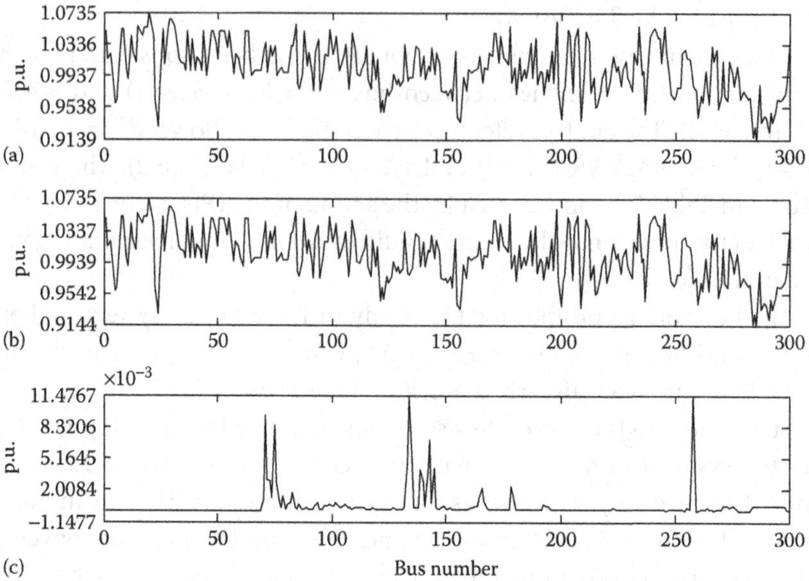

FIGURE 5.6 Bus voltage profile corresponding to the case study of Table 5.4. (a) Bus voltage magnitude without IPFC; (b) bus voltage magnitude with IPFC; (c) voltage magnitude difference.

Case 1: In this case, the violations of only the series injected voltage of converter 1 (V_{se1}^{Lim}) have been studied. Following the philosophy described in Section 5.5, this limit has been imposed by relaxing Q_{LINE} for line 1. The results for the 118-bus system are given in Table 5.5.

The bus voltage profile for this study with the 118-bus system is shown in Figure 5.7. From this figure, it is observed that the bus voltage profile does not change much when a single device limit constraint of the IPFC is enforced.

TABLE 5.5 Study of IEEE 118-Bus System with IPFC and V_{se}^{Lim} Constraint

Solution of IPFC Quantities with Limit on Series Injected Voltage of Converter 1 Only		
Specified converter 1 voltage limit: $V_{se1}^{Lim} = 0.111$ p.u.		
Quantity	**Converter 1**	**Converter 2**
V_{se} (p.u.)	**0.111**	0.096
θ_{se}	−113.59°	−116.37°
I_{se} (p.u.)	0.5102	0.809
P_{DC} (p.u.)	−0.0021	0.0021
P_{Line} (MW)	**50**	**80**
Q_{Line} (MVAR)	5.12	4.85
NI	10	

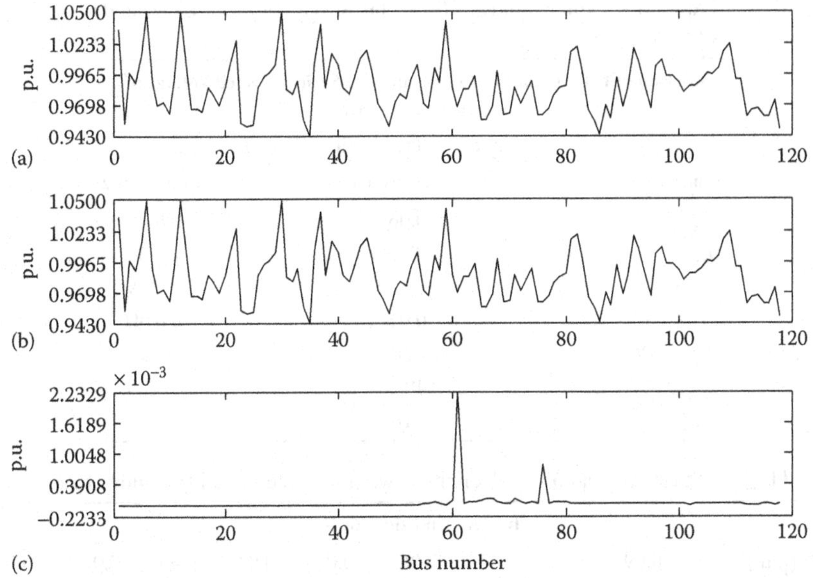

FIGURE 5.7 Bus voltage profile corresponding to the case study of Table 5.5. (a) Bus voltage magnitude without IPFC; (b) bus voltage magnitude with IPFC; (c) voltage magnitude difference.

In this particular case, it was observed that with the 300-bus system, adoption of the initial condition of $V_{se}^{0}\angle\theta_{se}^{0} = 0.1\angle-(\pi/2)$ yielded an unrealistic power flow solution (unrealistic values of complex bus voltages, converter currents, and power flows) for converter 2. The results are given in Table 5.6. From this table, it is observed that when the injected voltage of series converter 1 is limited to a very close value (0.064 p.u.) to that of the unconstrained value (0.065 p.u.), there occur impractically wide variations in the values of the voltages, line currents, dc link power transfer, and line reactive power flows pertaining to series converter 2. These unrealistic power flow solutions (corresponding to converter 2) are highlighted in bold cases in Table 5.6.

To investigate the convergence pattern and the final power flow solution *vis-à-vis* varying values of V_{se}^{0}, further case studies were again carried out. Some representative results (for converter 2) are presented in Table 5.7. Those with double asterisks (**) indicate unrealistic power flow solutions.

The realistic final power flow solutions (corresponding to $V_{se}^{0} = 0.04$ p.u.) for the 300-bus systems are shown in Table 5.8. The converged final values of the constrained variable along with the values of the control objectives are shown in bold cases. The bus voltage profiles for the final case study

TABLE 5.6 Initial Study of IEEE 300-Bus System with UPFC and V_{se}^{Lim} Constraint

Solution of IPFC Quantities with Limit on Series Injected Voltage of Converter 1 Only		
Specified converter 1 voltage limit: $V_{se1}^{Lim} = 0.064$ p.u.		
Quantity	**Converter 1**	**Converter 2**
V_{se} (p.u.)	0.064	0.5507
θ_{se}	−107.95°	−27.36°
I_{se} (p.u.)	0.6658	**2.339**
P_{DC} (p.u.)	−0.0141	0.0141
P_{Line} (MW)	50	50
Q_{Line} (MVAR)	48.38	**260.9**
NI	16	

TABLE 5.7 Effect of Variation of V_{se}^{0} on the Power Flow Solution for the Study of Table 5.6

	Initial Condition for V_{se}								
V_{se}^{0} (p.u.)	0.09	0.08	0.07	0.06	0.05	0.04	0.03	0.02	0.01
	Final Power Flow Solution for Converter 2								
V_{se2} (p.u.)	0.577	0.577	0.577	0.577	0.577	0.0767	0.0767	0.5507	
I_{se2} (p.u.)	2.443	2.443	2.443	2.443	2.443	0.5127	0.5127	2.339	
Q_{Line2} (MVAR)	274.7	274.7	274.7	274.7	274.7	4.55	4.55	260.9	div
P_{DC} (p.u.)	0.0205	0.0205	0.0205	0.0205	0.0205	0.0148	0.0148	0.0141	
NI	9 **	15 **	18 **	8 **	9 **	10	11	17 **	

TABLE 5.8 Final Study of IEEE 300-Bus System with UPFC and V_{se}^{Lim} Constraint

Solution of IPFC Quantities with Limit on Series Injected Voltage of Converter 1 Only		
Specified converter 1 voltage limit: $V_{se1}^{Lim} = 0.064$ p.u.		
Quantity	**Converter 1**	**Converter 2**
V_{se} (p.u.)	**0.064**	0.0767
θ_{se}	−106.15°	−121.61°
I_{se} (p.u.)	0.6648	0.5127
P_{DC} (p.u.)	−0.0148	0.0148
P_{Line} (MW)	**50**	**50**
Q_{Line} (MVAR)	48.21	4.55
NI	10 (with $V_{se}^{0} = 0.04$ p.u.)	

with the 300-bus system are shown in Figure 5.8. From this figure, it is again observed that the bus voltage profiles do not change much when a single device limit constraint of the IPFC is enforced.

Case 2: In this case, simultaneous limit violations of the series injected voltage of converter 1 (V_{se1}^{Lim}) and the dc link power transfer (P_{DC}^{Lim}) have

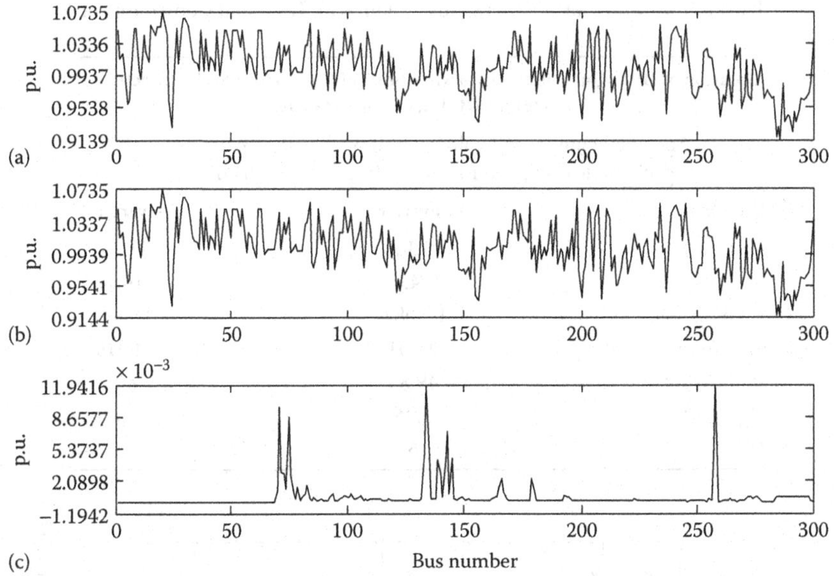

FIGURE 5.8 Bus voltage profile corresponding to the case study of Table 5.8. (a) Bus voltage magnitude without IPFC; (b) bus voltage magnitude with IPFC; (c) voltage magnitude difference.

been considered. The limits have been imposed by relaxing the active (P_{LINE}) and reactive power flow (Q_{LINE}) control objectives for the first transmission line, following the philosophy described in Section 5.5. The results for the 118-bus system are given in Table 5.9.

The bus voltage profiles for this study with the 118-bus system are shown in Figure 5.9. From this figure, it is observed that in the presence of IPFC, the bus voltage profiles do not change much when two simultaneous IPFC device limit constraints are enforced.

Again, in this case study, it was observed that with the 300-bus system, adoption of the initial condition of $V_{se}^0\angle\theta_{se}^0 = 0.1\angle-(\pi/2)$ yielded an unrealistic power flow solution for converter 1. The results are given in Table 5.10. From this table, the unrealistic power flow solutions (corresponding to converter 1) can be observed, which are highlighted in bold cases.

To investigate the convergence pattern and the final power flow solution *vis-à-vis* varying values of V_{se}^0, further case studies were again carried out. Some representative results (for converter 1) are presented in Table 5.11. Those with asterisks (**) indicate unrealistic power flow solutions.

TABLE 5.9 Study of IEEE 118-Bus System with IPFC and Constraints of V_{se}^{Lim} and P_{DC}^{Lim}

Solution of IPFC Quantities with Twin Limits of Converter 1 Series Injected Voltage and DC Link Power Transfer

Specified converter 1 voltage limit: $V_{se1}^{Lim} = 0.11$ p.u.
Specified dc link power transfer limit: $P_{DC}^{Lim} = -0.0019$ p.u.

Quantity	Converter 1	Converter 2
V_{se} (p.u.)	**0.11**	0.096
θ_{se}	–114.04°	–116.12°
I_{se} (p.u.)	0.5087	0.8089
P_{DC} (p.u.)	**–0.0019**	**0.0019**
P_{Line} (MW)	49.83	**80**
Q_{Line} (MVAR)	5.36	4.7
NI	6	

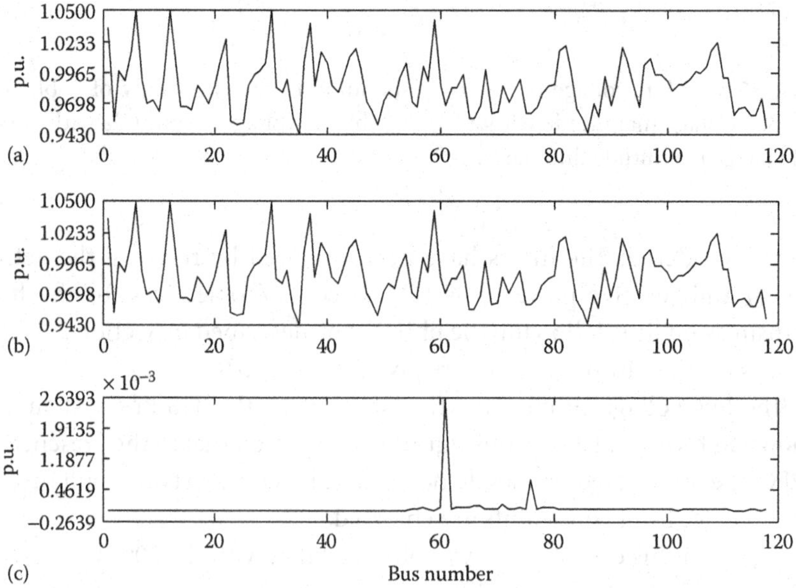

FIGURE 5.9 Bus voltage profile corresponding to the case study of Table 5.9. (a) Bus voltage magnitude without IPFC; (b) bus voltage magnitude with IPFC; (c) voltage magnitude difference.

The realistic final power flow solutions (corresponding to $V_{se}^{0} = 0.06$ p.u.) for the 300-bus systems are shown in Table 5.12. The converged final values of the constrained variable along with the values of the control objectives are shown in bold cases.

TABLE 5.10 Initial Study of IEEE 300-Bus System with IPFC and Constraints of V_{se}^{Lim} and P_{DC}^{Lim}

Solution of IPFC Quantities with Twin Limits of Converter 1 Series Injected Voltage and DC Link Power Transfer

Specified converter 1 voltage limit: $V_{se1}^{Lim} = 0.063$ p.u.
Specified dc link power transfer limit: $P_{DC}^{Lim} = -0.0143$ p.u.

Quantity	Converter 1	Converter 2
V_{se} (p.u.)	0.063	0.0763
θ_{se}	−10.04°	−119.63°
I_{se} (p.u.)	**0.2645**	0.5136
P_{DC} (p.u.)	−0.0143	0.0143
P_{Line} (MW)	**27.02**	50
Q_{Line} (MVAR)	**−6.3**	5.7
NI	7	

TABLE 5.11 Effect of Variation of V_{se}^0 on the Power Flow Solution for the Study of Table 5.10

Initial Condition for V_{se}									
V_{se}^0 (p.u.)	0.09	0.08	0.07	0.06	0.05	0.04	0.03	0.02	0.01
Final Power Flow Solution for Converter 1									
V_{se1} (p.u.)	0.063	0.063	0.063	0.063	0.063	0.063	0.063	0.063	
I_{se1} (p.u.)	0.2645	0.6623	0.6623	0.6623	0.2674	0.2674	0.6623	0.2674	
P_{Line1} (MW)	27.02	49.22	49.22	49.22	27.18	27.18	49.22	27.18	div
Q_{Line1} (MVAR)	−6.3	48.63	48.63	48.63	−6.96	−6.96	48.63	−6.96	
NI	10**	8	9	7	14**	13**	10	17**	

TABLE 5.12 Final Study of IEEE 300-Bus System with IPFC and Constraints of V_{se}^{Lim} and P_{DC}^{Lim}

Solution of IPFC Quantities with Twin Limits of Converter 1 Series Injected Voltage and DC Link Power Transfer

Specified converter 1 voltage limit: $V_{se1}^{Lim} = 0.063$ p.u.
Specified dc link power transfer limit: $P_{DC}^{Lim} = -0.0143$ p.u.

Quantity	Converter 1	Converter 2
V_{se} (p.u.)	**0.063**	0.0762
θ_{se}	−107.17°	−119.86°
I_{se} (p.u.)	0.6623	0.5133
P_{DC} (p.u.)	**−0.0143**	**0.0143**
P_{Line} (MW)	49.22	50
Q_{Line} (MVAR)	48.63	5.49
NI	7 (with $V_{se}^0 = 0.06$ p.u.)	

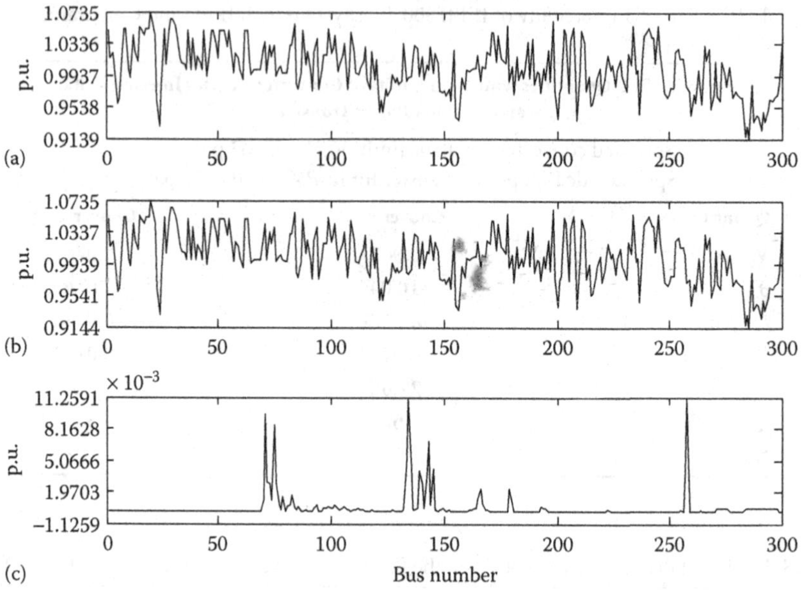

FIGURE 5.10 Bus voltage profile corresponding to the case study of Table 5.12. (a) Bus voltage magnitude without IPFC; (b) bus voltage magnitude with IPFC; (c) voltage magnitude difference.

The bus voltage profiles for this final case study with the 300-bus system are shown in Figure 5.10. From this figure, it is again observed that the bus voltage profiles do not change much when two simultaneous device limit constraints of the IPFC are enforced.

Case 3: In this case, simultaneous limit violations of the dc link power transfer (P_{DC}^{Lim}), the injected voltage of the series converter in line 1 (V_{se1}^{Lim}) and the line current (I_{se2}^{Lim}) in the second series converter (in line 2) have been considered. For enforcing these limits, both P_{LINE} and Q_{LINE} of line 1 as well as P_{LINE} of line 2 are relaxed following the discussion in Section 5.5. The results for the 118-bus system are presented in Table 5.13.

The bus voltage profiles for this case study with the 118-bus system are shown in Figure 5.11. From this figure, it is observed that in the presence of IPFC, the simultaneous enforcement of three IPFC device limit constraints causes very little change in the bus voltage profile.

However, again with the 300-bus system, adoption of the initial condition of $V_{se}^{0}\angle\theta_{se}^{0} = 0.1\angle-(\pi/2)$ yielded an unrealistic power flow solution for converter 1. The results are given in Table 5.14. From this table, the

TABLE 5.13 Study of IEEE 118-Bus System with IPFC and Constraints of V_{se}^{Lim}, I_{se}^{Lim}, and P_{DC}^{Lim}

Solution of IPFC Quantities with All Three Limits of DC Link Power Transfer, Series Injected Voltage of Converter 1, and Line Current of Converter 2

Specified dc link power transfer limit: $P_{DC}^{Lim} = -0.0018$ p.u.
Specified voltage limit of converter 1: $V_{se1}^{Lim} = 0.1$ p.u.
Specified line current limit of converter 2: $I_{se2}^{Lim} = 0.79$ p.u.

Quantity	Converter 1	Converter 2
V_{se} (p.u.)	**0.1**	0.088
θ_{se}	–115°	–116.69°
I_{se} (p.u.)	0.4908	**0.79**
P_{DC} (p.u.)	**–0.0018**	**0.0018**
P_{Line} (MW)	47.95	78.09
Q_{Line} (MVAR)	6.19	5.29
NI	6	

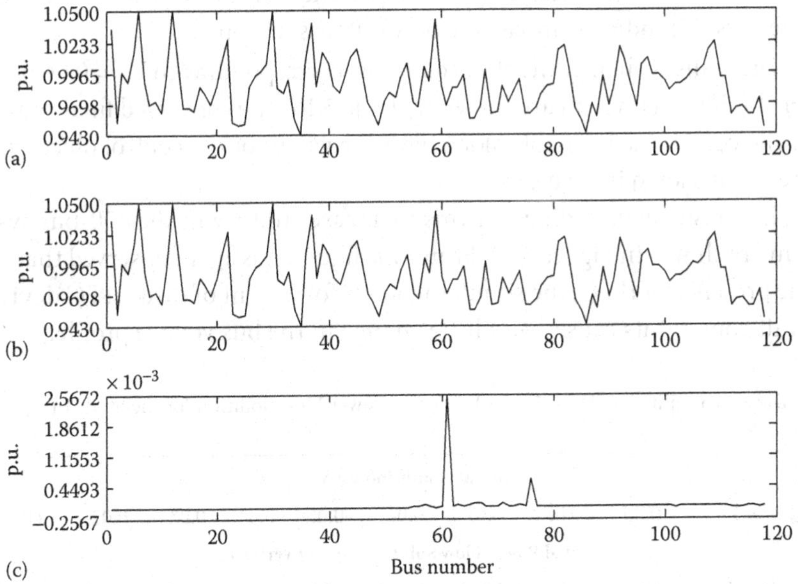

FIGURE 5.11 Bus voltage profile corresponding to the case study of Table 5.13. (a) Bus voltage magnitude without IPFC; (b) bus voltage magnitude with IPFC; (c) voltage magnitude difference.

TABLE 5.14 Initial Study of IEEE 300-Bus System with IPFC and V_{se}^{Lim}, I_{se}^{Lim}, and P_{DC}^{Lim} Constraints

Solution of IPFC Quantities with All Three Limits of DC Link Power Transfer, Series Injected Voltage of Converter 1, and Line Current of Converter 2		
Specified dc link power transfer limit: $P_{DC}^{Lim} = -0.0142$ p.u. Specified voltage limit of converter 1: $V_{se1}^{Lim} = 0.062$ p.u. Specified line current limit of converter 2: $I_{se2}^{Lim} = 0.51$ p.u.		
Quantity	**Converter 1**	**Converter 2**
V_{se} (p.u.)	0.062	0.0753
θ_{se}	−10.53°	−120.04°
I_{se} (p.u.)	**0.2642**	0.51
P_{DC} (p.u.)	−0.0142	0.0142
P_{Line} (MW)	**27.14**	49.66
Q_{Line} (MVAR)	**−5.59**	5.54
NI	10	

unrealistic power flow solutions (corresponding to converter 1) can be observed, which are highlighted in bold cases. Consequently, further case studies were again carried out *vis-à-vis* varying values of V_{se}^{0}. Some representative results (for converter 1) are presented in Table 5.15. Those with asterisks (**) indicate unrealistic power flow solutions.

The realistic final power flow solutions (corresponding to $V_{se}^{0} = 0.06$ p.u.) for the 300-bus system are shown in Table 5.16. The converged final values of the constrained variable along with the values of the control objectives are again shown in bold cases.

The bus voltage profiles for this final case study with the 300-bus system are shown in Figure 5.12. From this figure, it is again observed that in the presence of IPFC, the simultaneous enforcement of three IPFC device limit constraints causes very little change in the bus voltage profile.

TABLE 5.15 Effect of Variation of V_{se}^{0} on the Power Flow Solution for the Study of Table 5.14

	Initial Condition for V_{se}								
V_{se}^{0} (p.u.)	0.09	0.08	0.07	0.06	0.05	0.04	0.03	0.02	0.01
	Final Power Flow Solution for Converter 1								
V_{se1} (p.u.)	0.062	0.062	0.062	0.062	0.062		0.062	0.062	0.062
I_{se1} (p.u.)	0.2642	0.6556	0.6556	0.6556	0.2642		0.6556	0.2642	0.2642
P_{Line1} (MW)	27.14	48.83	48.83	48.83	27.14	div	48.83	27.14	27.14
Q_{Line1} (MVAR)	−5.59	48.04	48.04	48.04	−5.59		48.04	−5.59	−5.59
NI	10**	8	7	10	14**		8	9**	17**

TABLE 5.16 Final Study of IEEE 300-Bus System with IPFC and V_{se}^{Lim}, I_{se}^{Lim}, and P_{DC}^{Lim} Constraints

Solution of IPFC Quantities with All Three Limits of DC Link Power Transfer, Series Injected Voltage of Converter 1, and Line Current of Converter 2

Specified dc link power transfer limit: $P_{DC}^{Lim} = -0.0142$ p.u.
Specified voltage limit of converter 1: $V_{se1}^{Lim} = 0.062$ p.u.
Specified line current limit of converter 2: $I_{se2}^{Lim} = 0.51$ p.u.

Quantity	Converter 1	Converter 2
V_{se} (p.u.)	**0.062**	0.0753
θ_{se}	–106.64°	–120.21°
I_{se} (p.u.)	0.6556	**0.51**
P_{DC} (p.u.)	**–0.0142**	**0.0142**
P_{Line} (MW)	48.83	49.68
Q_{Line} (MVAR)	48.04	5.36
NI	7 (with $V_{se}^{0} = 0.07$ p.u.)	

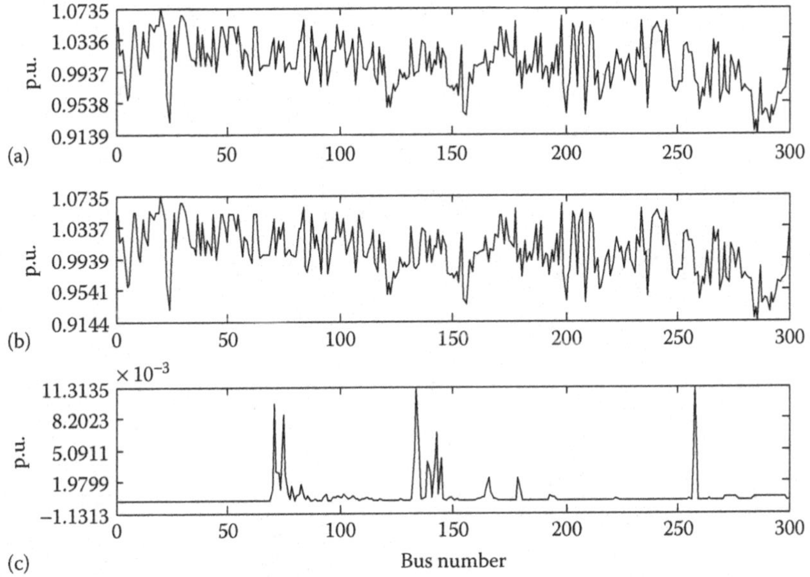

FIGURE 5.12 Bus voltage profile corresponding to the case study of Table 5.16. (a) Bus voltage magnitude without IPFC; (b) bus voltage magnitude with IPFC; (c) voltage magnitude difference.

5.8 SUMMARY

In this chapter, a Newton power flow model of the IPFC has been developed. The proposed method transforms an existing n-bus power system installed with an IPFC having p series converters into an equivalent $(n + p)$ bus system without any IPFC. Consequently, existing power flow and Jacobian codes can be reused in the proposed model, in conjunction with simple codes for matrix extraction. As a result, a substantial reduction in the complexity of the software codes can be achieved. The developed technique can also handle practical device limit constraints of the IPFC. Validity of the proposed method has been demonstrated on IEEE 118- and 300-bus systems with excellent convergence characteristics.

CHAPTER 6

Newton Power Flow Model of the Generalized Unified Power Flow Controller

6.1 INTRODUCTION

Similar to the interline power flow controller (IPFC), a generalized unified power flow controller (GUPFC) addresses the problem of compensating multiple transmission lines simultaneously. However, a GUPFC incorporates a shunt converter in addition to the two or more series converters employed by an IPFC. The simplest GUPFC configuration (employing two series converters and one shunt converter) possesses 5 degrees of freedom. Usually, these correspond to the active and reactive power flows in both the lines incorporating the series converters and the voltage of a sending end (SE) bus.

For proper utilization of GUPFCs in power system planning, operation, and control, power flow solutions of the networks incorporating GUPFC(s) are necessary. Some excellent research works were carried out in the literature [77,81] for developing efficient power-flow and optimal power flow algorithms for the GUPFC. In a way similar to the static synchronous series compensator, the unified power flow controller and the IPFC, it is observed from these works that the complexity of software codes is increased manifold when a GUPFC is modeled in an existing Newton–Raphson power flow algorithm. In fact, the GUPFC possesses multiple series converters. Contributions from these converters necessitate modifications in the existing power injection equations of the concerned buses (SE and receiving end [RE]) in all the transmission lines involved. Further, an entirely new expression has to be written for the GUPFC real power. Moreover, new codes have to be written to compute each of the multiple Jacobian subblocks exclusively related to the series and shunt converters of the GUPFC. As a result, the complexity of software codes for incorporating a GUPFC in a Newton–Raphson power flow code is greatly enhanced due to these factors. The problem aggravates as the number of GUPFCs in a system increases.

To reduce the complexities of the software codes for implementing Newton power flow algorithm, in this chapter, a novel approach for a GUPFC power flow model is proposed [84]. By this modeling approach, an existing power system installed with GUPFCs is transformed to an equivalent augmented network without any GUPFC. This results in a substantial reduction in the programming complexity because of the following reasons:

1. In the proposed model, the power injections for the buses concerned can be computed using existing power flow codes. This is because they no longer contain contributions from any series converter of the GUPFC.
2. In the proposed model, the active power flow of the GUPFC itself equals the sum of bus active power injections of additional power-flow buses. Thus, existing power flow codes can be used to compute them.
3. Only three Jacobian subblocks need be evaluated in the proposed model. Two of these subblocks can be evaluated using existing Jacobian codes directly, whereas the third can be computed with very minor modifications of the existing Jacobian codes.

6.2 GUPFC MODEL FOR NEWTON POWER FLOW ANALYSIS

Figure 6.1 shows an n-bus power system network in which p series converters along with the (single) shunt converter of a GUPFC are connected. Without loss of generality, it is assumed that the series converters are connected between the buses i–j, $(i + 1)$–$(j + 1)$, and so on, up to buses

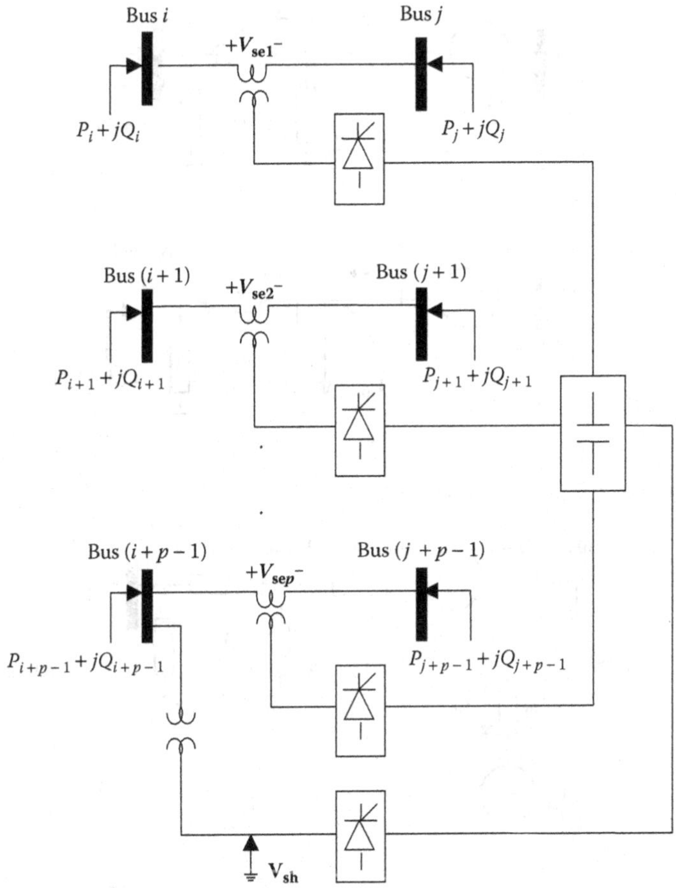

FIGURE 6.1 GUPFC with one shunt and p series converters connected to a power system.

$(i+p-1)$–$(j+p-1)$, whereas the shunt converter is connected to bus $(i+p-1)$. It is also further assumed that the gth converter $(1 \le g \le p)$ is connected at the SE, that is, at $(i+g-1)$th bus of the corresponding transmission line. The equivalent circuit of Figure 6.1 is shown in Figure 6.2.

In Figure 6.2, the GUPFC is represented by p (series) voltage sources and one shunt voltage source. The gth series voltage source $\mathbf{V}_{\mathbf{se}g}$ (not shown) (representing the gth series converter) is in series with the impedance $\mathbf{Z}_{\mathbf{se}g}$ (representing the impedance of the coupling transformer of the gth series converter) and is connected in series with the gth transmission line (which is represented by its equivalent pi circuit). The shunt voltage source $\mathbf{V}_{\mathbf{sh}}$ represents the shunt converter, whereas $\mathbf{Z}_{\mathbf{sh}}$ represents the

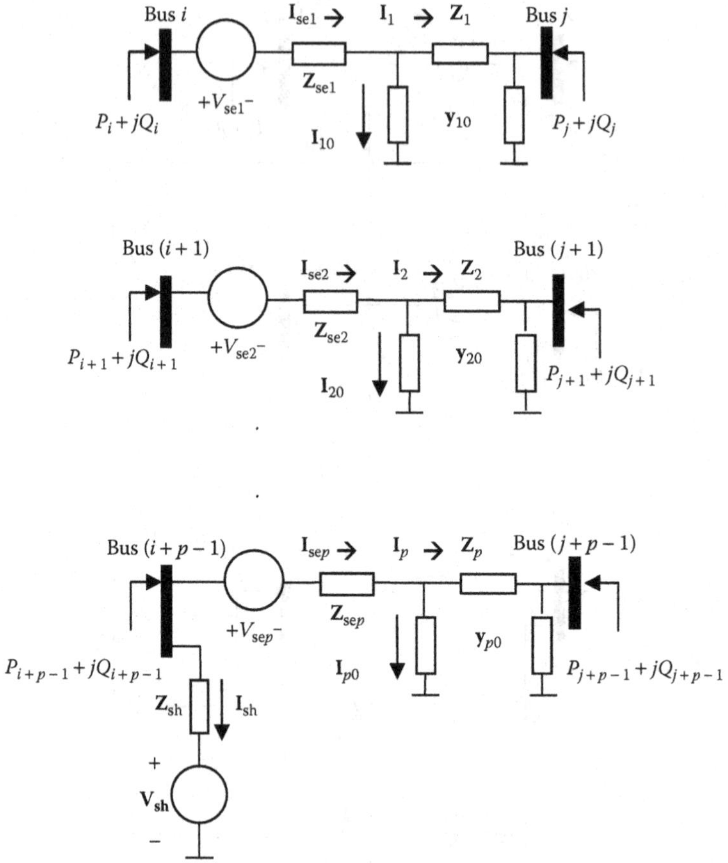

FIGURE 6.2 Equivalent circuit of GUPFC-incorporated power system network.

impedance of the coupling transformer of the shunt converter. The total current through the gth ($1 \le g \le p$) series coupling transformer is $\mathbf{I}_{seg}$, which consists of two parts: (1) $\mathbf{I}_g$ flowing flowing into the transmission line and (2) $\mathbf{I}_{g0}$, the line charging current.

Now, let

$$\mathbf{Z}_g = \mathrm{R}_g + \mathrm{j}\mathrm{X}_g, \mathbf{y}_g = \frac{\mathbf{1}}{\mathbf{Z}_g}, \text{ and } \mathbf{y}_{seg} = \frac{\mathbf{1}}{\mathbf{Z}_{seg}} \tag{6.1}$$

Also,

$$\mathbf{Y}_{ij} = \mathbf{Y}_{ji} = -\mathbf{y}_{ij} = -\mathbf{y}_{ji} \tag{6.2}$$

From Figure 6.2, it can be shown that for the first converter

$$\mathbf{I}_{se1} = \alpha_1 \mathbf{V}_i - \beta_1 \mathbf{V}_j - \alpha_1 \mathbf{V}_{se1} \tag{6.3}$$

where:

$$\alpha_1 = \frac{\mathbf{y}_{se1}(\mathbf{y}_1 + \mathbf{y}_{10})}{\mathbf{y}_1 + \mathbf{y}_{10} + \mathbf{y}_{se1}} \quad \text{and} \quad \beta_1 = \frac{\mathbf{y}_{se1}\mathbf{y}_1}{\mathbf{y}_1 + \mathbf{y}_{10} + \mathbf{y}_{se1}} \tag{6.4}$$

Again, from Figure 6.2, the net injected current at bus i is

$$\mathbf{I}_i = \left[\mathbf{Y}_{ii}^{\text{old}} - (\mathbf{y}_1 + \mathbf{y}_{10})\right]\mathbf{V}_i + \sum_{q=1, q \neq i, q \neq j}^{n} \mathbf{Y}_{iq}\mathbf{V}_q + \mathbf{I}_{se1} \tag{6.5}$$

In the above equation, $\mathbf{Y}_{ii}^{\text{old}} = \sum_{q=1,q\neq i,q\neq j}^{n} \mathbf{y}_{iq} + \mathbf{y}_{i0}^{p} + \mathbf{y}_1 + \mathbf{y}_{10}$ is the self-admittance of bus i for the existing n-bus system without any GUPFC connected and $\mathbf{y}_{i0}^{p}$ accounts for the shunt capacitances of all transmission lines connected to bus i, except the line in branch (i–j). From Equations 6.3 and 6.5, the net injected current at bus i with GUPFC becomes

$$\mathbf{I}_i = \sum_{q=1}^{n+1} \mathbf{Y}_{iq}\mathbf{V}_q \tag{6.6}$$

where $\mathbf{Y}_{ii} = \mathbf{Y}_{ii}^{\text{old}} - (\mathbf{y}_1 + \mathbf{y}_{10}) + \alpha_1$ is the new value of self-admittance for the ith bus with GUPFC and

$$\mathbf{V}_{n+1} = \mathbf{V}_{se1}, \mathbf{Y}_{i(n+1)} = -\alpha_1, \text{and } \mathbf{Y}_{ij} = -\beta_1 \tag{6.7}$$

Similarly, the net injected current at bus j can be written as

$$\mathbf{I}_j = \sum_{q=1}^{n+1} \mathbf{Y}_{jq}\mathbf{V}_q \tag{6.8}$$

where $\mathbf{Y}_{jj} = \mathbf{Y}_{jj}^{\text{old}} - \mathbf{y}_1 + \gamma_1$ is the new value of self-admittance for bus j with GUPFC and

$$\gamma_1 = \frac{\mathbf{y}_1(\mathbf{y}_{se1} + \mathbf{y}_{10})}{\mathbf{y}_1 + \mathbf{y}_{10} + \mathbf{y}_{se1}}, \mathbf{Y}_{j(n+1)} = \beta_1, \text{and } \mathbf{Y}_{ji} = -\beta_1 \tag{6.9}$$

Thus, the effect of the first series converter of the GUPFC is equivalent to an additional (n + 1)th bus without any GUPFC. Proceeding in a similar way, it can be shown that the effect of incorporation of the remaining

$(p-1)$ series converters can be treated as equivalent to addition of $(p-1)$ more buses $(n+2)$ up to $(n+p)$ to the existing n-bus system, provided

$$\mathbf{V}_{n+2}=\mathbf{V}_{se2},\mathbf{V}_{n+3}=\mathbf{V}_{se3},\ldots,\mathbf{V}_{n+p}=\mathbf{V}_{sep} \tag{6.10}$$

with

$$\mathbf{Y}_{(i+1)(n+2)}=-\alpha_2,\mathbf{Y}_{(i+2)(n+3)}=-\alpha_3,\ldots,\mathbf{Y}_{(i+p-1)(n+p)}=-\alpha_p \tag{6.11}$$

$$\mathbf{Y}_{(i+1)(j+1)}=-\beta_2,\mathbf{Y}_{(i+2)(j+2)}=-\beta_3,\ldots,\mathbf{Y}_{(i+p-1)(j+p-1)}=-\beta_p \tag{6.12}$$

$$\mathbf{Y}_{(j+1)(n+2)}=\beta_2,\mathbf{Y}_{(j+2)(n+3)}=\beta_3,\ldots,\mathbf{Y}_{(j+p-1)(n+p)}=\beta_p \tag{6.13}$$

where, for the transmission line incorporating the pth converter,

$$\alpha_p=\frac{y_{sep}(y_p+y_{p0})}{y_p+y_{p0}+y_{sep}} \quad \text{and} \quad \beta_p=\frac{y_{sep}y_p}{y_p+y_{p0}+y_{sep}} \tag{6.14}$$

Again, from Figure 6.2, it is observed that bus $(i+p-1)$ is connected to the pth series converter as well as the shunt converter. Now, the net injected current at bus $(i+p-1)$ can be written as

$$\begin{aligned}\mathbf{I}_{i+p-1}&=\left[\mathbf{Y}^{\text{old}}_{(i+p-1)i+p-1)}-(\mathbf{y}_p+\mathbf{y}_{p0})\right]\mathbf{V}_{i+p-1}\\&+\sum_{q=1,\,q\neq(i+p-1),\,q\neq(j+p-1)}^{n}\mathbf{Y}_{(i+p-1)q}\mathbf{V}_q+\mathbf{I}_{sep}+\mathbf{I}_{sh}\end{aligned} \tag{6.15}$$

Also

$$\mathbf{I}_{sep}=\alpha_p\mathbf{V}_{i+p-1}-\beta_p\mathbf{V}_{j+p-1}-\alpha_p\mathbf{V}_{sep} \tag{6.16}$$

and

$$\mathbf{I}_{sh}=\mathbf{y}_{sh}(\mathbf{V}_{i+p-1}-\mathbf{V}_{sh}) \tag{6.17}$$

Now, using Equations 6.16 and 6.17 in Equation 6.15,

$$\begin{aligned}\mathbf{I}_{i+p-1}&=\left[\mathbf{Y}^{\text{old}}_{(i+p-1)(i+p-1)}-(\mathbf{y}_p+\mathbf{y}_{p0})+\alpha_p+\mathbf{y}_{sh}\right]\mathbf{V}_{i+p-1}\\&+\sum_{q=1,\,q\neq(i+p-1)}^{n+p+1}\mathbf{Y}_{(i+p-1)q}\mathbf{V}_q\end{aligned} \tag{6.18}$$

or

$$\mathbf{I}_{i+p-1}=\sum_{q=1}^{n+p+1}\mathbf{Y}_{(i+p-1)q}\mathbf{V}_q \tag{6.19}$$

where $\mathbf{Y}_{(i+p-1)(i+p-1)} = \mathbf{Y}^{\text{old}}_{(i+p-1)(i+p-1)} - (\mathbf{y}_p + \mathbf{y}_{p0}) + \alpha_p + \mathbf{y}_{\text{sh}}$ is the new value of self-admittance for the $(i + p - 1)$th bus with GUPFC with

$$\mathbf{V}_{n+p+1} = \mathbf{V}_{\text{sh}} \quad \text{and} \quad \mathbf{Y}_{(i+p-1)(n+p+1)} = -\mathbf{y}_{\text{sh}} \tag{6.20}$$

Thus, the effect of incorporation of the shunt converter can be treated as equivalent to addition of another fictitious bus $(n + p + 1)$ to the existing n-bus system. Consequently, incorporation of the GUPFC (consisting of p series converters and one shunt converter) transforms the original n-bus power system to an $(n + p + 1)$-bus system without any GUPFC.

Now, the net injected current at the first fictitious power-flow bus [bus $(n + 1)$] equals the current flowing into the transmission system from this bus, and can be written using Equation 6.3 as

$$\mathbf{I}_{n+1} = -\mathbf{I}_{\text{se}1} = \alpha_1 \mathbf{V}_{\text{se}1} - \alpha_1 \mathbf{V}_i + \beta_1 \mathbf{V}_j = \sum_{q=1}^{n+p+1} \mathbf{Y}_{(n+1)q} \mathbf{V}_q \tag{6.21}$$

where:

$$\mathbf{V}_{n+1} = \mathbf{V}_{\text{se}1}, \mathbf{Y}_{(n+1)(n+1)} = \alpha_1, \mathbf{Y}_{(n+1)i} = -\alpha_1, \text{and } \mathbf{Y}_{(n+1)j} = \beta_1 \tag{6.22}$$

Proceeding in a similar way, the net injected currents at the remaining p fictitious power-flow buses [buses $(n + 2)$ up to $(n + p + 1)$] can also be written as

$$\mathbf{I}_{n+2} = \sum_{q=1}^{n+p+1} \mathbf{Y}_{(n+2)q} \mathbf{V}_q, \text{and so on, up to } \mathbf{I}_{n+p+1} = \sum_{q=1}^{n+p+1} \mathbf{Y}_{(n+p+1)q} \mathbf{V}_q \tag{6.23}$$

with

$$\begin{gathered} \mathbf{Y}_{(n+2)(n+2)} = \alpha_2, \text{and so on, up to } \mathbf{Y}_{(n+p)(n+p)} = \alpha_p \\ \text{with } \mathbf{Y}_{(n+p+1)(n+p+1)} = \mathbf{y}_{\text{sh}} \end{gathered} \tag{6.24}$$

$$\begin{gathered} \mathbf{Y}_{(n+2)(i+1)} = -\alpha_2, \text{and so on, up to } \mathbf{Y}_{(n+p)(i+p-1)} = -\alpha_p \\ \text{with } \mathbf{Y}_{(n+p+1)(i+p-1)} = -\mathbf{y}_{\text{sh}} \end{gathered} \tag{6.25}$$

$$\begin{gathered} \mathbf{Y}_{(n+2)(j+1)} = \beta_2, \text{and so on, up to } \mathbf{Y}_{(n+p)(j+p-1)} = \beta_p \\ \text{with } \mathbf{Y}_{(n+p+1)(j+p-1)} = 0 \end{gathered} \tag{6.26}$$

$$\mathbf{V}_{n+2} = \mathbf{V}_{\text{se}2}, \text{and so on, up to } \mathbf{V}_{n+p} = \mathbf{V}_{\text{se}p} \text{ with } \mathbf{V}_{n+p+1} = \mathbf{V}_{\text{sh}} \tag{6.27}$$

6.3 POWER FLOW EQUATIONS IN PROPOSED GUPFC MODEL

From Figure 6.1, it is observed that with a GUPFC-connected system, four possibilities exist for any SE or RE bus. These are as follows:

1. The bus is not connected to any GUPFC converter.
2. The bus is connected to a GUPFC series converter only as an SE bus.
3. The bus is connected to a GUPFC series converter only as an RE bus.
4. The bus is connected to both a series converter and a shunt converter of the GUPFC as an SE bus.

With existing GUPFC models, the expression of the active power injection at any SE or RE bus a can be written (using Equations 6.5 and 6.15) as

$$P_a = \sum_{q=1}^{n} V_a V_q Y_{aq} \cos\left(\theta_a - \theta_q - \varphi_{yaq}\right),\ a \le n \tag{6.28}$$

if no GUPFC is connected to bus a;

$$\begin{aligned} P_a = &\sum_{q=1, q\neq b}^{n} V_a V_q Y_{aq} \cos\left(\theta_a - \theta_q - \varphi_{yaq}\right) - V_a V_b \beta_c \cos\left(\theta_a - \theta_b - \varphi_{y\beta c}\right) \\ &- V_a V_{\text{se}\,c} \alpha_c \cos\left(\theta_a - \theta_{\text{se}\,c} - \varphi_{y\alpha c}\right) \end{aligned} \tag{6.29}$$

if series converter c is connected in branch a–b with a as SE bus;

$$\begin{aligned} P_a = &\sum_{q=1, q\neq b}^{n} V_a V_q Y_{aq} \cos\left(\theta_a - \theta_q - \varphi_{yaq}\right) - V_a V_b \beta_c \cos\left(\theta_a - \theta_b - \varphi_{y\beta c}\right) \\ &+ V_a V_{\text{se}\,c} \beta_c \cos\left(\theta_a - \theta_{\text{se}\,c} - \varphi_{y\beta c}\right) \end{aligned} \tag{6.30}$$

if series converter c is connected in branch a–b with a as RE bus;

$$\begin{aligned} P_a = &\sum_{q=1, q\neq b}^{n} V_a V_q Y_{aq} \cos\left(\theta_a - \theta_q - \varphi_{yaq}\right) \\ &- V_a V_b \beta_c \cos\left(\theta_a - \theta_b - \varphi_{y\beta c}\right) \\ &- V_a V_{\text{se}\,c} \alpha_c \cos\left(\theta_a - \theta_{\text{se}\,c} - \varphi_{y\alpha c}\right) \\ &- V_a V_{\text{sh}\,d} y_{\text{sh}\,d} \cos\left(\theta_a - \theta_{\text{sh}\,d} - \varphi_{y\text{sh}\,d}\right) \end{aligned} \tag{6.31}$$

if both the series converter c (connected in branch a–b) and the shunt converter d are connected to bus a as SE bus.

Thus, from the above equation, it is observed that with existing GUPFC models, additional terms due to contributions from the voltage sources representing the multiple series converters and the shunt converter of the GUPFC are present in the expression of the bus active power injection(s). This causes the existing power flow codes to be modified. This is also true for the bus reactive power injection(s).

Now, it has already been established in Section 6.2 that in the proposed GUPFC model, there are a total of $(n + p + 1)$ buses, with $(p + 1)$ fictitious power-flow buses. Therefore, the net injected current at any of the p SE buses [buses i, $(i + 1)$, and so on, up to $(i + p - 1)$] or RE buses [buses j, $(j + 1)$, and so on, up to $(j + p - 1)$] can be written as

$$\mathbf{I}_h = \sum_{q=1}^{n+p+1} \mathbf{Y}_{hq}\mathbf{V}_q \quad i \le h \le (i+p-1) \quad \text{or} \quad j \le h \le (j+p-1) \qquad (6.32)$$

Thus, in the proposed model, the active power injection equation at any SE or RE bus a $(a \le n + p + 1)$ can be written using Equation 6.32 as

$$P_a = \sum_{q=1}^{n+p+1} V_a V_q Y_{aq} \cos\left(\theta_a - \theta_q - \varphi_{yaq}\right) \quad a \le (n+p+1) \qquad (6.33)$$

Similarly, the expression for the reactive power at any bus a (SE or RE) can be written as

$$Q_a = \sum_{q=1}^{n+p+1} V_a V_q Y_{aq} \sin\left(\theta_a - \theta_q - \varphi_{yaq}\right) \quad a \le (n+p+1) \qquad (6.34)$$

From Equations 6.33 and 6.34, it can be observed that both the active and reactive power injections at any bus can be computed using existing power flow codes.

Now, let us consider the real power delivered by a GUPFC with two series converters (for simplicity, the simplest GUPFC configuration with only two series converters is assumed) and one shunt converter. The two series converters e and f are in series with the branches between buses a–b and c–d, respectively. The shunt converter g is connected to bus c. The real power delivered by the GUPFC can be written using Equations 6.3 and 6.17 as

$$P_{\text{GUPFC}} = \text{Re}\left[\mathbf{V}_{\text{se}e}(-\mathbf{I}^*_{\text{se}e}) + \mathbf{V}_{\text{se}f}(-\mathbf{I}^*_{\text{se}f}) + \mathbf{V}_{\text{sh}g}(-\mathbf{I}^*_{\text{sh}g})\right] \tag{6.35}$$

$$\begin{aligned} P_{\text{GUPFC}} = {} & V^2_{\text{se}e}\alpha_e\cos\varphi_{y\alpha e} + V_{\text{se}e}V_b\beta_e\cos\left(\theta_{\text{se}e} - \theta_b - \varphi_{y\beta e}\right) \\ & - V_{\text{se}e}V_a\alpha_e\cos\left(\theta_{\text{se}e} - \theta_a - \varphi_{y\alpha e}\right) \\ & + V_{\text{se}f}V_d\beta_f\cos\left(\theta_{\text{se}f} - \theta_d - \varphi_{y\beta f}\right) \\ & - V_{\text{se}f}V_c\alpha_f\cos\left(\theta_{\text{se}f} - \theta_c - \varphi_{y\alpha f}\right) \\ & + V^2_{\text{se}f}\alpha_f\cos\varphi_{y\alpha f} + V^2_{\text{sh}g}y_{\text{sh}g}\cos\varphi_{y\text{sh}g} \\ & - V_{\text{sh}g}V_c y_{\text{sh}g}\cos\left(\theta_{\text{sh}g} - \theta_c - \varphi_{y\text{sh}g}\right) \end{aligned} \tag{6.36}$$

From Equation 6.36, it is observed that the GUPFC real power expression cannot be computed using existing power flow codes. It will require fresh codes for its implementation.

However, in the proposed model, GUPFC series converters e and f along with shunt converter g are transformed to power-flow buses $(n + 1)$, $(n + 2)$, and $(n + 3)$, respectively, and hence, using Equation 6.35,

$$\begin{aligned} P_{\text{GUPFC}} & = \text{Re}\left[\mathbf{V}_{\text{se}e}(-\mathbf{I}^*_{\text{se}e}) + \mathbf{V}_{\text{se}f}(-\mathbf{I}^*_{\text{se}f}) + \mathbf{V}_{\text{sh}g}(-\mathbf{I}^*_{\text{sh}g})\right] \\ & = \text{Re}\left[\mathbf{V}_{n+1}(\mathbf{I}^*_{n+1}) + \mathbf{V}_{n+2}(\mathbf{I}^*_{n+2}) + \mathbf{V}_{n+3}(\mathbf{I}^*_{n+3})\right] \\ & = \text{Re}\left(\mathbf{S}_{n+1} + \mathbf{S}_{n+2} + \mathbf{S}_{n+3}\right) = P_{n+1} + P_{n+2} + P_{n+3} \end{aligned} \tag{6.37}$$

Hence, the sum of the active power injections of three power-flow buses (computed from Equation 6.33 using existing power flow codes) yields the GUPFC real power. It can be shown that if the GUPFC considered has p series converters along with a shunt converter, the real power delivered by it can be computed from the sum of the active power injections of $(p + 1)$ power-flow buses using existing codes. In such a case, Equation 6.37 becomes

$$\begin{aligned} P_{\text{GUPFC}} & = \text{Re}\left[\mathbf{V}_{\text{se}1}(-\mathbf{I}^*_{\text{se}1}) + \mathbf{V}_{\text{se}2}(-\mathbf{I}^*_{\text{se}2}) + \cdots + \mathbf{V}_{\text{se}p}(-\mathbf{I}^*_{\text{se}p}) + \mathbf{V}_{\text{sh}}(-\mathbf{I}^*_{\text{sh}})\right] \\ & = P_{n+1} + P_{n+2} + \cdots + P_{n+p+1} \end{aligned} \tag{6.38}$$

Similar to the bus power injections, the expressions corresponding to the various control objectives can also be evaluated in the proposed model using existing codes. Usually, for a GUPFC with p series converters and a shunt converter, the line active and reactive power flows in the p transmission lines incorporating the series converters are chosen as the control objectives for the p series converters, whereas the voltage of the bus connected to the shunt converter is chosen as the control objective for the shunt converter. In this chapter too, the same control objectives have been considered.

Now, with existing GUPFC models, the expression for the active power flow in the line connected between buses a and b and incorporating the cth ($1 \leq c \leq p$) series converter can be written using Equation 6.3 as

$$P_{\text{LINE}\,c} = P_{ab} = \text{Re}\left(\mathbf{V}_a \mathbf{I}_{\mathbf{se}c}^*\right)$$

or

$$\begin{aligned} P_{ab} &= V_a^2 \alpha_c \cos\varphi_{y\alpha c} - V_a V_b \beta_c \cos\left(\theta_a - \theta_b - \varphi_{y\beta c}\right) \\ &\quad - V_a V_{\text{se}\,c} \alpha_c \cos\left(\theta_a - \theta_{\text{se}\,c} - \varphi_{y\alpha c}\right) \end{aligned} \tag{6.39}$$

Similarly, the expression for the line reactive power flow can be written as

$$\begin{aligned} Q_{ab} &= -V_a^2 \alpha_c \sin\varphi_{y\alpha c} - V_a V_b \beta_c \sin\left(\theta_a - \theta_b - \varphi_{y\beta c}\right) \\ &\quad - V_a V_{\text{se}\,c} \alpha_c \sin\left(\theta_a - \theta_{\text{se}\,c} - \varphi_{y\alpha c}\right) \end{aligned} \tag{6.40}$$

Both Equations 6.39 and 6.40 have new terms involving the GUPFC series converter and the modified admittances due to the series coupling transformer of the converter. These necessitate modifications in the existing power flow codes.

In the proposed model, using Equation 6.21, these expressions become

$$P_{\text{LINE}\,c} = P_{ab} = \text{Re}\left(\mathbf{V}_a \mathbf{I}_{\mathbf{se}c}^*\right) = \text{Re}\left[\mathbf{V}_a\left(-\mathbf{I}_{n+c}^*\right)\right]$$

or

$$P_{ab} = -\sum_{q=1}^{n+p+1} V_a V_q Y_{(n+c)q} \cos\left(\theta_a - \theta_q - \varphi_{y(n+c)q}\right) \tag{6.41}$$

The above equation is written because of the fact that the cth series converter of the GUPFC is transformed into the $(n + c)$th power-flow bus in the proposed model. In a similar way,

$$Q_{ab} = -\sum_{q=1}^{n+p+1} V_a V_q Y_{(n+c)q} \sin\left(\theta_a - \theta_q - \varphi_{y(n+c)q}\right) \tag{6.42}$$

From Equations 6.41 and 6.42, it can be observed that in the proposed model, both the line active and reactive power flows can be computed using very minor modifications of the existing power flow codes.

6.4 IMPLEMENTATION IN NEWTON POWER FLOW ANALYSIS

If the number of voltage-controlled buses is $(m-1)$, the power-flow problem for an n-bus system incorporated with x GUPFCs each having p series converters and one shunt converter can be formulated as follows:

Solve θ, $\mathbf{V}$, $\theta_{\mathbf{se}}$, $\mathbf{V}_{\mathbf{se}}$, $\theta_{\mathbf{sh}}$, and $\mathbf{V}_{\mathbf{sh}}$

Specified $\mathbf{P}$, $\mathbf{Q}$, $\mathbf{P}_{\mathbf{G}}$, $\mathbf{V}_{\mathbf{B}}$, $\mathbf{P}_{\mathbf{L}}$, and $\mathbf{Q}_{\mathbf{L}}$

where:

$$z = \text{number of series converters} = px \tag{6.43}$$

Also,

$$\text{Number of shunt converters} = x \tag{6.44}$$

$$\text{Total number of converters} = w = (p+1)x \tag{6.45}$$

$$\theta = [\theta_2 \ldots \theta_n]^{\mathrm{T}} \quad \mathbf{V} = [V_{m+1} \ldots V_n]^{\mathrm{T}} \tag{6.46}$$

$$\theta_{\mathbf{se}} = [\theta_{\mathrm{se}1} \ldots \theta_{\mathrm{se}z}]^{\mathrm{T}} \quad \mathbf{V}_{\mathbf{se}} = [V_{\mathrm{se}1} \ldots V_{\mathrm{se}z}]^{\mathrm{T}} \tag{6.47}$$

$$\theta_{\mathbf{sh}} = [\theta_{\mathrm{sh}1} \ldots \theta_{\mathrm{sh}x}]^{\mathrm{T}} \quad \mathbf{V}_{\mathbf{sh}} = [V_{\mathrm{sh}1} \ldots V_{\mathrm{sh}x}]^{\mathrm{T}} \tag{6.48}$$

$$\mathbf{P} = [P_2 \ldots P_n]^{\mathrm{T}} \quad \mathbf{Q} = [Q_{m+1} \ldots Q_n]^{\mathrm{T}} \tag{6.49}$$

$$\mathbf{P}_{\mathbf{G}} = [P_{\mathrm{GUPFC}1} \ldots P_{\mathrm{GUPFC}x}]^{\mathrm{T}} \quad \mathbf{V}_{\mathbf{B}} = [V_{\mathrm{BUS}1} \ldots V_{\mathrm{BUS}x}]^{\mathrm{T}} \tag{6.50}$$

$$\mathbf{P}_{\mathbf{L}} = [P_{\mathrm{LINE}1} \ldots P_{\mathrm{LINE}z}]^{\mathrm{T}} \quad \mathbf{Q}_{\mathbf{L}} = [Q_{\mathrm{LINE}1} \ldots Q_{\mathrm{LINE}z}]^{\mathrm{T}} \tag{6.51}$$

In Equations 6.50 and 6.51, $\mathbf{P}_{\mathbf{G}}$, $\mathbf{V}_{\mathbf{B}}$, $\mathbf{P}_{\mathbf{L}}$, and $\mathbf{Q}_{\mathbf{L}}$ represent the vectors of the specified real powers for the x GUPFCs, the SE bus voltages (at which the shunt converters of the x GUPFCs are connected), and the active and reactive power flows of the z transmission lines, respectively. In Equations 6.46 through 6.51, it is assumed that without any loss of generality, there are m generators connected at the first m buses of the system with bus 1 being the slack bus. Thus, the basic power-flow equation for the Newton power flow solution is represented as

$$\begin{bmatrix} \frac{\partial P}{\partial \theta} & \frac{\partial P}{\partial V} & \frac{\partial P}{\partial \theta_{se}} & \frac{\partial P}{\partial V_{se}} & \frac{\partial P}{\partial \theta_{sh}} & \frac{\partial P}{\partial V_{sh}} \\ \frac{\partial Q}{\partial \theta} & \frac{\partial Q}{\partial V} & \frac{\partial Q}{\partial \theta_{se}} & \frac{\partial Q}{\partial V_{se}} & \frac{\partial Q}{\partial \theta_{sh}} & \frac{\partial Q}{\partial V_{sh}} \\ \frac{\partial P_G}{\partial \theta} & \frac{\partial P_G}{\partial V} & \frac{\partial P_G}{\partial \theta_{se}} & \frac{\partial P_G}{\partial V_{se}} & \frac{\partial P_G}{\partial \theta_{sh}} & \frac{\partial P_G}{\partial V_{sh}} \\ \frac{\partial V_B}{\partial \theta} & \frac{\partial V_B}{\partial V} & \frac{\partial V_B}{\partial \theta_{se}} & \frac{\partial V_B}{\partial V_{se}} & \frac{\partial V_B}{\partial \theta_{sh}} & \frac{\partial V_B}{\partial V_{sh}} \\ \frac{\partial P_L}{\partial \theta} & \frac{\partial P_L}{\partial V} & \frac{\partial P_L}{\partial \theta_{se}} & \frac{\partial P_L}{\partial V_{se}} & \frac{\partial P_L}{\partial \theta_{sh}} & \frac{\partial P_L}{\partial V_{sh}} \\ \frac{\partial Q_L}{\partial \theta} & \frac{\partial Q_L}{\partial V} & \frac{\partial Q_L}{\partial \theta_{se}} & \frac{\partial Q_L}{\partial V_{se}} & \frac{\partial Q_L}{\partial \theta_{sh}} & \frac{\partial Q_L}{\partial V_{sh}} \end{bmatrix} \begin{bmatrix} \Delta\theta \\ \Delta V \\ \Delta\theta_{se} \\ \Delta V_{se} \\ \Delta\theta_{sh} \\ \Delta V_{sh} \end{bmatrix} = \begin{bmatrix} \Delta P \\ \Delta Q \\ \Delta P_G \\ \Delta V_B \\ \Delta P_L \\ \Delta Q_L \end{bmatrix} \tag{6.52}$$

or

$$\begin{bmatrix} \multicolumn{2}{c}{\multirow{2}{*}{J_{old}}} & J_1 & J_2 & J_3 & J_4 \\ & & J_5 & J_6 & J_7 & J_8 \\ J_9 & J_{10} & J_{11} & J_{12} & J_{13} & J_{14} \\ 0 & J_{15} & 0 & 0 & 0 & 0 \\ J_{16} & J_{17} & J_{18} & J_{19} & J_{20} & J_{21} \\ J_{22} & J_{23} & J_{24} & J_{25} & J_{26} & J_{27} \end{bmatrix} \begin{bmatrix} \Delta\theta \\ \Delta V \\ \Delta\theta_{se} \\ \Delta V_{se} \\ \Delta\theta_{sh} \\ \Delta V_{sh} \end{bmatrix} = \begin{bmatrix} \Delta P \\ \Delta Q \\ \Delta P_G \\ \Delta V_B \\ \Delta P_L \\ \Delta Q_L \end{bmatrix} \tag{6.53}$$

In the above equation, J_{old} is the conventional power-flow Jacobian subblock corresponding to the angle and voltage magnitude variables of the *n*-buses. The other Jacobian submatrices can be identified easily from Equation 6.53.

Now, in the proposed model, there are $(n + w)$ buses. Thus, the quantities to be solved for power-flow are θ^{new} and V^{new}, where

$$\theta^{new} = [\theta_2 \ldots \theta_{n+w}]^T \text{ and } V^{new} = [V_{m+1} \ldots V_{n+w}]^T \tag{6.54}$$

Thus, Equation 6.53 is transformed in the proposed model as

$$\begin{bmatrix} JX1 \\ JX2 \\ 0 \quad JX3 \\ JX4 \end{bmatrix} \begin{bmatrix} \Delta\theta^{new} \\ \Delta V^{new} \end{bmatrix} = \begin{bmatrix} \Delta PQ \\ \Delta P_G \\ \Delta V_B \\ \Delta PQ_L \end{bmatrix} \tag{6.55}$$

where:

$$\Delta\mathbf{PQ} = [\Delta\mathbf{P}^{T}\ \Delta\mathbf{Q}^{T}]^{T} \quad \Delta\mathbf{PQ}_{L} = [\Delta\mathbf{P}_{L}^{T}\ \Delta\mathbf{Q}_{L}^{T}]^{T} \tag{6.56}$$

Also, **JX1**, **JX2**, **JX3**, and **JX4** can be identified easily from Equation 6.55.

The elements of the matrix **JX3** are either unity or zero, depending on whether an element of the vector $\mathbf{V}_{B}$ is also an element of the vector $\mathbf{V}^{new}$ or not. Thus, **JX3** is a constant matrix known *a priori* and does not need to be computed.

Now, it can be shown that in Equation 6.55,

1. The matrices **JX1** and **JX2** can be computed using existing Jacobian codes.
2. The matrix **JX4** can be computed using very minor modifications of the existing Jacobian codes.

The justification of the above two statements are shown as follows:

Let us first define for the proposed $(n + w)$ bus system,

$$\mathbf{P}^{new} = [P_{2} \ldots P_{n+w}]^{T} \tag{6.57}$$

Subsequently, a new Jacobian matrix is computed as

$$\mathbf{J}^{new} = \begin{bmatrix} \dfrac{\partial \mathbf{P}^{new}}{\partial \boldsymbol{\theta}^{new}} & \dfrac{\partial \mathbf{P}^{new}}{\partial \mathbf{V}^{new}} \\ \hdashline \dfrac{\partial \mathbf{Q}}{\partial \boldsymbol{\theta}^{new}} & \dfrac{\partial \mathbf{Q}}{\partial \mathbf{V}^{new}} \end{bmatrix} = \begin{bmatrix} \mathbf{J}^{A} \\ \hdashline \mathbf{J}^{B} \end{bmatrix} \tag{6.58}$$

From Equations 6.33 and 6.34 of the proposed model, it can be observed that the expressions for P_{i} $(2 \le i \le n+w)$ and Q_{i} $(m+1 \le i \le n)$ in $\mathbf{P}^{new}$ and $\mathbf{Q}$, respectively (of Equation 6.58), can be computed using existing power flow codes. Hence, the matrix $\mathbf{J}^{new}$ can be computed with existing codes for calculating the Jacobian matrix.

From Equations 6.46, 6.49, and 6.57,

$$\frac{\partial \mathbf{P}^{new}}{\partial \boldsymbol{\theta}^{new}} = \begin{bmatrix} \dfrac{\partial \mathbf{P}}{\partial \boldsymbol{\theta}^{new}} \\ \hdashline \mathbf{JX5} \end{bmatrix} \tag{6.59}$$

where:

$$\mathbf{JX5} = \left[\frac{\partial P_{n+1}}{\partial \theta^{\text{new}}} \quad \cdots \quad \frac{\partial P_{n+p+1}}{\partial \theta^{\text{new}}} \;\vdots\; \cdots \quad \cdots \;\vdots\; \frac{\partial P_{n+w-p}}{\partial \theta^{\text{new}}} \quad \cdots \quad \frac{\partial P_{n+w}}{\partial \theta^{\text{new}}} \right]^{\mathrm{T}} \tag{6.60}$$

From Equation 6.59, it is observed that the subblock $\partial \mathbf{P}/\partial \boldsymbol{\theta}^{\text{new}}$ is contained within $\partial \mathbf{P}^{\text{new}}/\partial \boldsymbol{\theta}^{\text{new}}$. In a similar way, it can be shown that $\partial \mathbf{P}/\partial \mathbf{V}^{\text{new}}$ is contained within $\partial \mathbf{P}^{\text{new}}/\partial \mathbf{V}^{\text{new}}$. Hence, once the matrix $\mathbf{J}^{\text{new}}$ is computed (using existing codes), the matrix

$$\mathbf{JX1} = \begin{bmatrix} \partial \mathbf{P}/\partial \boldsymbol{\theta}^{\text{new}} & \partial \mathbf{P}/\partial \mathbf{V}^{\text{new}} \\ \hline \multicolumn{2}{c}{\mathbf{J}^{B}} \end{bmatrix}$$

can be very easily extracted from the matrix $\mathbf{J}^{\text{new}}$ using simple matrix extraction codes only. Hence, no fresh codes need to be written for computing **JX1**.

Now, using Equation 6.38 in Equation 6.60,

$$\frac{\partial}{\partial \theta^{\text{new}}}\left[P_{n+1} + P_{n+2} + \cdots + P_{n+p+1}\right] = \frac{\partial P_{\text{GUPFC1}}}{\partial \theta^{\text{new}}} \tag{6.61}$$

Thus, in Equation 6.60, the sum of the first $(p + 1)$ rows of **JX5** equals $\partial P_{\text{GUPFC1}}/\partial \boldsymbol{\theta}^{\text{new}}$ (the first element of the Jacobian subblock $\partial \mathbf{P}_G/\partial \boldsymbol{\theta}^{\text{new}}$). Similarly, the sum of the last $(p + 1)$ rows yields the last element of $\partial \mathbf{P}_G/\partial \boldsymbol{\theta}^{\text{new}}$. Therefore, $\partial \mathbf{P}_G/\partial \boldsymbol{\theta}^{\text{new}}$ can be easily extracted from $\partial \mathbf{P}^{\text{new}}/\partial \boldsymbol{\theta}^{\text{new}}$. In a similar way, it can be shown that $\partial \mathbf{P}_G/\partial \mathbf{V}^{\text{new}}$ can also be easily extracted from $\partial \mathbf{P}^{\text{new}}/\partial \mathbf{V}^{\text{new}}$. Hence the matrix $\mathbf{JX2} = \left[\partial \mathbf{P}_G/\partial \boldsymbol{\theta}^{\text{new}} \quad \partial \mathbf{P}_G/\partial \mathbf{V}^{\text{new}} \right]$ need not be computed—it can be formed from the matrix $\mathbf{J}^{\text{A}}$ of $\mathbf{J}^{\text{new}}$ by matrix extraction codes (in conjunction with codes for simple matrix row addition). Hence, both **JX1** and **JX2** need not be computed and can be extracted from $\mathbf{J}^{\text{new}}$ (subsequent to its computation using existing Jacobian codes).

Subsequently, it is shown that in the proposed model, the matrix

$$\mathbf{JX4} = \begin{bmatrix} \partial \mathbf{P}_{\text{L}}/\partial \boldsymbol{\theta}^{\text{new}} & \partial \mathbf{P}_{\text{L}}/\partial \mathbf{V}^{\text{new}} \\ \partial \mathbf{Q}_{\text{L}}/\partial \boldsymbol{\theta}^{\text{new}} & \partial \mathbf{Q}_{\text{L}}/\partial \mathbf{V}^{\text{new}} \end{bmatrix}$$

can be computed using very minor modifications of the existing Jacobian codes.

Let c $(1 \le c \le w)$ denote the GUPFC converter number. Now because an arbitrary GUPFC converter can be either a series or a shunt one, two expressions are possible for the converter number. Corresponding to the ath $(1 \le a \le p)$ series converter of the bth $(1 \le b \le x)$ GUPFC, $c=(b-1)(p+1)+a$. Similarly, corresponding to the shunt converter of the bth GUPFC, $c=b(p+1)$. In the proposed model, the cth GUPFC converter is transformed to the $(n+c)$th fictitious power-flow bus. Hence, the expression for the active power flow in the line connected between any arbitrary SE and RE buses g and h, respectively, and incorporating the series converter c can be written (using Equation 6.41) as

$$P_{\mathrm{LINE}\,c} = P_{gh} = -\sum_{q=1}^{n+w} V_g V_q Y_{(n+c)q} \cos\left(\theta_g - \theta_q - \varphi_{y(n+c)q}\right) \tag{6.62}$$

In a similar way, the reactive power flow can be written as

$$Q_{\mathrm{LINE}\,c} = Q_{gh} = -\sum_{q=1}^{n+w} V_g V_q Y_{(n+c)q} \sin\left(\theta_g - \theta_q - \varphi_{y(n+c)q}\right) \tag{6.63}$$

From Equations 6.62 and 6.63, it is observed that in the proposed model, both the line active $P_{\mathrm{LINE}\,i}$ $(1 \le i \le z)$ and reactive power flows $Q_{\mathrm{LINE}\,i}$ $(1 \le i \le z)$ can be computed using very minor modifications of the existing power flow codes. Consequently, it can be shown that all the subblocks of **JX4** can be computed using very minor modifications of the existing Jacobian codes, unlike with existing GUPFC models. This reduces the complexity of software codes substantially.

6.5 ACCOMMODATION OF GUPFC DEVICE LIMIT CONSTRAINTS

In this chapter, five major device limit constraints [82,83] of the GUPFC have been considered, and they are listed as follows:

1. The series-injected voltage magnitude $V_{\mathrm{se}}^{\mathrm{Lim}}$
2. The line current through the series converter $I_{\mathrm{se}}^{\mathrm{Lim}}$
3. The real power transfer through the dc link $P_{\mathrm{DC}}^{\mathrm{Lim}}$
4. The shunt converter current $I_{\mathrm{sh}}^{\mathrm{Lim}}$
5. The bus voltage on line side of the series converter V_m^{Lim}

The device limit constraints have been accommodated by the principle that whenever a particular constraint limit is violated, it is kept at its specified limit, whereas a control objective is relaxed. Mathematically, this signifies the replacement of the Jacobian elements pertaining to the control objective by those of the constraint violated. The control strategies to incorporate the above five limits are detailed in the text that follows. For simplicity, we consider a single GUPFC with two series converters (installed in two different transmission lines) and a shunt converter. The original power flow control specifications considered are $P_{\text{LINE 1}}$ and $Q_{\text{LINE 1}}$ (for line 1), $P_{\text{LINE 2}}$ and $Q_{\text{LINE 2}}$ (for line 2), and V_{BUS} (for SE bus of line 2). It is to be noted that the choice of the SE bus voltage (of line 2) as a control objective is purely arbitrary. The SE bus voltage of line 1 could also be chosen.

1. In this case,

$$V_{\text{se}} = V_{\text{se}}^{\text{Lim}} \tag{6.64}$$

   If $V_{\text{se 1}}^{\text{Lim}}$ is violated, V_{n+1} is preset at the limit $V_{\text{se 1}}^{\text{Lim}}$, and either the line active ($P_{\text{LINE 1}}$) or the reactive power flow ($Q_{\text{LINE 1}}$) control objective is relaxed for line 1. The corresponding relaxed active or reactive power mismatch is replaced by $\Delta V_{n+1} = V_{\text{se 1}}^{\text{Lim}} - V_{n+1}$. The Jacobian elements are changed accordingly. If $V_{\text{se 2}}^{\text{Lim}}$ is violated, V_{n+2} is fixed at the limit, and the control objective in vogue [line active ($P_{\text{LINE 2}}$) or reactive power flow ($Q_{\text{LINE 2}}$)] is relaxed for line 2 with its mismatch replaced by $\Delta V_{n+2} = V_{\text{se 2}}^{\text{Lim}} - V_{n+2}$. Again, the Jacobian elements are changed accordingly. As already discussed in Section 6.4, the corresponding row of the matrix **JX3** would have all elements equal to zero except the entry pertaining to V_{se}, which would be unity.

2. If $I_{\text{se 1}}^{\text{Lim}}$ is violated, I_{n+1} is preset at the limit $I_{\text{se 1}}^{\text{Lim}}$, and either the line active ($P_{\text{LINE 1}}$) or reactive power flow ($Q_{\text{LINE 1}}$) control objective is relaxed for line 1. Thus, the corresponding mismatch is replaced by $\Delta I_{n+1} = I_{\text{se 1}}^{\text{Lim}} - I_{n+1}$, where, for transmission line 1 (connected between SE and RE buses a and b, respectively), the expression for I_{n+1} can be written using Equations 6.21 and 6.22 as

$$I_{n+1} = \left|\mathbf{I}_{n+1}\right| = \left[\mathrm{e}_1 + \mathrm{e}_2 + \mathrm{e}_3 + \mathrm{e}_4\right]^{1/2} \tag{6.65}$$

where:

$$e_1 = Y_{(n+1)a}^2 V_a^2 + Y_{(n+1)b}^2 V_b^2 + Y_{(n+1)(n+1)}^2 V_{n+1}^2$$

$$e_2 = 2Y_{(n+1)a} Y_{(n+1)b} V_a V_b \cos\left(\theta_a - \theta_b + \varphi_{y(n+1)a} - \varphi_{y(n+1)b}\right)$$

$$e_3 = 2Y_{(n+1)a} Y_{(n+1)(n+1)} V_a V_{n+1} \cos\left(\theta_a - \theta_{n+1} + \varphi_{y(n+1)a} - \varphi_{y(n+1)(n+1)}\right)$$

$$e_4 = 2Y_{(n+1)b} Y_{(n+1)(n+1)} V_b V_{n+1} \cos\left(\theta_b - \theta_{n+1} + \varphi_{y(n+1)b} - \varphi_{y(n+1)(n+1)}\right)$$

The Jacobian elements are changed accordingly. If $I_{se\,2}^{Lim}$ is violated, I_{n+2} is preset at the limit $I_{se\,2}^{Lim}$, and the line active ($P_{LINE\,2}$) or reactive power flow ($Q_{LINE\,2}$) control objective is relaxed for line 2, with its mismatch and Jacobian elements changed accordingly. If, however, both $V_{se\,1}^{Lim}$ and $I_{se\,1}^{Lim}$ are violated, leniency is exercised on both $P_{LINE\,1}$ and $Q_{LINE\,1}$ control objectives, which are replaced by $\Delta V_{n+1} = V_{se\,1}^{Lim} - V_{n+1}$ and $\Delta I_{n+1} = I_{se\,1}^{Lim} - I_{n+1}$. The corresponding Jacobian elements are also replaced. The derivations of the expressions of e_1, e_2, e_3, and e_4 are given in the Appendix.

3. If P_{DC}^{Lim} is violated, P_{DC} is preset at the limit P_{DC}^{Lim}, and either the P_{LINE} or the Q_{LINE} control objective is relaxed. Thus, the corresponding line power flow (active or reactive) mismatch is replaced by $\Delta P_{DC} = P_{DC}^{Lim} - P_{DC}$, where, for transmission line 1 (connected between SE and RE buses a and b, respectively), the expression for P_{DC} can be written using Equations 6.21 and 6.22 as

$$\begin{aligned} P_{DC} &= \mathrm{Re}\left[\mathbf{V}_{se}\left(-\mathbf{I}_{se}^{*}\right)\right] = \mathrm{Re}\left(\mathbf{V}_{n+1}\mathbf{I}_{n+1}^{*}\right) \\ &= V_{n+1} V_a Y_{(n+1)a} \cos\left(\theta_{n+1} - \theta_a - \varphi_{y(n+1)a}\right) \\ &\quad + V_{n+1} V_b Y_{(n+1)b} \cos\left(\theta_{n+1} - \theta_b - \varphi_{y(n+1)b}\right) \\ &\quad + V_{n+1}^2 Y_{(n+1)(n+1)} \cos\varphi_{y(n+1)(n+1)} \end{aligned} \tag{6.66}$$

The Jacobian elements are changed accordingly.

If, however, all three quantities P_{DC}^{Lim}, $I_{se\,1}^{Lim}$, and $I_{se\,2}^{Lim}$ (or say, P_{DC}^{Lim}, V_{se}^{Lim}, and I_{se}^{Lim}) are violated, all the control objectives, that is, the line active and reactive power flow of the first line along with the line active (reactive) power flow of the second line, are relaxed. The corresponding mismatches are replaced, along with the modification of the corresponding Jacobian elements.

4. If $I_{\text{sh}}^{\text{Lim}}$ is violated, I_{n+3} (because the shunt converter is transformed to $[n+3]$th power-flow bus for a GUPFC with two series and one shunt converters) is preset at the limit $I_{\text{sh}}^{\text{Lim}}$, and the SE bus voltage control objective (for line 2) is relaxed, with the voltage mismatch replaced by $\Delta I_{n+3} = I_{\text{sh}}^{\text{Lim}} - I_{n+3}$, where, for transmission line 2 (connected between SE and RE buses c and d, respectively), the expression for I_{n+3} can be written using Equations 6.23 and 6.24 as

$$I_{n+3} = \left|\mathbf{I}_{n+3}\right| = [f_1 + f_2]^{1/2} \tag{6.67}$$

where:

$$f_1 = Y_{(n+3)c}^2 V_c^2 + Y_{(n+3)(n+3)}^2 V_{n+3}^2$$

$$f_2 = 2Y_{(n+3)c} Y_{(n+3)(n+3)} V_c V_{n+3} \cos\left(\theta_c - \theta_{n+3} + \varphi_{y(n+3)c} - \varphi_{y(n+3)(n+3)}\right)$$

The Jacobian elements are changed accordingly. The derivations of the expressions of f_1 and f_2 are given in the Appendix.

5. If V_{m1}^{Lim} (bus voltage on the line side of the first series converter) is violated, V_{m1} is preset at the limit V_{m1}^{Lim}, and either the line active ($P_{\text{LINE 1}}$) or reactive power flow ($Q_{\text{LINE 1}}$) control objective is relaxed for line 1. The corresponding relaxed active or reactive power mismatch is replaced by $\Delta V_{m1} = V_{m1}^{\text{Lim}} - V_{m1}$, where, corresponding to line 1 (connected between SE and RE buses a and b, respectively), it can be shown (using Figure 6.2) that

$$\mathbf{V}_{m1} = \frac{\mathbf{y}_{\text{se1}}\mathbf{V}_a + \mathbf{y}_1\mathbf{V}_b - \mathbf{y}_{\text{se1}}\mathbf{V}_{\text{se1}}}{\mathbf{y}_1 + \mathbf{y}_{10} + \mathbf{y}_{\text{se1}}} \tag{6.68}$$

where:

$$V_{m1} = \left|\mathbf{V}_{m1}\right| = [h_1 + h_2 + h_3 + h_4]^{1/2} \tag{6.69}$$

where:

$$h_1 = c_1^2 V_a^2 + c_2^2 V_b^2 + c_1^2 V_{n+1}^2$$

$$h_2 = 2c_1 c_2 V_a V_b \cos\left(\theta_a - \theta_b + \varphi_{c1} - \varphi_{c2}\right)$$

$$h_3 = -2c_1^2 V_a V_{n+1} \cos\left(\theta_a - \theta_{n+1}\right)$$

$$h_4 = -2c_1 c_2 V_b V_{n+1} \cos\left(\theta_b - \theta_{n+1} + \varphi_{c2} - \varphi_{c1}\right)$$

The Jacobian elements are changed accordingly. The derivations of the expressions of h_1, h_2, h_3, and h_4 are given in the Appendix.

If V_{m2}^{Lim} is violated, V_{m2} is preset at the limit V_{m2}^{Lim}, and either the line active ($P_{\text{LINE }2}$) or reactive power flow ($Q_{\text{LINE }2}$) control objective is relaxed for line 2. The corresponding relaxed active or reactive power mismatch is replaced by $\Delta V_{m2} = V_{m2}^{\text{Lim}} - V_{m2}$. The Jacobian elements are replaced accordingly.

6.6 SELECTION OF INITIAL CONDITIONS

In this chapter, the initial conditions for the series voltage source(s) were chosen as $V_{\text{se}}^0 \angle \theta_{\text{se}}^0 = 0.1 \angle -(\pi/2)$ p.u. following suggestions of [49], whereas those for the shunt voltage source were chosen as $V_{\text{sh}}^0 \angle \theta_{\text{sh}}^0 = 1.0 \angle 0^0$ following [60]. However, while enforcing the limits of shunt converter current, it was observed that adoption of this initial condition for the shunt voltage source makes the shunt converter current magnitude zero. As a consequence, the Jacobians of the shunt converter current magnitude were rendered indeterminate (using Equation 6.68). This is shown in the Appendix. Modifying the shunt source initial condition to $V_{\text{sh}}^0 \angle \theta_{\text{sh}}^0 = 1.0 \angle -(\pi/9)$ solves this problem, without any observed detrimental effect on the convergence.

6.7 CASE STUDIES AND RESULTS

The proposed method was applied to the IEEE 300-bus system to validate its feasibility. In this test system, GUPFCs with multiple series converters and one shunt converter were included, and studies have been carried out for (1) converters without any device limit constraints and (2) practical converters with device limit constraints. In all the case studies, a convergence tolerance of 10^{-12} p.u. has been chosen. Although a large number of case studies confirmed the validity of the model, a few sets of representative results are presented below. As in Chapters 3 through 5, in all the subsequent tables, the symbol NI denotes the number of iterations taken by the algorithm.

6.7.1 Studies of GUPFCs without Any Device Limit Constraints

Case 1: In this case, a single GUPFC with three series converters and one shunt converter has been considered in the 300-bus system. The series converters are incorporated on the transmission line branches between buses 3-7 (converter 1), 3-19 (converter 2), and 3-150 (converter 3).

TABLE 6.1 Study of IEEE 300-Bus System with Single GUPFC Having Four Converters (No Device Limit Constraints)

Solution of Base Case Power Flow (without Any GUPFC) NI = 7			
Parameter	**Line 1**	**Line 2**	**Line 3**
P_{LINE} (MW)	260	138.75	95.06
Q_{LINE} (MVAR)	120.55	11.19	138.26
V_3 (p.u.)	0.9969		

Unconstrained Solution of GUPFC Quantities NI = 7			
	Series Converters		
Quantity	**Converter 1**	**Converter 2**	**Converter 3**
V_{se} (p.u.)	0.2746	0.1625	0.1075
θ_{se}	–80.73°	–87.03°	–90.25°
I_{se} (p.u.)	2.6751	1.4884	1.107
P_{Line} (MW)	**270**	**150**	**100**
Q_{Line} (MVAR)	**10**	**10**	**50**
	Shunt Converter		
V_{sh} (p.u.)	**θ_{sh}**	**I_{sh} (p.u.)**	**V_3 (p.u.)**
1.0184	6.95°	0.134	1.01

The shunt converter is connected to bus number 3. The control references chosen are as follows: $P_{3\text{-}7}^{SP} = 270$ MW and $Q_{3\text{-}7}^{SP} = 10$ MVAR (line 1), $P_{3\text{-}19}^{SP} = 150$ MW and $Q_{3\text{-}19}^{SP} = 10$ MVAR (line 2), $P_{3\text{-}150}^{SP} = 100$ MW and $Q_{3\text{-}19}^{SP} = 50$ MVAR (line 3), and $V_3^{SP} = 1.01$ p.u. The results are shown in Table 6.1. For ready reference, the power-flow solution of the 300-bus system without any GUPFC is also shown in the table. The converged final values of the control objectives are shown in bold cases. From the table, it can be observed that in the presence of the GUPFC, values of the line active power flows can be maintained or enhanced, even for reduced levels of line reactive power flows.

The bus voltage profiles for this study without and with GUPFC are shown in Figure 6.3a and b, respectively. Further, the difference between the bus voltage magnitudes with and without GUPFC is shown in Figure 6.3c. From this figure, it is observed that in the presence of GUPFC, there is not much change in the bus voltage profile.

Case 2: In this case, two GUPFCs, each having two series converters and one shunt converter, have been considered in the 300-bus system. For the first GUPFC, the series converters are incorporated on the transmission line

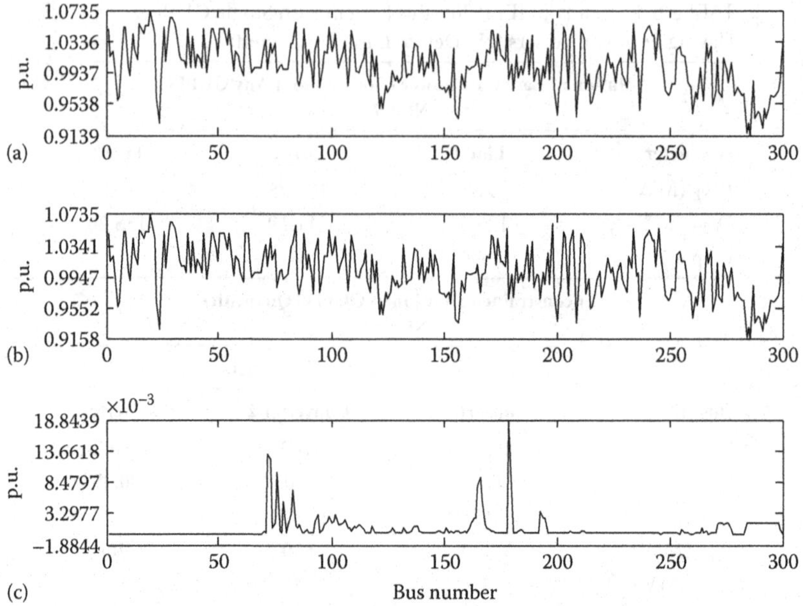

FIGURE 6.3 Bus voltage profile corresponding to the case study of Table 6.1. (a) Bus voltage magnitude without GUPFC; (b) bus voltage magnitude with GUPFC; (c) voltage magnitude difference.

branches between buses 2-3 (converter 1) and 86-102 (converter 2). The shunt converter is connected to bus number 86. The control references chosen are as follows: $P_{2\text{-}3}^{SP} = 50$ MW and $Q_{2\text{-}3}^{SP} = 10$ MVAR (line 1), $P_{86\text{-}102}^{SP} = 50$ MW and $Q_{86\text{-}102}^{SP} = 0$ (line 2), and $V_{86}^{SP} = 1.02$ p.u. For the second GUPFC, the series converters are incorporated on the transmission line branches between buses 35-72 (converter 1) and 35-77 (converter 2). The shunt converter is connected to bus number 35. The control references chosen are as follows: $P_{35\text{-}72}^{SP} = 50$ MW and $Q_{72\text{-}35}^{SP} = 5$ MVAR (line 1), $P_{35\text{-}77}^{SP} = 50$ MW and $Q_{35\text{-}77}^{SP} = 0$ (line 2), and $V_{35}^{SP} = 0.98$ p.u. The results are shown in Table 6.2. The converged final values of the control objectives are again shown in bold cases.

The bus voltage profiles for this study without and with GUPFCs are shown in Figure 6.4. From this figure, it is observed that in the presence of GUPFCs, there is very little change in the bus voltage profiles.

It is important to note that the choices of $Q_{86\text{-}102}^{SP} = 0$ and $Q_{35\text{-}77}^{SP} = 0$ for the second transmission lines (line 2) of both the GUPFCs are purely arbitrary. It only shows that desirable levels of line active power flows can be maintained by the GUPFC even when the line reactive power flows

TABLE 6.2 First Study of IEEE 300-Bus System with Two GUPFCs (No Device Limit Constraints)

Solution of Base Case Power Flow (without Any GUPFC)
NI = 6

	Lines Corresponding to Converter 1		Lines Corresponding to Converter 2	
Parameter	**Line 1**	**Line 2**	**Line 1**	**Line 2**
P_{LINE} (MW)	36.03	37.95	48.7	–47.69
Q_{LINE} (MVAR)	73.23	–19.93	1.87	–10.3
V_{BUS} (SE) (p.u.)	V_{86} = 0.991		V_{35} = 0.9755	

Unconstrained Solution of GUPFC Quantities
NI = 7

Series Converters

	GUPFC 1 (In Double-Line System 1)		GUPFC 2 (In Double-Line System 2)	
Quantity	**Converter 1**	**Converter 2**	**Converter 1**	**Converter 2**
V_{se} (p.u.)	0.0683	0.0711	0.0918	0.1225
θ_{se}	–52.92°	–96.65°	–110.88°	–123.74°
I_{se} (p.u.)	0.4867	0.4902	0.5127	0.5102
P_{Line} (MW)	**50**	**50**	**50**	**50**
Q_{Line} (MVAR)	**10**	**0**	**–5**	**0**

Shunt Converters

GUPFC 1				GUPFC 2			
V_{sh} **(p.u.)**	θ_{sh}	I_{sh} **(p.u.)**	V_{86} **(p.u.)**	V_{sh} **(p.u.)**	θ_{sh}	**Ish (p.u.)**	V_{35} **(p.u.)**
1.0817	–14.58°	0.6175	1.02	1.0568	–27.13°	0.7685	0.98

are enforced to zero. Subsequently, both the values of the line reactive power flows ($Q_{86\text{-}102}^{SP}$ and $Q_{35\text{-}77}^{SP}$) are modified to –10 and –5 MVAR, respectively. All other control objectives are maintained identical to the values given above (in the first part of case 2). The results are given in Table 6.3. It is observed that the proposed algorithm again converges to the modified control objectives with the same number of iterations. This demonstrates the versatility of the proposed technique.

The bus voltage profiles for this study without and with GUPFCs are shown in Figure 6.5. From this figure, it is observed that in the presence of GUPFCs, there is very little change in the bus voltage profile.

6.7.2 Studies of GUPFCs with Device Limit Constraints

In these case studies, various device limit constraints have been considered for GUPFCs incorporated in the 300-bus test system. As already

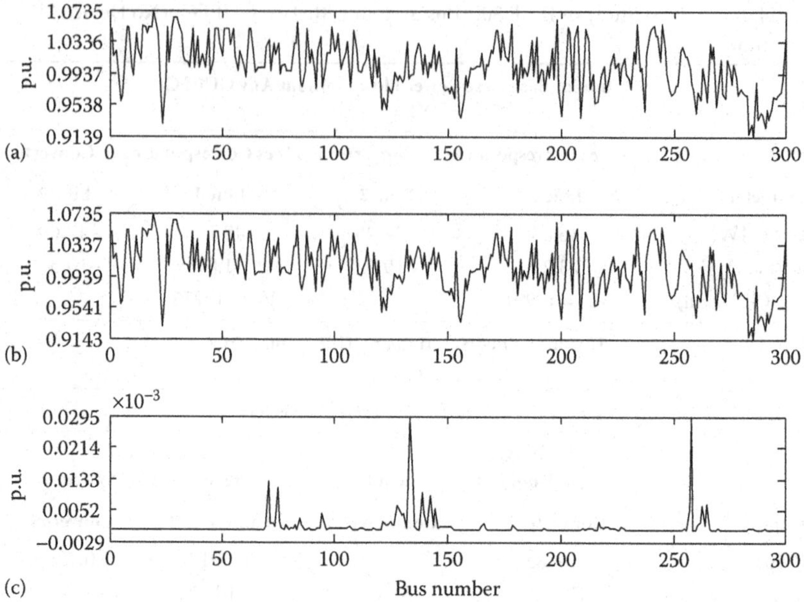

FIGURE 6.4 Bus voltage profile corresponding to the case study of Table 6.2. (a) Bus voltage magnitude without GUPFC; (b) bus voltage magnitude with GUPFC; (c) voltage magnitude difference.

TABLE 6.3 Second Study of IEEE 300-Bus System with Two GUPFCs (No Device Limit Constraints)

Unconstrained Solution of GUPFC Quantities
NI = 7

	Series Converters			
	GUPFC 1 (In Double-Line System 1)		**GUPFC 2 (In Double-Line System 2)**	
Quantity	**Converter 1**	**Converter 2**	**Converter 1**	**Converter 2**
V_{se} (p.u.)	0.0683	0.0769	0.0921	0.1227
θ_{se}	−52.92°	−83.63°	−110.55°	−120.26°
I_{se} (p.u.)	0.4867	0.4999	0.5127	0.5127
P_{Line} (MW)	**50**	**50**	**50**	**50**
Q_{Line} (MVAR)	**10**	**−10**	**−5**	**−5**

Shunt Converters

GUPFC 1 (In Double-line System 1)				**GUPFC 2 (In Double-Line System 2)**			
V_{sh} **(p.u.)**	θ_{sh}	I_{sh} **(p.u.)**	V_{86} **(p.u.)**	V_{sh} **(p.u.)**	θ_{sh}	I_{sh} **(p.u.)**	V_{35} **(p.u.)**
1.072	−14.57°	0.5208	1.02	1.0519	−27.15°	0.7191	0.98

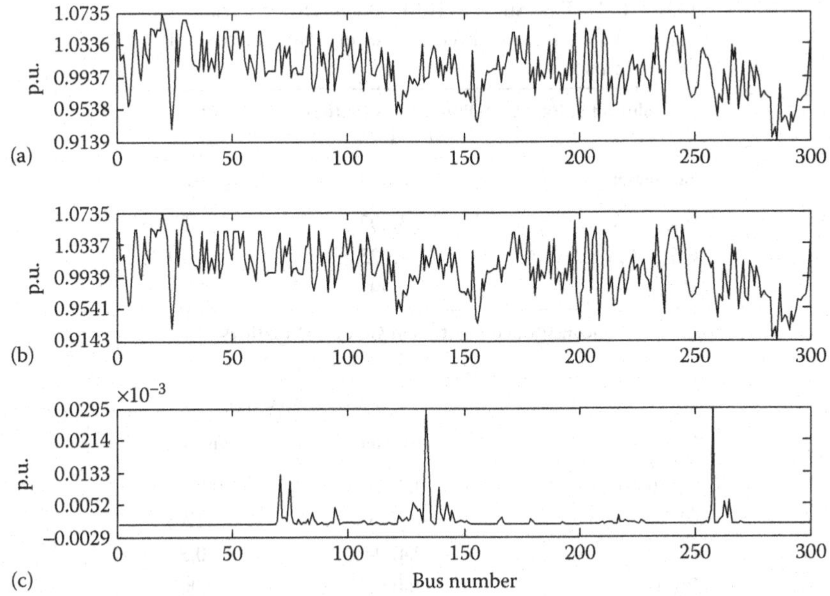

FIGURE 6.5 Bus voltage profile corresponding to the case study of Table 6.3. (a) Bus voltage magnitude without GUPFC; (b) bus voltage magnitude with GUPFC; (c) voltage magnitude difference.

mentioned in Section 6.5, five major device constraint limits have been considered. In the 300-bus test system, case studies were carried out for implementation of these limit constraints in five different ways: (1) limit violation of a single constraint, (2) limit violations of two different constraints simultaneously, (3) limit violations of all three different constraints simultaneously, (4) limit violations of four different constraints simultaneously, and (5) limit violations of all five different constraints simultaneously. For simplicity, in the 300-bus test system, a GUPFC with two series converters and a shunt converter (minimum possible configuration) is considered. The power flow solutions for this GUPFC without any device limit constraint are obtained in the 300-bus test system. The details of the GUPFC series and shunt converters considered are as follows.

Case 3: As already mentioned, in the 300-bus test system, a GUPFC with two series converters and a shunt converter has been considered. The series converters are incorporated in the transmission line branches between buses 2-3 (converter 1) and 86-102 (converter 2). The control references chosen are as follows: $P_{2\text{-}3}^{SP} = 50$ MW and $Q_{2\text{-}3}^{SP} = 5$ MVAR (line 1), $P_{86\text{-}102}^{SP} =$ 50 MW and $Q_{86\text{-}102}^{SP} = 0$ (line 2), and $V_{86}^{SP} = 1.0$ p.u. The results are shown

TABLE 6.4 First Study of IEEE 300-Bus System with Single GUPFC having Three Converters (No Device Limit Constraints)

Solution of Base Case Power Flow (without Any GUPFC)
NI = 6

Parameter	**Line 1**	**Line 2**
P_{LINE} (MW)	36.03	37.95
Q_{LINE} (MVAR)	73.23	−19.93
V_{86} (p.u.)	0.991	

Unconstrained Solution of GUPFC Quantities
NI = 7

	Series Converters	
Quantity	**Converter 1**	**Converter 2**
V_{se} (p.u.)	0.0727	0.0719
θ_{se}	−47.45°	−112.6°
I_{se} (p.u.)	0.4794	0.5
P_{DC} (p.u.)	−0.0228	0.005
P_{Line} (MW)	**50**	**50**
Q_{Line} (MVAR)	**5**	**0**
V_m (p.u.)	1.001	1.0097

Shunt Converter

V_{sh} (p.u.)	θ_{sh}	I_{sh} (p.u.)	V_{86} (p.u.)	P_{DC} (p.u.)
1.0338	−14.54°	0.3383	**1.0**	0.0178

in Table 6.4. From this table, it can be observed that in the presence of the GUPFC, the enhanced level of active power flow can be maintained in line 1, even though the reactive power flow in the line is reduced. It may again be noted that the choice of $Q_{86-102}^{SP} = 0$ for line 2 is purely arbitrary and that the proposed technique is applicable equally well for any other suitably chosen nonzero value of Q_{86-102}^{SP}. This is demonstrated by modifying the control objective to $Q_{86-102}^{SP} = -1$ MVAR. All other control objective values are kept identical to those in Table 6.4. The results are given in Table 6.5. It is observed from this table that the power flow solution again converges to the modified value of control objectives (shown in bold cases). The bus voltage profiles for these two cases (both without and with GUPFCs) are shown in Figures 6.6 and 6.7, respectively. From these figures, it is observed that in the presence of GUPFCs, the bus voltage profiles do not change much.

Subsequent to the power flow solutions obtained (Table 6.5) without any device limit constraints, the various constraint limits of the GUPFC

TABLE 6.5 Second Study of IEEE 300-Bus System with Single GUPFC Having Three Converters (No Device Limit Constraints)

Unconstrained Solution of GUPFC Quantities NI = 8		
	Series Converters	
Quantity	Converter 1	Converter 2
V_{se} (p.u.)	0.0727	0.0718
θ_{se}	–47.45°	–111.16°
I_{se} (p.u.)	0.4794	0.5001
P_{DC} (p.u.)	–0.0228	0.0048
P_{Line} (MW)	**50**	**50**
Q_{Line} (MVAR)	**5**	**–1**
V_m (p.u.)	1.001	1.0089

Shunt Converter				
V_{sh} (p.u.)	θ_{sh}	I_{sh} (p.u.)	V_{86} (p.u.)	P_{DC} (p.u.)
1.0328	–14.54°	0.3284	**1.0**	**0.018**

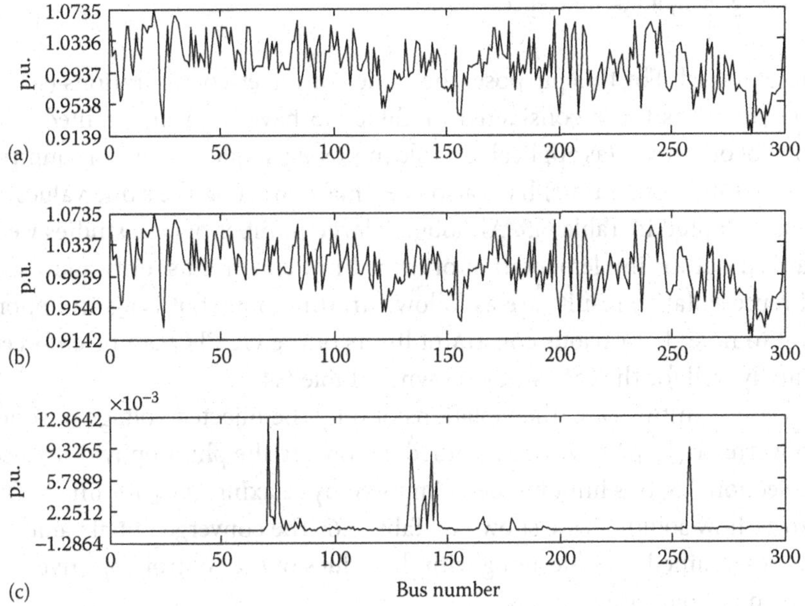

FIGURE 6.6 Bus voltage profile corresponding to the case study of Table 6.4. (a) Bus voltage magnitude without GUPFC; (b) bus voltage magnitude with GUPFC; (c) voltage magnitude difference.

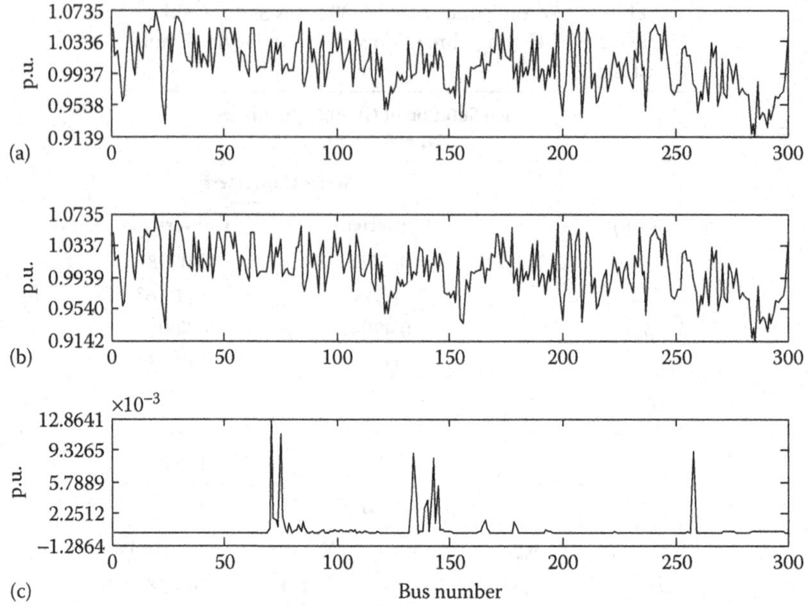

FIGURE 6.7 Bus voltage profile corresponding to the case study of Table 6.5. (a) Bus voltage magnitude without GUPFC; (b) bus voltage magnitude with GUPFC; (c) voltage magnitude difference.

are enforced. For this purpose, the same converter configurations (series and shunt) as those considered in Table 6.5 have been maintained. The control objectives (as applicable while imposing single, double, or multiple device limit constraints) have also been maintained at the same values as those adopted in Table 6.5. Although a large number of case studies were carried out for implementation of these limit constraints, only a few sets of representative results are as follows. In this context, it is again important to note that various constraint limits of the GUPFC can be enforced equally well for the case study shown in Table 6.4.

Case 4: In this case, the violations of only the injected voltage of series converter 1 (V_{se1}^{Lim}) have been studied. Following the philosophy described in Section 6.5, this limit has been imposed by relaxing Q_{LINE} for line 1. The power flow solutions are shown in Table 6.6. The converged final values of the constrained variable along with the values of the control objectives are shown in bold cases.

The bus voltage profile for the sixth case study without and with GUPFC is shown in Figure 6.8. From this figure, it is again observed that in the presence of GUPFC, there is very little change in the bus voltage profile.

TABLE 6.6 Study of IEEE 300-Bus System with GUPFC Constraint Limit of V_{se1}^{Lim}

Solution of GUPFC Quantities with Limit on Series Injected Voltage of Converter 1 Only NI = 7		
Series Converters		
Specified converter 1 voltage limit: $V_{se1}^{Lim} = 0.07$ p.u.		
Quantity	**Converter 1**	**Converter 2**
V_{se} (p.u.)	**0.07**	0.0718
θ_{se}	−50.61°	−111.16°
I_{se} (p.u.)	0.4833	0.5001
P_{DC} (p.u.)	−0.0223	0.0048
P_{Line} (MW)	**50**	**50**
Q_{Line} (MVAR)	8.06	**−1**
V_m (p.u.)	1.0027	1.0089

Shunt Converter

V_{sh} (p.u.)	θ_{sh}	I_{sh} (p.u.)	V_{86} (p.u.)	P_{DC} (p.u.)
1.0328	−14.54°	0.3283	**1.0**	0.0175

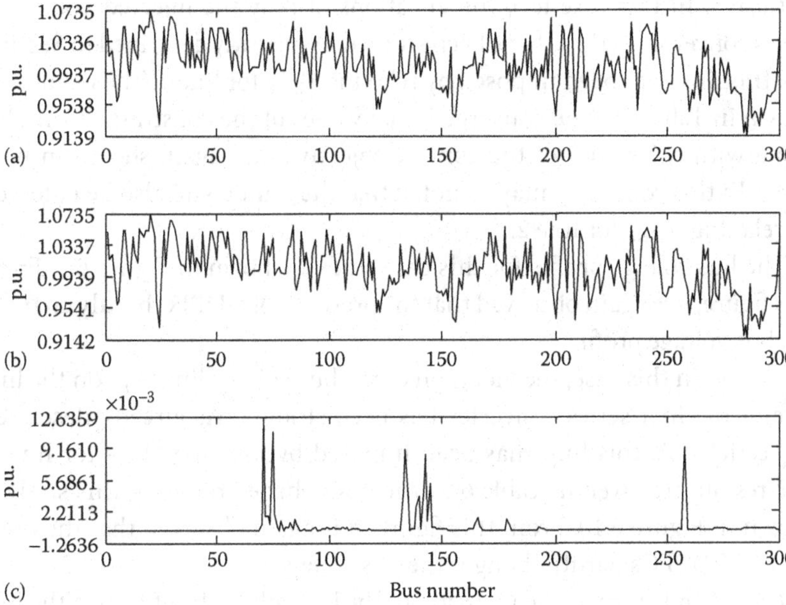

FIGURE 6.8 Bus voltage profile corresponding to the case study of Table 6.6. (a) Bus voltage magnitude without GUPFC; (b) bus voltage magnitude with GUPFC; (c) voltage magnitude difference.

TABLE 6.7 Study of IEEE 300-Bus System with GUPFC Constraint Limit of V_{se2}^{Lim}

Solution of GUPFC Quantities with Limit on Series Injected Voltage of Converter 2 Only NI = 7				
Series Converters				
Specified converter 2 voltage limit: $V_{se2}^{Lim} = 0.07$ p.u.				
Quantity		**Converter 1**		**Converter 2**
V_{se} (p.u.)		0.0727		**0.068**
θ_{se}		–47.44°		–111.27°
I_{se} (p.u.)		0.4794		0.4877
P_{DC} (p.u.)		–0.0228		0.0045
P_{Line} (MW)		**50**		48.76
Q_{Line} (MVAR)		**5**		**–1**
V_m (p.u.)		1.001		1.0086
Shunt Converter				
V_{sh} **(p.u.)**	θ_{sh}	I_{sh} **(p.u.)**	V_{86} **(p.u.)**	P_{DC} **(p.u.)**
1.0326	–14.48°	0.3263	**1.0**	0.0183

Case 5: In this case too, the violations of only the injected voltage of series converter 2 (V_{se2}^{Lim}) have been studied. As already described in Section 6.5, this limit has been imposed by relaxing P_{LINE} for line 2. The results are shown in Table 6.7. The converged final values of the constrained variable along with the values of the control objectives are again shown in bold cases. In this context, it may be noted that the limit could also be enforced by relaxing Q_{LINE} for line 2.

The bus voltage profile for this case study is shown in Figure 6.9. From this figure, it is again observed that the presence of GUPFC hardly changes the bus voltage profile.

Case 6: In this case, the violation of the bus voltage limit V_{m1} on the line side of the first series converter has been studied. As already described in Section 6.5, this limit has been imposed by relaxing Q_{LINE} for line 1. The results are given in Table 6.8. The bus voltage profile for this study is shown in Figure 6.10. From this figure, it is again observed that the presence of GUPFCs hardly changes the bus voltage profile.

Case 7: In this case, the violation of the bus voltage limit V_{m2} on the line side of the second series converter has been studied. This limit has been imposed by relaxing Q_{LINE} for line 2. The results are given in Table 6.9. The bus voltage profile for this case is shown in Figure 6.11. From this figure,

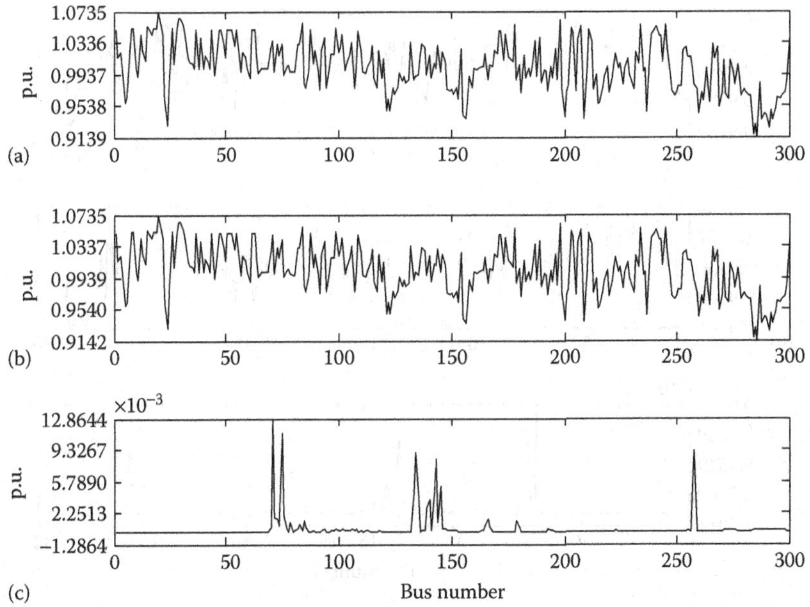

FIGURE 6.9 Bus voltage profile corresponding to the case study of Table 6.7. (a) Bus voltage magnitude without GUPFC; (b) bus voltage magnitude with GUPFC; (c) voltage magnitude difference.

TABLE 6.8 Study of IEEE 300-Bus System with GUPFC Constraint Limit of V_{m1}^{Lim}

Solution of GUPFC Quantities with Bus Voltage Limit on the Line Side of Series Converter 1 Only
NI = 7

Series Converters

Specified bus voltage limit on the line side of series converter 1: $V_{m1}^{\text{Lim}} = 1.0$ p.u.

Quantity	**Converter 1**	**Converter 2**
V_{se} (p.u.)	0.0745	0.0717
θ_{se}	–45.54°	–111.16°
I_{se} (p.u.)	0.4779	0.5001
P_{DC} (p.u.)	–0.0232	0.0048
P_{Line} (MW)	**50**	**50**
Q_{Line} (MVAR)	3.03	**–1**
V_m (p.u.)	**1.0**	1.0089

Shunt Converter

V_{sh} (p.u.)	**θ_{sh}**	**I_{sh} (p.u.)**	**V_{86} (p.u.)**	**P_{DC} (p.u.)**
1.0328	–14.53°	0.3285	**1.0**	0.0184

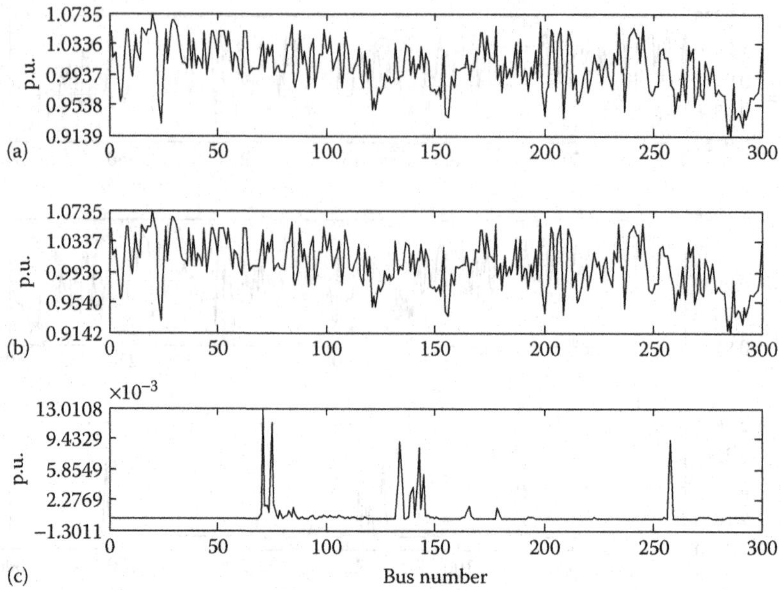

FIGURE 6.10 Bus voltage profile corresponding to the case study of Table 6.8. (a) Bus voltage magnitude without GUPFC; (b) bus voltage magnitude with GUPFC; (c) voltage magnitude difference.

TABLE 6.9 Study of IEEE 300-Bus System with GUPFC Constraint Limit of V_{m2}^{Lim}

Solution of GUPFC Quantities with Bus Voltage Limit on the Line Side of Series Converter 2 Only NI = 9		
Series Converters		
Specified bus voltage limit on the line side of series converter 2: $V_{m2}^{\mathrm{Lim}} = 1.0$ p.u.		
Quantity	**Converter 1**	**Converter 2**
V_{se} (p.u.)	0.0727	0.0738
θ_{se}	−47.44°	−95.38°
I_{se} (p.u.)	0.4794	0.5144
P_{DC} (p.u.)	−0.0228	0.0029
P_{Line} (MW)	**50**	**50**
Q_{Line} (MVAR)	**5**	**−12**
V_m (p.u.)	1.001	**1.0**

Shunt Converter

V_{sh} (p.u.)	θ_{sh}	I_{sh} (p.u.)	V_{86} (p.u.)	P_{DC} (p.u.)
1.0219	−14.52°	0.2193	**1.0**	0.0199

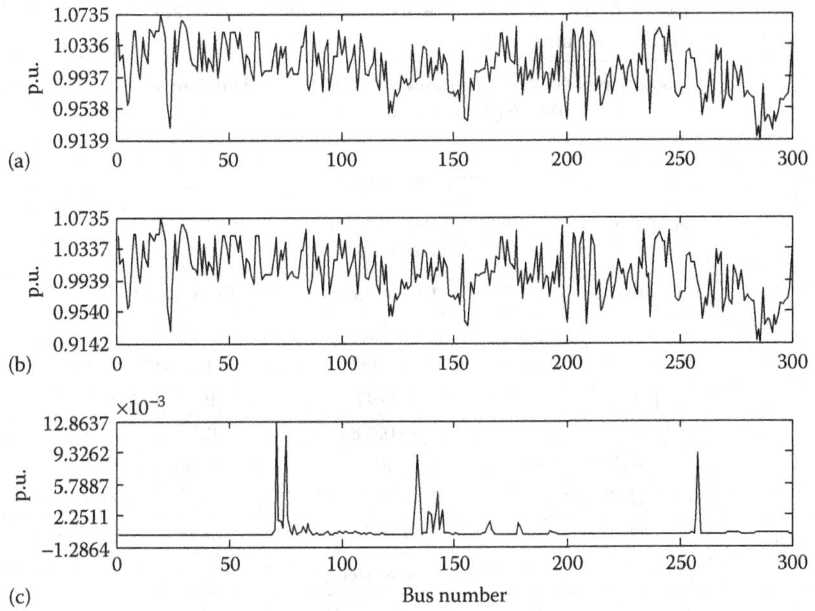

FIGURE 6.11 Bus voltage profile corresponding to the case study of Table 6.9. (a) Bus voltage magnitude without GUPFC; (b) bus voltage magnitude with GUPFC; (c) voltage magnitude difference.

it is again observed that the presence of GUPFC hardly changes the bus voltage profile.

Case 8: In this case, the violation of only the dc link power transfer of series converter 2 (P_{DC2}^{Lim}) have been studied. Following the philosophy described in Section 6.5, this limit has been imposed by relaxing Q_{LINE} for line 2. The results are given in Table 6.10. The bus voltage profile for this study is shown in Figure 6.12. From this figure, it is again observed that the bus voltage profile does not change much in the presence of GUPFC.

Case 9: In this case, simultaneous limit violations of the series injected voltage (V_{se1}^{Lim}) and the dc link power transfer (P_{DC1}^{Lim}) of converter 1 have been considered. The limits have been imposed by relaxing the active (P_{LINE}) and reactive power flow (Q_{LINE}) control objectives for the first transmission line, following the philosophy described in Section 6.5. The results are given in Table 6.11.

From this table, it can be observed from the power flow solution that the value of the reactive power flow Q_{LINE1} for the first transmission line (corresponding series converter 1) shows an impractically wide variation (26.7 MVAR) from its unconstrained value (5 MVAR) even for a very small variation in

TABLE 6.10 Study of IEEE 300-Bus System with GUPFC Constraint Limit of P_{DC2}^{Lim}

Solution of GUPFC Quantities with Limit of DC Link Power Transfer for Series Converter 2 Only NI = 11				
Series Converters				
Specified dc link power transfer limit of series converter 2: $P_{DC2}^{Lim} = 0.0046$ p.u.				
Quantity		**Converter 1**		**Converter 2**
V_{se} (p.u.)		0.0727		0.0717
θ_{se}		–47.45°		–109.61°
I_{se} (p.u.)		0.4794		0.5004
P_{DC} (p.u.)		–0.0228		**0.0046**
P_{Line} (MW)		**50**		**50**
Q_{Line} (MVAR)		**5**		–2.07
V_m (p.u.)		1.001		1.008
Shunt Converter				
V_{sh} (p.u.)	**θ_{sh}**	**I_{sh} (p.u.)**	**V_{86} (p.u.)**	**P_{DC} (p.u.)**
1.0318	–14.53°	0.3178	**1.0**	0.0182

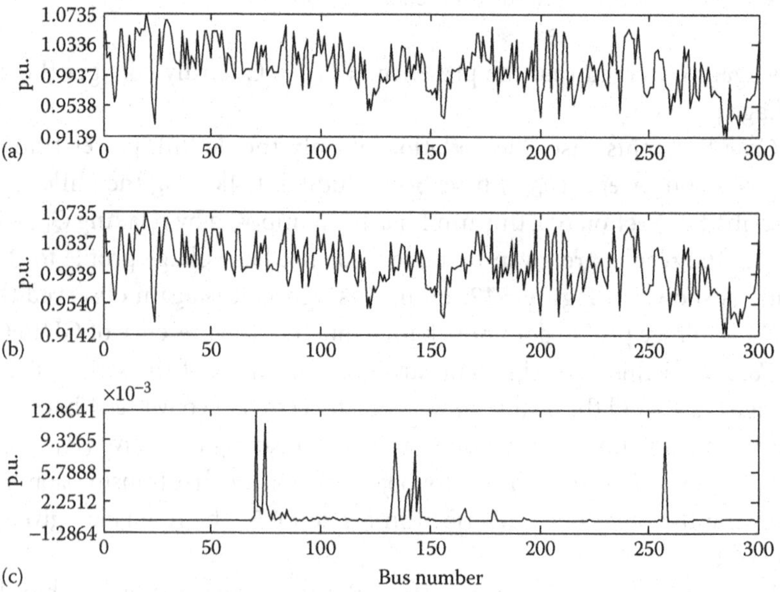

FIGURE 6.12 Bus voltage profile corresponding to the case study of Table 6.10. (a) Bus voltage magnitude without GUPFC; (b) bus voltage magnitude with GUPFC; (c) voltage magnitude difference.

TABLE 6.11 Initial Study of IEEE 300-Bus System with GUPFC Constraint Limits of V_{se1}^{Lim} and P_{DC1}^{Lim}

Solution of GUPFC Quantities with Twin Limits of Series Injected Voltage and DC Link Power Transfer for Converter 1
NI = 9

Series Converters

Specified voltage limit of series converter 1: $V_{se1}^{Lim} = 0.0725$ p.u.
Specified dc link power transfer limit of series converter 1:
$P_{DC1}^{Lim} = -0.0227$ p.u.

Quantity	Converter 1	Converter 2
V_{se} (p.u.)	0.0725	0.0717
θ_{se}	–76.48°	–111.15°
I_{se} (p.u.)	**0.605**	0.5001
P_{DC} (p.u.)	–0.0227	0.0048
P_{Line} (MW)	**57.4**	50
Q_{Line} (MVAR)	**26.7**	–1
V_m (p.u.)	1.0127	1.0089

Shunt Converter

V_{sh} (p.u.)	θ_{sh}	I_{sh} (p.u.)	V_{86} (p.u.)	P_{DC} (p.u.)
1.0327	–14.52°	0.3272	1.0	0.0179

the values of V_{se1}^{Lim}(0.0725 p.u.) and P_{DC1}^{Lim} (–0.0227 p.u.) from their unconstrained values (0.0727 and –0.0228 p.u., respectively). The same is also true for the values of P_{LINE1} and I_{se1}. Thus, the power flow solution obtained is not a realistic one (unrealistic values of complex bus voltages, converter currents, and power flows). These unrealistic power flow solutions (corresponding to series converter 1) are highlighted in bold cases in Table 6.11.

To investigate the convergence pattern and the final power flow solution *vis-à-vis* varying values of V_{se}^{0}, further case studies were again carried out. Some representative convergence patterns are presented in Table 6.12. Those with double asterisks (**) indicate unrealistic power flow solutions.

TABLE 6.12 Effect of Variation of V_{se}^{0} on the Power Flow Solution for the Study of Table 6.11

	Initial Condition for V_{se}^{0}								
V_{se}^{0} (p.u.)	0.09	0.08	0.07	0.06	0.05	0.04	0.03	0.02	0.01
	Final Power Flow Solution for Converter 2								
I_{se2} (p.u.)	0.605	0.4757	0.605	0.605	0.4757	0.605	0.605	0.4757	0.4757
P_{Line1} (MW)	57.4	49.65	57.4	57.4	49.65	57.4	57.4	49.65	49.65
Q_{Line1} (MVAR)	26.7	4.65	26.7	26.7	4.65	26.7	26.7	4.65	4.65
NI	7**	9	10**	9**	11	9**	8**	10	10

TABLE 6.13 Final Study of IEEE 300-Bus System with GUPFC Constraint Limits of V_{se1}^{Lim} and P_{DC1}^{Lim}

Solution of GUPFC Quantities with Twin Limits of Series Injected Voltage and DC Link Power Transfer for Converter 1
NI = 9 (with V_{se}^{0} = 0.08 p.u.)

Series Converters

Specified voltage limit of series converter 1: V_{se1}^{Lim} = 0.0725 p.u.
Specified dc link power transfer limit of series converter 1: P_{DC1}^{Lim} = −0.0227 p.u.

Quantity	Converter 1	Converter 2
V_{se} (p.u.)	**0.0725**	0.0718
θ_{se}	−46.81°	−111.16°
I_{se} (p.u.)	0.4757	0.5001
P_{DC} (p.u.)	**−0.0227**	0.0048
P_{Line} (MW)	49.65	**50**
Q_{Line} (MVAR)	4.65	**−1**
V_m (p.u.)	1.0009	1.0089

Shunt Converter

V_{sh} (p.u.)	θ_{sh}	I_{sh} (p.u.)	V_{86} (p.u.)	P_{DC} (p.u.)
1.0328	−14.54°	0.3284	**1.0**	0.0179

The realistic final power flow solutions (corresponding to V_{se}^{0} = 0.08 p.u) are shown in Table 6.13. The converged final values of the constrained variable along with the values of the control objectives are shown in bold cases.

The bus voltage profile for this study is shown in Figure 6.13. From this figure, it is again observed that the presence of GUPFCs hardly changes the bus voltage profile.

Case 10: In this case, simultaneous limit violations of the series injected voltage (V_{se2}^{Lim}) and the dc link power transfer (P_{DC2}^{Lim}) of converter 2 have been considered. The limits have been imposed by relaxing the active (P_{LINE}) and reactive power flow (Q_{LINE}) control objectives for the second transmission line. The results are given in Table 6.14.

The bus voltage profile for this study is shown in Figure 6.14. From this figure, it is again observed that the bus voltage profile does not change much in the presence of GUPFC.

Case 11: In this case, simultaneous limit violations of three quantities, that is, the dc link power transfer (P_{DC1}^{Lim}) and the injected voltage (V_{se1}^{Lim}) of the first series converter (in line 1) along with the line current (I_{se2}^{Lim}) in the second series converter (in line 2), have been considered. For enforcing these

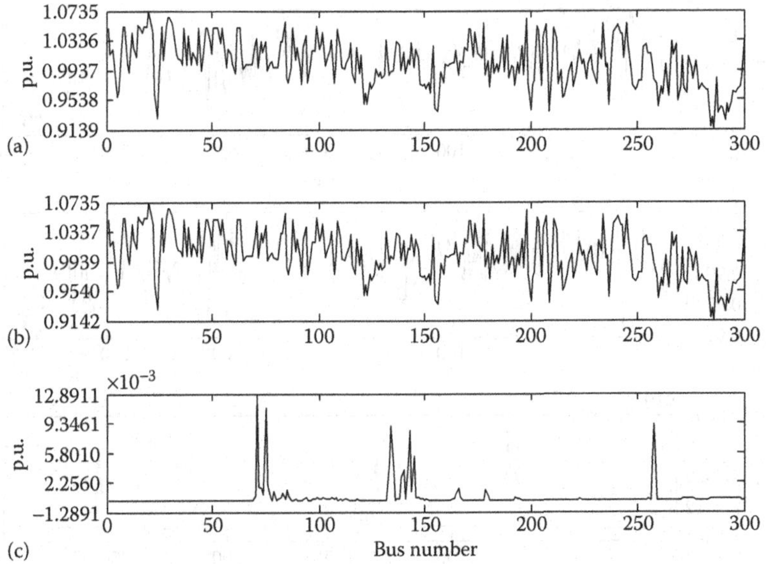

FIGURE 6.13 Bus voltage profile corresponding to the case study of Table 6.13. (a) Bus voltage magnitude without GUPFC; (b) bus voltage magnitude with GUPFC; (c) voltage magnitude difference.

TABLE 6.14 Study of IEEE 300-Bus System with GUPFC Constraint Limits of V_{se2}^{Lim} and P_{DC2}^{Lim}

Solution of GUPFC Quantities with Twin Limits of Series Injected Voltage and DC Link Power Transfer for Converter 2 NI = 7		
Series Converters		
Specified voltage limit of series converter 2: $V_{Se2}^{Lim} = 0.07$ p.u. Specified dc link power transfer limit of series converter 2: $P_{DC2}^{Lim} = 0.0047$ p.u.		
Quantity	**Converter 1**	**Converter 2**
V_{se} (p.u.)	0.0727	**0.07**
θ_{se}	−47.44°	−111.36°
I_{se} (p.u.)	0.4794	0.4942
P_{DC} (p.u.)	−0.0228	**0.0047**
P_{Line} (MW)	**50**	49.42
Q_{Line} (MVAR)	**5**	−0.9
V_m (p.u.)	1.001	1.0089

Shunt Converter

V_{sh} (p.u.)	θ_{sh}	I_{sh} (p.u.)	V_{86} (p.u.)	P_{DC} (p.u.)
1.0328	−14.51°	0.3284	**1.0**	0.0181

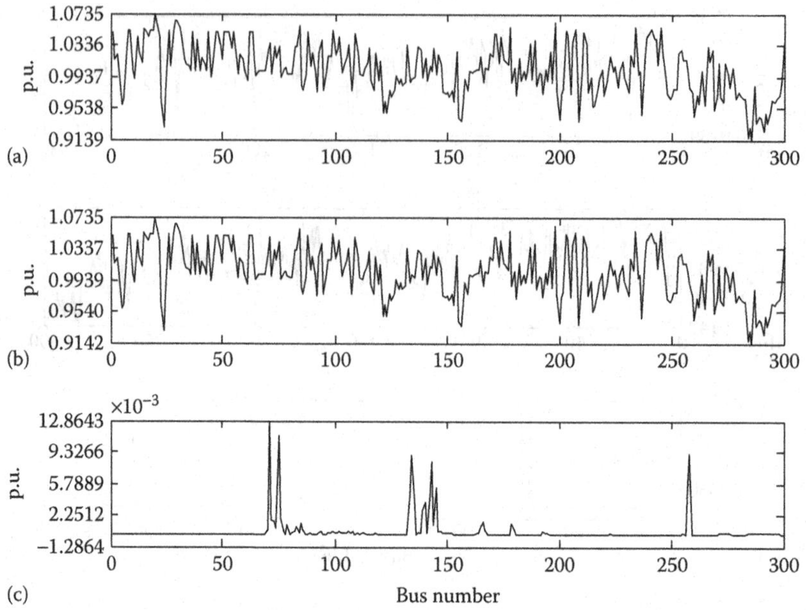

FIGURE 6.14 Bus voltage profile corresponding to the case study of Table 6.14. (a) Bus voltage magnitude without GUPFC; (b) bus voltage magnitude with GUPFC; (c) voltage magnitude difference.

limits, both P_{LINE} and Q_{LINE} of line 1 as well as P_{LINE} of line 2 are relaxed following the discussion in Section 6.5. The results are presented in Table 6.15.

From this table, it can be observed from the power flow solution that the value of the reactive power flow Q_{LINE1} for the first transmission line (corresponding series converter 1) shows an impractically wide variation (26.28 MVAR) from its unconstrained value (5 MVAR) even for a very small variation in the values of $V_{\text{se1}}^{\text{Lim}}$(0.072 p.u.) and $P_{\text{DC1}}^{\text{Lim}}$ (–0.0226 p.u.) from their unconstrained values (0.0727 and –0.0228 p.u., respectively). The same is also true for the values of P_{LINE1} and I_{se1}. Thus, the power flow solution obtained is not a realistic one. These unrealistic power flow solutions (corresponding to series converter 1) are highlighted in bold cases in Table 6.15.

To investigate the convergence pattern and the final power flow solution *vis-à-vis* varying values of V_{se}^0, further case studies were again carried out. Some representative convergence patterns are presented in Table 6.16. Those with double asterisks (**) indicate unrealistic power flow solutions.

The realistic power flow solution (corresponding to $V_{\text{se}}^0 = 0.08$ p.u.) is shown in Table 6.17. The converged final values of the constrained variable

TABLE 6.15 Initial Study of IEEE 300-Bus System with GUPFC Constraint Limits of V_{se1}^{Lim}, P_{DC1}^{Lim}, and I_{se2}^{Lim}

Solution of GUPFC Quantities with Three Simultaneous Limits of DC Link Power Transfer and Injected Voltage of Series Converter 1 along with Line Current of Series Converter 2 NI = 7		
Series Converters		
Specified dc link power transfer limit: $P_{DC1}^{Lim} = -0.0226$ p.u. Specified voltage limit of converter 1: $V_{se1}^{Lim} = 0.072$ p.u. Specified line current limit of converter 2: $I_{se2}^{Lim} = 0.48$ p.u.		
Quantity	**Converter 1**	**Converter 2**
V_{se} (p.u.)	0.072	0.0656
θ_{se}	−75.9°	−111.33°
I_{se} (p.u.)	**0.6003**	0.48
P_{DC} (p.u.)	−0.0226	0.0044
P_{Line} (MW)	**57.06**	47.99
Q_{Line} (MVAR)	**26.28**	−1
V_m (p.u.)	1.0124	1.0085

Shunt Converter

V_{sh} (p.u.)	θ_{sh}	I_{sh} (p.u.)	V_{86} (p.u.)	P_{DC} (p.u.)
1.0323	−14.43°	0.3238	**1.0**	0.0182

TABLE 6.16 Effect of Variation of V_{se}^{0} on the Power Flow Solution for the Study of Table 6.15

	Initial Condition for V_{se}								
V_{se}^{0} (p.u.)	0.09	0.08	0.07	0.06	0.05	0.04	0.03	0.02	0.01
	Final Power Flow Solution for Converter 2								
I_{se1} (p.u.)	0.6003	0.4761	0.6003	0.6003	0.6003	0.6003	0.6003	0.4761	0.6003
P_{Line1} (MW)	57.06	49.63	57.06	57.06	57.06	57.06	57.06	49.63	57.06
Q_{Line1} (MVAR)	26.28	5.19	26.28	26.28	26.28	26.28	26.28	5.19	26.28
NI	7**	7	9**	10**	10**	9**	9**	9	11**

along with the values of the control objectives are shown in bold cases. In this context, it is important to note that the limit of I_{se2}^{Lim} could be enforced equally well by relaxing Q_{LINE2} (instead of P_{LINE2}).

The bus voltage profile for this study is shown in Figure 6.15. From this figure, it is again observed that the bus voltage profile does not change much in the presence of GUPFC.

Case 12: In this case too, simultaneous limit violations of three more quantities, that is, the dc link power transfer (P_{DC2}^{Lim}) and the injected voltage (V_{se2}^{Lim}) of the second series converter (in line 2) along with the line

TABLE 6.17 Final Study of IEEE 300-Bus System with GUPFC Constraint Limits of V_{se1}^{Lim}, P_{DC1}^{Lim}, and I_{se2}^{Lim}

Solution of GUPFC Quantities with Three Simultaneous Limits of DC Link Power Transfer and Injected Voltage of Series Converter 1 along with Line Current of Series Converter 2
NI = 7 (with V_{se}^{0} = 0.08 p.u.)

Series Converters

Specified dc link power transfer limit: $P_{DC1}^{Lim} = -0.0226$ p.u.
Specified voltage limit of converter 1: $V_{se1}^{Lim} = 0.072$ p.u.
Specified line current limit of converter 2: $I_{se2}^{Lim} = 0.48$ p.u.

Quantity	**Converter 1**	**Converter 2**
V_{se} (p.u.)	**0.072**	0.0657
θ_{se}	−47.33°	−111.34°
I_{se} (p.u.)	0.4761	**0.48**
P_{DC} (p.u.)	**−0.0226**	0.0044
P_{Line} (MW)	49.63	47.99
Q_{Line} (MVAR)	5.19	**−1**
V_m (p.u.)	1.0011	1.0085

Shunt Converter

V_{sh} (p.u.)	θ_{sh}	I_{sh} (p.u.)	V_{86} (p.u.)	P_{DC} (p.u.)
1.0325	−14.45°	0.325	**1.0**	0.0182

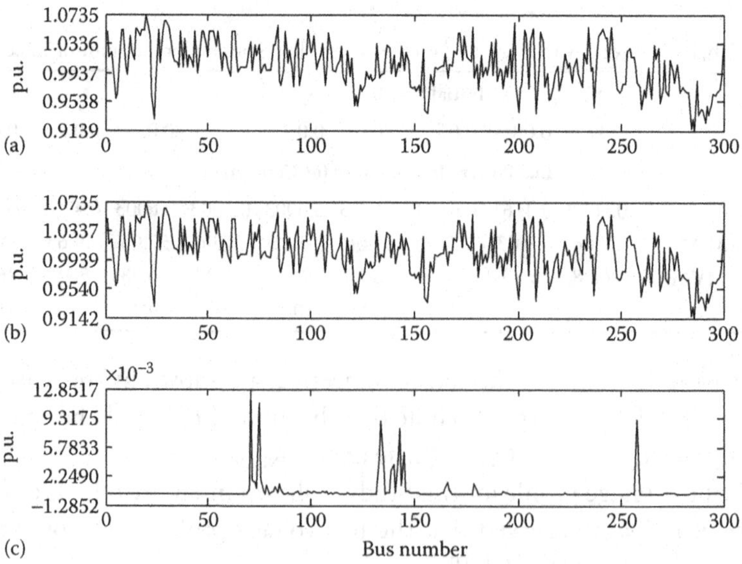

FIGURE 6.15 Bus voltage profile corresponding to the case study of Table 6.17. (a) Bus voltage magnitude without GUPFC; (b) bus voltage magnitude with GUPFC; (c) voltage magnitude difference.

current (I_{se1}^{Lim}) in the first series converter (in line 1), have been considered. For enforcing these limits, both P_{LINE} and Q_{LINE} of line 2 as well as Q_{LINE} of line 1 are relaxed. The results are presented in Table 6.18.

The bus voltage profile for this study is shown in Figure 6.16. From this figure, it is again observed that in the presence of GUPFC, the bus voltage profile shows little change.

Case 13: In this case, simultaneous limit violations of four different quantities, that is, the dc link power transfer (P_{DC1}^{Lim}) and the injected voltage (V_{se1}^{Lim}) of the first series converter (in line 1), and the line current (I_{se2}^{Lim}) and the bus voltage on the line side V_{m2} of the second series converter (in line 2), have been considered. For enforcing these limits, both P_{LINE} and Q_{LINE} of line 1 along with P_{LINE} and Q_{LINE} of line 2 are relaxed, following the discussion in Section 6.5. The results are presented in Table 6.19.

From this table, it can be observed from the power flow solution that the value of the reactive power flow Q_{LINE1} for the first transmission line (corresponding series converter 1) shows an impractically wide variation

TABLE 6.18 Study of IEEE 300-Bus System with GUPFC Constraint Limits of V_{se1}^{Lim}, P_{DC1}^{Lim}, and I_{se2}^{Lim}

Solution of GUPFC Quantities with Three Simultaneous Limits of DC Link Power Transfer and Injected Voltage of Series Converter 2 along with Line Current of Series Converter 1
NI = 10

Series Converters

Specified dc link power transfer limit: $P_{DC2}^{Lim} = 0.0047$ p.u.
Specified voltage limit of converter 2: $V_{se2}^{Lim} = 0.071$ p.u.
Specified line current limit of converter 1: $I_{se1}^{Lim} = 0.4792$ p.u.

Quantity	**Converter 1**	**Converter 2**
V_{se} (p.u.)	0.0729	**0.071**
θ_{se}	–47.2°	–110.8°
I_{se} (p.u.)	**0.4792**	0.4977
P_{DC} (p.u.)	–0.0229	**0.0047**
P_{Line} (MW)	**50**	49.75
Q_{Line} (MVAR)	4.75	–1.26
V_m (p.u.)	1.0009	1.0086

Shunt Converter

V_{sh} (p.u.)	θ_{sh}	I_{sh} (p.u.)	V_{86} (p.u.)	P_{DC} (p.u.)
1.0325	–14.45°	0.3254	**1.0**	0.0182

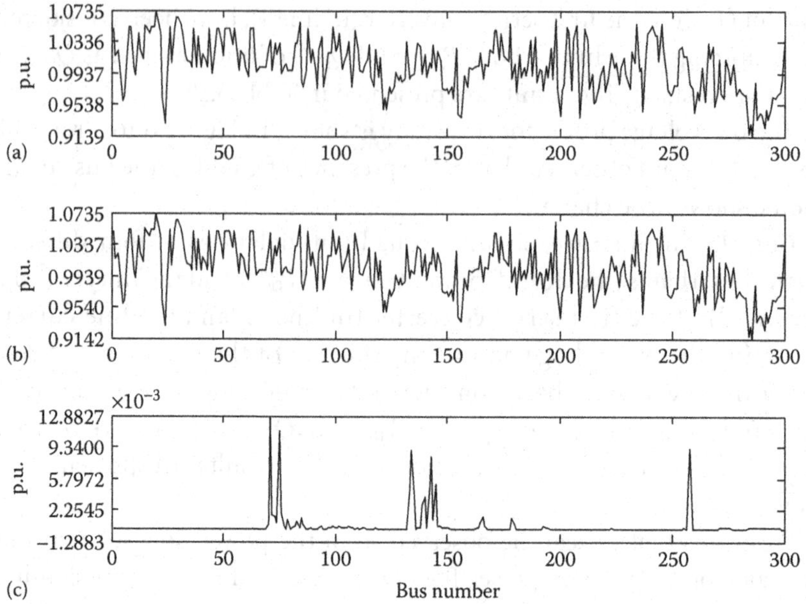

FIGURE 6.16 Bus voltage profile corresponding to the case study of Table 6.18. (a) Bus voltage magnitude without GUPFC; (b) bus voltage magnitude with GUPFC; (c) voltage magnitude difference.

(26.29 MVAR) from its unconstrained value (5 MVAR) even for a very small variation in the values of V_{se1}^{Lim} (0.072 p.u.) and P_{DC1}^{Lim} (−0.0226 p.u.) from their unconstrained values (0.0727 and −0.0228 p.u., respectively). The same is also true for the values of I_{se1}, P_{LINE1}, and V_{m1}. Thus, the power flow solution obtained is not a realistic one. These unrealistic power flow solutions (corresponding to series converter 1) are highlighted in bold cases in Table 6.19.

To investigate the convergence pattern and the final power flow solution *vis-à-vis* varying values of V_{se}^{0}, further case studies were again carried out. Some representative convergence patterns are presented in Table 6.20. Those with double asterisks (**) indicate unrealistic power flow solutions.

The realistic power flow solution (corresponding to V_{se}^{0} = 0.08 p.u.) is shown in Table 6.21. The converged final values of the constrained variable along with the values of the control objectives are shown in bold cases.

The bus voltage profile for this study is shown in Figure 6.17. From this figure, it is again observed that the bus voltage profile does not change much in the presence of GUPFC.

Case 14: In this case too, simultaneous limit violations of four more quantities, that is, the dc link power transfer (P_{DC2}^{Lim}) and the injected voltage (V_{se2}^{Lim}) of

TABLE 6.19 Study of IEEE 300-Bus System with GUPFC Constraint Limits of V_{se1}^{Lim}, P_{DC1}^{Lim}, I_{se2}^{Lim}, and V_{m2}^{Lim}

Solution of GUPFC Quantities with Four Simultaneous Limits of DC Link Power Transfer and Injected Voltage of Series Converter 1, and the Line Current and the Line Side Bus Voltage of Series Converter 2 NI = 7				
Series Converters				
Specified dc link power transfer limit: $P_{DC1}^{Lim} = -0.0226$ p.u. Specified voltage limit of converter 1: $V_{se1}^{Lim} = 0.072$ p.u. Specified line current limit of converter 2: $I_{se2}^{Lim} = 0.48$ p.u. Specified line side bus voltage limit of converter 2: $V_{m2}^{Lim} = 1.0$ p.u.				
Quantity		**Converter 1**		**Converter 2**
V_{se} (p.u.)		0.072		0.0634
θ_{se}		–75.9°		–94.62°
I_{se} (p.u.)		**0.6003**		0.48
P_{DC} (p.u.)		–0.0226		0.0019
P_{Line} (MW)		**57.06**		46.68
Q_{Line} (MVAR)		**26.29**		–11.17
V_m (p.u.)		**1.0124**		1.0
Shunt Converter				
V_{sh} (p.u.)	**θ_{sh}**	**I_{sh} (p.u.)**	**V_{86} (p.u.)**	**P_{DC} (p.u.)**
1.0221	–14.35°	0.2214	1.0	0.0207

TABLE 6.20 Effect of Variation of V_{se}^0 on the Power Flow Solution for the Study of Table 6.19

	Initial Condition for V_{se}								
V_{se}^0 (p.u.)	0.09	0.08	0.07	0.06	0.05	0.04	0.03	0.02	0.01
	Final Power Flow Solution for Converter 2								
I_{se1} (p.u.)	0.6003	0.4762	0.6003	0.6003	0.6003	0.6003	0.6003	0.4762	0.6003
P_{Line1} (MW)	57.06	49.63	57.06	57.06	57.06	57.06	57.06	49.63	57.06
Q_{Line1} (MVAR)	26.29	5.2	26.29	26.29	26.29	26.29	26.29	5.2	26.29
V_{m1} (p.u.)	1.0124	1.0011	1.0124	1.0124	1.0124	1.0124	1.0124	1.0011	1.0124
NI	7**	7	9**	9**	10**	9**	10**	9	15**

the second series converter (in line 2) along with the line current (I_{se1}^{Lim}) and the bus voltage on the line side V_{m1} of the first series converter (in line 1), have been considered. For enforcing these limits, both P_{LINE} and Q_{LINE} of line 2 along with P_{LINE} and Q_{LINE} of line 1 are relaxed. The results are presented in Table 6.22.

The bus voltage profile for this study is shown in Figure 6.18. From this figure, it is again observed that in the presence of GUPFC, the bus voltage profile shows little change.

TABLE 6.21 Final Study of IEEE 300-Bus System with GUPFC Constraint Limits of V_{se1}^{Lim}, P_{DC1}^{Lim}, I_{se2}^{Lim}, and V_{m2}^{Lim}

Solution of GUPFC Quantities with Four Simultaneous Limits of DC Link Power Transfer and Injected Voltage of Series Converter 1, and the Line Current and the Line Side Bus Voltage of Series Converter 2
NI = 7 (with $\mathbf{V_{se}^{0}}$ = 0.08 p.u.)

Series Converters

Specified dc link power transfer limit: $P_{DC1}^{Lim} = -0.0226$ p.u.
Specified voltage limit of converter 1: $V_{se1}^{Lim} = 0.072$ p.u.
Specified line current limit of converter 2: $I_{se2}^{Lim} = 0.48$ p.u.
Specified line side bus voltage limit of converter 2: $V_{m2}^{Lim} = 1.0$ p.u.

Quantity	Converter 1	Converter 2
V_{se} (p.u.)	**0.072**	0.0635
θ_{se}	–47.34°	–94.65°
I_{se} (p.u.)	0.4762	**0.48**
P_{DC} (p.u.)	**–0.0226**	0.0019
P_{Line} (MW)	49.63	46.68
Q_{Line} (MVAR)	5.2	–11.17
V_m (p.u.)	1.0011	**1.0**

Shunt Converter

V_{sh} (p.u.)	θ_{sh}	I_{sh} (p.u.)	V_{86} (p.u.)	P_{DC} (p.u.)
1.0222	–14.37°	0.2226	**1.0**	0.0207

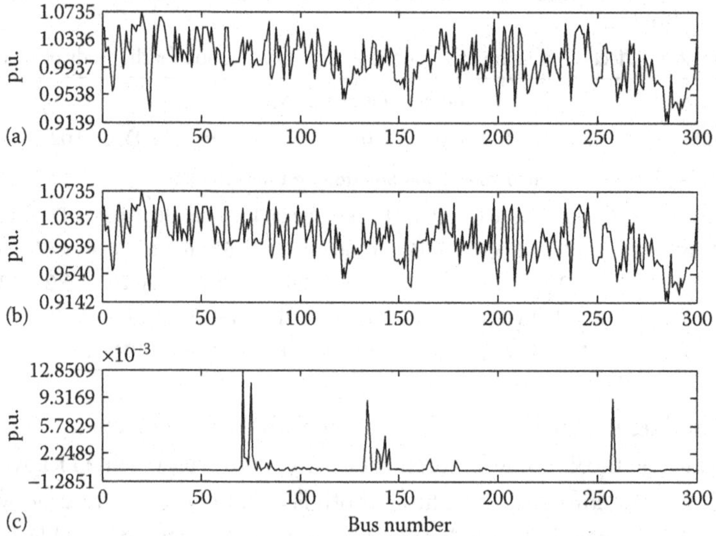

FIGURE 6.17 Bus voltage profile corresponding to the case study of Table 6.21. (a) Bus voltage magnitude without GUPFC; (b) bus voltage magnitude with GUPFC; (c) voltage magnitude difference.

TABLE 6.22 Study of IEEE 300-Bus System with GUPFC Constraint Limits of V_{se2}^{Lim}, P_{DC2}^{Lim}, I_{se1}^{Lim}, and V_{m1}^{Lim}

Solution of GUPFC Quantities with Four Simultaneous Limits of DC Link Power Transfer and Injected Voltage of Series Converter 2, and the Line Current and Line Side Bus Voltage of Series Converter 1
NI = 7

Series Converters

Specified dc link power transfer limit: $P_{DC2}^{Lim} = 0.0047$ p.u.
Specified voltage limit of converter 2: $V_{se2}^{Lim} = 0.071$ p.u.
Specified line current limit of converter 1: $I_{se1}^{Lim} = 0.479$ p.u.
Specified line side bus voltage limit of converter 1: $V_{m1}^{Lim} = 1.0$ p.u.

Quantity	Converter 1	Converter 2
V_{se} (p.u.)	0.0747	**0.071**
θ_{se}	–45.63°	–110.79°
I_{se} (p.u.)	**0.479**	0.4978
P_{DC} (p.u.)	–0.0232	**0.0047**
P_{Line} (MW)	50.12	49.76
Q_{Line} (MVAR)	3.03	–1.27
V_m (p.u.)	**1.0**	1.0086

Shunt Converter

V_{sh} (p.u.)	θ_{sh}	I_{sh} (p.u.)	V_{86} (p.u.)	P_{DC} (p.u.)
1.0325	–14.52°	0.3254	**1.0**	0.0185

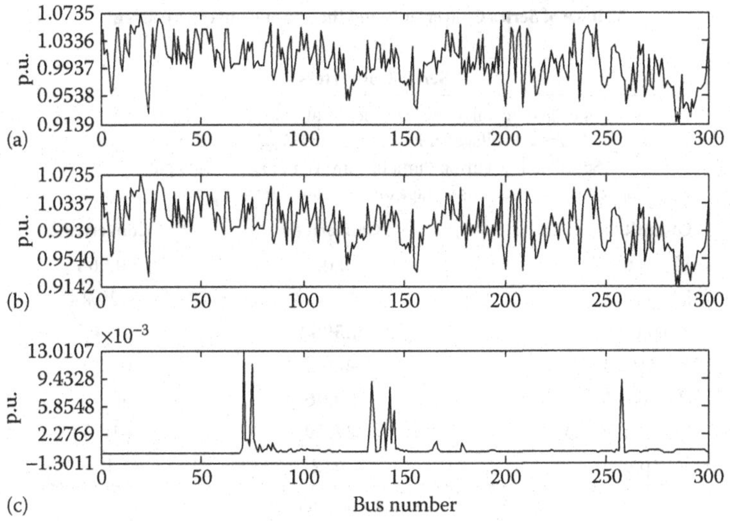

FIGURE 6.18 Bus voltage profile corresponding to the case study of Table 6.22. (a) Bus voltage magnitude without GUPFC; (b) bus voltage magnitude with GUPFC; (c) voltage magnitude difference.

Case 15: In this case, simultaneous limit violations of all five quantities, that is, the dc link power transfer (P_{DC1}^{Lim}) and the injected voltage (V_{se1}^{Lim}) of the first series converter (in line 1), the line current (I_{se2}^{Lim}) and the bus voltage on the line side V_{m2}^{Lim} of the second series converter (in line 2), and the shunt converter current (I_{sh}^{Lim}), have been considered. For enforcing these limits, both P_{LINE} and Q_{LINE} of line 1 as well as line 2 and the SE bus voltage control objective (V_{86}) are relaxed, following the discussion in Section 6.5. The results are presented in Table 6.23.

From this table, it can be observed from the power flow solution that the value of the reactive power flow Q_{LINE1} for the first transmission line (corresponding series converter 1) shows an impractically wide variation (27.39 MVAR) from its unconstrained value (5 MVAR) even for a very small variation in the values of V_{se1}^{Lim} (0.07 p.u.) and P_{DC1}^{Lim} (–0.022 p.u.) from their unconstrained values (0.0727 and –0.0228 p.u., respectively). The same is also true for the values of I_{se1}, P_{LINE1}, and V_{m1}. Thus, the power flow solution obtained is not a realistic one. These unrealistic power flow solutions (corresponding to series converter 1) are highlighted in bold cases in Table 6.23.

TABLE 6.23 Initial Study of IEEE 300-Bus System with GUPFC Constraint Limits of V_{se1}^{Lim}, P_{DC1}^{Lim}, I_{se2}^{Lim}, V_{m2}^{Lim}, and I_{sh}^{Lim}

Solution of GUPFC Quantities with All Five Limits of DC Link Power Transfer and Injected Voltage of Series Converter 1, the Line Current and Line Side Bus Voltage of Series Converter 2, and the Shunt Converter Current

NI = 7

Series Converters

Specified dc link power transfer limit: $P_{DC1}^{Lim} = -0.022$ p.u.
Specified voltage limit of converter 1: $V_{se1}^{Lim} = 0.07$ p.u.
Specified line current limit of converter 2: $I_{se2}^{Lim} = 0.49$ p.u.
Specified line side bus voltage limit of converter 2: $V_{m2}^{Lim} = 1.0$ p.u.

Quantity	**Converter 1**	**Converter 2**
V_{se} (p.u.)	0.07	0.068
θ_{se}	–76.87°	–129.24°
I_{se} (p.u.)	**0.5954**	0.49
P_{DC} (p.u.)	–0.022	0.0203
P_{Line} (MW)	**55.96**	45.95
Q_{Line} (MVAR)	**27.39**	–10.33
V_m (p.u.)	**1.0127**	1.0

Shunt Converter

Specified shunt converter current limits: $I_{sh}^{Lim} = 0.32$ p.u.

V_{sh} (p.u.)	**θ_{sh}**	**I_{sh} (p.u.)**	**V_{86} (p.u.)**	**P_{DC} (p.u.)**
0.9291	–14.42°	0.32	0.9611	0.0017

TABLE 6.24 Effect of Variation of V_{se}^0 on the Power Flow Solution for the Study of Table 6.23

Initial Condition for V_{se}									
V_{se}^0 (p.u.)	0.09	0.08	0.07	0.06	0.05	0.04	0.03	0.02	0.01
Final Power Flow Solution for Converter 2									
I_{se1} (p.u.)	0.5954	0.4629	0.4629	0.5954	0.4629	0.5954	0.5954	0.5954	0.5954
P_{Line1} (MW)	55.96	48.21	48.21	55.96	48.21	55.96	55.96	55.96	55.96
Q_{Line1} (MVAR)	27.39	5.43	5.43	27.39	5.43	27.39	27.39	27.39	27.39
V_{m1} (p.u.)	1.0127	1.001	1.001	1.0127	1.001	1.0127	1.0127	1.0127	1.0127
NI	8**	7	8	9**	9	9**	9**	9**	25**

To investigate the convergence pattern and the final power flow solution *vis-à-vis* varying values of V_{se}^0, further case studies were again carried out. Some representative convergence patterns are presented in Table 6.24. Those with double asterisks (**) indicate unrealistic power flow solutions.

The realistic power flow solution (corresponding to $V_{se}^0 = 0.08$ p.u.) is shown in Table 6.25. The converged final values of the constrained variable along with the values of the control objectives are shown in bold cases. The bus voltage profile for this study is shown in Figure 6.19. From this figure,

TABLE 6.25 Final Study of IEEE 300-Bus System with GUPFC Constraint Limits of V_{se1}^{Lim}, P_{DC1}^{Lim}, I_{se2}^{Lim}, V_{m2}^{Lim}, and I_{sh}^{Lim}

NI = 7 (with $V_{se}^0 = 0.08$ p.u.)		
Series Converters		
Specified dc link power transfer limit: $P_{DC1}^{Lim} = 0.022$ p.u. Specified voltage limit of converter 1: $V_{se1}^{Lim} = 0.07$ p.u. Specified line current limit of converter 2: $I_{se2}^{Lim} = 0.49$ p.u. Specified line side bus voltage limit of converter 2: $V_{m2}^{Lim} = 1.0$ p.u.		
Quantity	**Converter 1**	**Converter 2**
V_{se} (p.u.)	**0.07**	0.0681
θ_{se}	−46.21°	−129.31°
I_{se} (p.u.)	0.4629	**0.49**
P_{DC} (p.u.)	**−0.022**	0.0203
P_{Line} (MW)	48.21	45.95
Q_{Line} (MVAR)	5.43	−10.32
V_m (p.u.)	1.001	**1.0**

Shunt Converter

Specified shunt converter current limits: $I_{sh}^{Lim} = 0.32$ p.u.

V_{sh} (p.u.)	θ_{sh}	I_{sh} (p.u.)	V_{86} (p.u.)	P_{DC} (p.u.)
0.929	−14.43°	**0.32**	0.961	0.0017

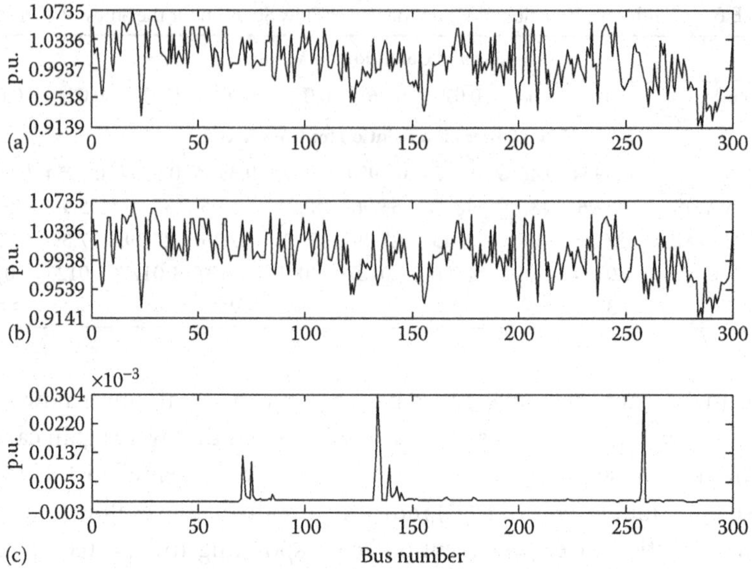

FIGURE 6.19 Bus voltage profile corresponding to the case study of Table 6.25. (a) Bus voltage magnitude without GUPFC; (b) bus voltage magnitude with GUPFC; (c) voltage magnitude difference.

TABLE 6.26 Study of IEEE 300-Bus System with GUPFC Constraint Limits of V_{se2}^{Lim}, P_{DC2}^{Lim}, I_{se1}^{Lim}, V_{m1}^{Lim}, and I_{sh}^{Lim}

NI = 7				
Series Converters				
Specified dc link power transfer limit: $P_{DC2}^{Lim} = 0.0047$ p.u. Specified voltage limit of converter 2: $V_{se2}^{Lim} = 0.07$ p.u. Specified line current limit of converter 1: $I_{se1}^{Lim} = 0.47$ p.u. Specified line side bus voltage limit of converter 1: $V_{m1}^{Lim} = 1.0$ p.u.				
Quantity	**Converter 1**	**Converter 2**		
V_{se} (p.u.)	0.0732	**0.07**		
θ_{se}	−45.01°	−74.76°		
I_{se} (p.u.)	**0.47**	0.5692		
P_{DC} (p.u.)	−0.0228	**0.0047**		
P_{Line} (MW)	49.15	45.01		
Q_{Line} (MVAR)	3.31	−33.2		
V_m (p.u.)	**1.0**	0.9814		
Shunt Converter				
Specified shunt converter current limits: $I_{sh}^{Lim} = 0.25$ p.u.				
V_{sh} (p.u.)	**θ_{sh}**	**I_{sh} (p.u.)**	**V_{86} (p.u.)**	**P_{DC} (p.u.)**
0.9576	−14.28°	**0.25**	0.9825	0.0181

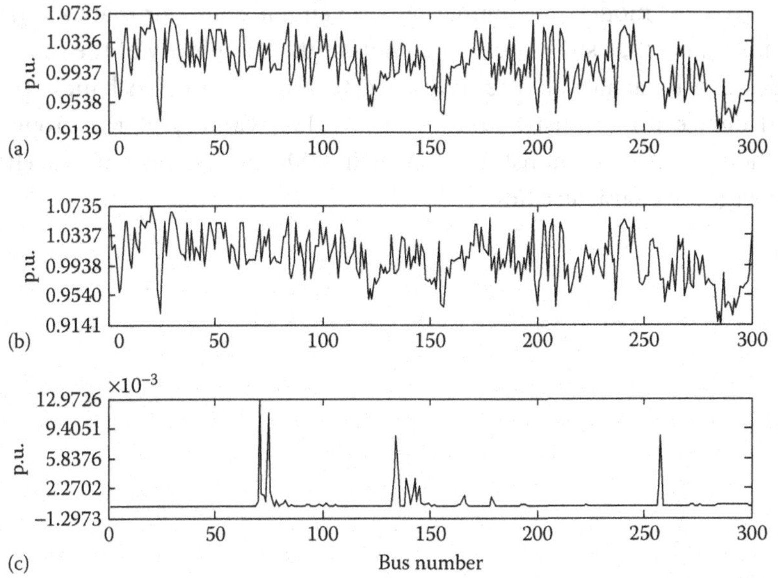

FIGURE 6.20 Bus voltage profile corresponding to the case study of Table 6.26. (a) Bus voltage magnitude without GUPFC; (b) bus voltage magnitude with GUPFC; (c) voltage magnitude difference.

it is again observed that in the presence of GUPFC, the bus voltage profile shows little change.

Case 16: In this case, simultaneous limit violations of all five quantities, that is, the dc link power transfer (P_{DC2}^{Lim}) and the injected voltage (V_{se2}^{Lim}) of the second series converter (in line 2), the line current (I_{se1}^{Lim}) and the bus voltage on the line side V_{m1} of the first series converter (in line 1), and the shunt converter current (I_{sh}^{Lim}), have been considered. For enforcing these limits, both P_{LINE} and Q_{LINE} of line 1 as well as line 2 and the SE bus voltage control objective (V_{86}) are relaxed. The results are presented in Table 6.26.

The bus voltage profile for the sixteenth case study is shown in Figure 6.20. From this figure, it is again observed that in the presence of GUPFC, the bus voltage profile shows little change.

6.8 SUMMARY

In this chapter, a Newton power flow model of the GUPFC has been developed. The proposed method transforms an existing n-bus power system installed with a GUPFC having p series converters and a shunt converter into an equivalent $(n + p + 1)$-bus system without any GUPFC. Consequently, existing power flow and Jacobian codes can be reused in the proposed model, in conjunction with simple codes for matrix extraction. As a result, a substantial reduction in the complexity of the software codes can be achieved. The developed technique can also handle practical device limit constraints of the GUPFC. Validity of the proposed method has been demonstrated on IEEE 300-bus systems with excellent convergence characteristics.

CHAPTER 7

Newton Power Flow Model of the Static Compensator

7.1 INTRODUCTION

The static compensator (STATCOM) has been the earliest among all the voltage-sourced converter-based flexible AC transmission system (FACTS) controllers. It is also the one that is installed in maximum numbers by most utilities worldwide. The first STATCOM, rated ±100 MVA, was commissioned at the Sullivan Substation by the Tennessee Valley Authority in 1995.

For maximum utilization of STATCOMs in power system planning, operation, and control, power flow solution of the network containing them is a fundamental requirement. Some excellent research works carried out in the literature are presented in [44–48] for developing efficient power-flow algorithms for the STATCOM. In all these works, it is observed that the voltage source(s) representing the shunt converter(s) of the STATCOM(s) contribute(s) new terms to (1) the expressions for the power injections at the concerned buses, (2) the real power of the STATCOM(s), and (3) the associated Jacobian blocks of (1) and (2) above. These new terms increase the complexity of software codes manifold.

To reduce the complexities of the software codes for incorporating a STATCOM in an existing Newton power flow algorithm, in this chapter, a novel modeling approach [84,88] is proposed. By this modeling approach, an existing power system installed with a STATCOM is transformed to an equivalent augmented network without any STATCOM. This results in a substantial reduction in the programming complexity because of the following reasons:

1. The power injections for the buses concerned can be computed in the proposed model using existing power flow codes, as it does not contain any STATCOM.
2. In the proposed model, existing power flow codes can be used to compute the active power flow of the STATCOM itself, which equals the bus active power injection at an additional power-flow bus.
3. Only two Jacobian subblocks need to be evaluated in the proposed model. Both of these can be evaluated using existing Jacobian codes directly.

This substantially reduces the complexity of software codes for modeling a STATCOM in an existing Newton–Raphson algorithm.

Moreover, it is also observed that none of the previously published works on the power flow modeling of a STATCOM directly explores the feasibility of decoupling. In this chapter, subsequent to the development of the proposed Newton power flow model, a decoupled power flow model of a STATCOM is proposed [84,88]. The subblocks of the system Jacobian matrix are rendered constant matrices with all elements known *a priori*. This results in a drastic reduction of the programming complexity. The computational time per iteration is also substantially reduced.

Furthermore, the proposed Newton–Raphson and decoupled models can also handle multiple control functions of the STATCOMs such as control of bus voltage, line active power, and line reactive power, similar to other models already reported in the literature. These models can also account for various device limit constraints of the STATCOM. The developed model is described in Sections 7.2 through 7.5.

7.2 STATCOM MODEL FOR NEWTON POWER FLOW ANALYSIS

Figure 7.1 shows an n-bus power system network in which a STATCOM is connected to bus j through a coupling transformer. The equivalent circuit of Figure 7.1 is shown in Figure 7.2, in which the STATCOM is represented

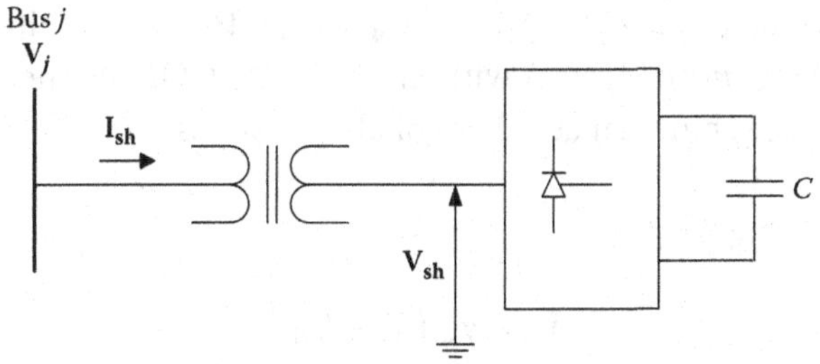

FIGURE 7.1 A STATCOM connected to bus j of an existing n-bus system.

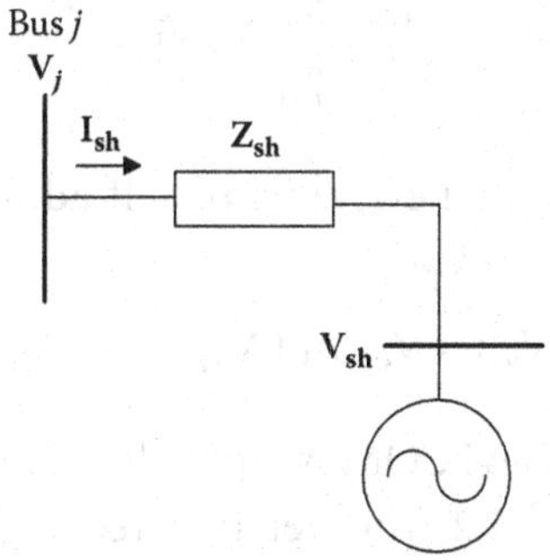

FIGURE 7.2 Equivalent circuit of STATCOM-incorporated power system network.

by a voltage source $\mathbf{V}_{sh}$, connected to node j via coupling transformer impedance $\mathbf{Z}_{sh}$. In Figure 7.1, the current flowing from bus j to the voltage source $\mathbf{V}_{sh}$ representing the STATCOM shunt converter (through the coupling transformer) is $\mathbf{I}_{sh}$.

Now, let us define

$$\mathbf{Z}_{sh} = R_{sh} + jX_{sh}, \; \mathbf{y}_{sh} = \frac{\mathbf{1}}{\mathbf{Z}_{sh}} \tag{7.1}$$

Again, from Figure 7.2, the net injected current at bus j is

$$\mathbf{I}_j = \mathbf{Y}_{jj}^{old}\mathbf{V}_j + \sum_{k=1,k\neq j}^{n} \mathbf{Y}_{jk}\mathbf{V}_k + \mathbf{I}_{sh} \tag{7.2}$$

In the above equation, $\mathbf{Y}_{jj}^{\text{old}} = \sum_{k=1,k\neq j}^{n} \mathbf{y}_{jk} + \mathbf{y}_{j0}$ is the self-admittance of bus j for the original n-bus system without any STATCOM connected and $\mathbf{y}_{j0}$ accounts for the shunt capacitances of all transmission lines connected to bus j.

Also, from Figure 7.2,

$$\mathbf{I}_{\text{sh}} = \mathbf{y}_{\text{sh}}\left(\mathbf{V}_j - \mathbf{V}_{\text{sh}}\right) \tag{7.3}$$

From Equations 7.2 and 7.3, the net injected current at bus j with STATCOM can be written as

$$\mathbf{I}_j = \sum_{k=1}^{n+1} \mathbf{Y}_{jk}\mathbf{V}_k \tag{7.4}$$

where $\mathbf{Y}_{ij} = \mathbf{Y}_{jj}^{\text{old}} + \mathbf{y}_{\text{sh}}$ is the new value of self-admittance for the jth bus with STATCOM and

$$\mathbf{V}_{n+1} = \mathbf{V}_{\text{sh}} \text{ and } \mathbf{Y}_{j(n+1)} = -\mathbf{y}_{\text{sh}} \tag{7.5}$$

Thus, the effect of the STATCOM is equivalent to an additional $(n + 1)$th bus. Now, from Figure 7.2, the net injected current at this fictitious $(n + 1)$th bus equals the current flowing into the transmission system from this bus and is

$$\mathbf{I}_{n+1} = -\mathbf{I}_{\text{sh}} = \mathbf{y}_{\text{sh}}\mathbf{V}_{\text{sh}} - \mathbf{y}_{\text{sh}}\mathbf{V}_j = \sum_{k=1}^{n+1} \mathbf{Y}_{(n+1)k}\mathbf{V}_k \tag{7.6}$$

with

$$\mathbf{Y}_{(n+1)(n+1)} = \mathbf{y}_{\text{sh}}, \mathbf{Y}_{(n+1)j} = -\mathbf{y}_{\text{sh}} \text{ as } \mathbf{V}_{n+1} = \mathbf{V}_{\text{sh}} \tag{7.7}$$

In general, in the proposed model, an existing n-bus system installed with p STATCOMs is transformed into an $(n + p)$ bus system without any STATCOM, and the expression for the net injected current at any bus g (including the STATCOM buses) of the network would be

$$\mathbf{I}_g = \sum_{k=1}^{n+p} \mathbf{Y}_{gk}\mathbf{V}_k \quad 1 \leq g \leq (n+p) \tag{7.8}$$

Also, the mth $(1 \leq m \leq p)$ STATCOM connected to bus h is transformed to the $(n + m)$th power-flow bus, and the net injected current at this mth fictitious bus is given by

$$\mathbf{I}_{n+m} = \sum_{k=1}^{n+p} \mathbf{Y}_{(n+m)k} \mathbf{V}_k \tag{7.9}$$

with

$$\mathbf{Y}_{(n+m)(n+m)} = \mathbf{y}_{\text{sh}\,m}, \mathbf{Y}_{(n+m)h} = -\mathbf{y}_{\text{sh}\,m} \text{ as } \mathbf{V}_{n+m} = \mathbf{V}_{\text{sh}\,m} \tag{7.10}$$

7.3 POWER FLOW EQUATIONS IN THE PROPOSED STATCOM MODEL

With existing STATCOM models, the net active power injection at any bus a $[1 \le a \le (n+p)]$ in an n-bus system installed with p STATCOMs can be written (using Equation 7.2) as

$$P_a = \sum_{k=1}^{n} V_a V_k Y_{ak} \cos(\theta_a - \theta_k - \varphi_{yak}), \quad a \le n \tag{7.11}$$

if no STATCOM is connected to bus a;

$$P_a = \sum_{k=1}^{n} V_a V_k Y_{ak} \cos\left(\theta_a - \theta_k - \varphi_{yak}\right) - V_a V_{\text{sh}\,b} y_{\text{sh}\,b} \cos\left(\theta_a - \theta_{\text{sh}\,b} - \varphi_{y\text{sh}\,b}\right) \tag{7.12}$$

if STATCOM b is connected to bus a.

Thus, from the above equation, it is observed that with existing STATCOM models, an additional term due to the contribution from the voltage source representing the shunt converter of the STATCOM is present in the expression of the bus active power injection(s). This causes the existing power flow codes to be modified. This is also true for the bus reactive power injection(s).

Now, in the proposed model, it is observed from Equations 7.8 and 7.9 that the net injected current at bus h (h can be any sending end [SE], receiving end [RE], or any of the p fictitious power-flow buses representing the STATCOM voltage sources) can be written as

$$\mathbf{I}_h = \sum_{q=1}^{n+p} \mathbf{V}_{hq} \mathbf{V}_q \quad 1 \le h \le (n+p) \tag{7.13}$$

Thus, in the proposed model, the active power injection equation at any SE or RE bus a ($a \le n + p$) can be written using Equation 7.13 as

$$P_a = \sum_{q=1}^{n+p} V_a V_q Y_{aq} \cos\left(\theta_a - \theta_q - \varphi_{yaq}\right) \quad a \le (n+p) \tag{7.14}$$

Similarly, the expression for the reactive power at any bus a (SE or RE) can be written as

$$Q_a = \sum_{q=1}^{n+p} V_a V_q Y_{aq} \sin\left(\theta_a - \theta_q - \varphi_{yaq}\right) \quad a \le (n+p) \tag{7.15}$$

From Equations 7.14 and 7.15, it can be observed that both the active and reactive power injections at any bus can be computed using existing power flow codes.

Now, for an n-bus system installed with p STATCOMs, the expression for the real power delivered by any STATCOM g $(1 \le g \le p)$ connected to any bus h $(1 \le h \le n)$ can be written using Equation 7.3 as

$$\mathbf{P}_{\mathrm{STAT}\,g} = \mathrm{Re}\left[\mathbf{V}_{\mathbf{sh}\,g}\left(-\mathbf{I}^{*}_{\mathbf{sh}\,g}\right)\right] \tag{7.16}$$

$$\mathbf{P}_{\mathrm{STAT}\,g} = V^2_{\mathrm{sh}\,g}\, y_{\mathrm{sh}\,g} \cos\varphi_{y\mathrm{sh}\,g} - V_{\mathrm{sh}\,g} V_h\, y_{\mathrm{sh}\,g} \cos\left(\theta_{\mathrm{sh}\,g} - \theta_h - \varphi_{y\mathrm{sh}\,g}\right) \tag{7.17}$$

From the above equation, it is observed that the STATCOM real power expression cannot be computed using existing power flow codes. It will require fresh codes for its implementation.

However, in the proposed model, STATCOM g is transformed to $(n+g)$th power-flow bus, and hence, using Equation 7.9 or 7.13,

$$\mathbf{P}_{\mathrm{STAT}\,g} = \mathrm{Re}\left[\mathbf{V}_{\mathbf{sh}\,g}\left(-\mathbf{I}^{*}_{\mathbf{sh}\,g}\right)\right] = \mathrm{Re}\left(\mathbf{V}_{n+g}\mathbf{I}^{*}_{n+g}\right) = \mathrm{Re}\left(\mathbf{S}_{n+g}\right) = P_{n+g} \tag{7.18}$$

Hence, the real power delivered by STATCOM g equals the active power injection at the $(n+g)$th power-flow bus, which can be computed using the existing power flow codes (from Equation 7.14).

7.4 IMPLEMENTATION IN NEWTON POWER FLOW ANALYSIS

If the number of voltage controlled buses is $(m-1)$, the power-flow problem for an n-bus system installed with p STATCOMs can be formulated as follows.

Solve θ, $\mathbf{V}$, $\theta_{\mathbf{sh}}$, and $\mathbf{V}_{\mathbf{sh}}$

Specified $\mathbf{P}$, $\mathbf{Q}$, $\mathbf{P}_{\mathrm{STAT}}$, and $\mathbf{V}_{\mathbf{BUS}}$

where:

$$\boldsymbol{\theta}=\left[\theta_2\ldots\theta_n\right]^{\mathrm{T}},\mathbf{V}=\left[V_{m+1}\ldots V_n\right]^{\mathrm{T}} \tag{7.19}$$

$$\boldsymbol{\theta}_{\mathbf{sh}}=\left[\theta_{\mathrm{sh}1}\ldots\theta_{\mathrm{sh}\,p}\right]^{\mathrm{T}},\mathbf{V}_{\mathbf{sh}}=\left[V_{\mathrm{sh}1}\ldots V_{\mathrm{sh}\,p}\right]^{\mathrm{T}} \tag{7.20}$$

$$\mathbf{P}=\left[P_2\ldots P_n\right]^{\mathrm{T}},\mathbf{Q}=\left[Q_{m+1}\ldots Q_n\right]^{\mathrm{T}} \tag{7.21}$$

$$\mathbf{P}_{\mathbf{STAT}}=\left[P_{\mathrm{STAT}1}\ldots P_{\mathrm{STAT}\,p}\right]^{\mathrm{T}},\mathbf{V}_{\mathbf{BUS}}=\left[V_{\mathrm{BUS}1}\ldots V_{\mathrm{BUS}\,p}\right]^{\mathrm{T}} \tag{7.22}$$

In Equations 7.19 through 7.22, $\mathbf{P}_{\mathbf{STAT}}$ is the vector of specified real powers of the p STATCOMs, which have been taken to be zero following [44]. $\mathbf{V}_{\mathbf{BUS}}$ represents the vector of specified voltage magnitudes of p buses at which STATCOMs are connected. Also, in Equations 7.19 through 7.22, it is assumed that without any loss of generality, there are m generators connected at the first m buses of the system with bus 1 being the slack bus. Thus, the basic power-flow equation for the Newton power flow solution is represented as

$$\begin{bmatrix} \frac{\partial \mathbf{P}}{\partial \boldsymbol{\theta}} & \frac{\partial \mathbf{P}}{\partial \mathbf{V}} & \frac{\partial \mathbf{P}}{\partial \boldsymbol{\theta}_{\mathbf{sh}}} & \frac{\partial \mathbf{P}}{\partial \mathbf{V}_{\mathbf{sh}}} \\ \frac{\partial \mathbf{Q}}{\partial \boldsymbol{\theta}} & \frac{\partial \mathbf{Q}}{\partial \mathbf{V}} & \frac{\partial \mathbf{Q}}{\partial \boldsymbol{\theta}_{\mathbf{sh}}} & \frac{\partial \mathbf{Q}}{\partial \mathbf{V}_{\mathbf{sh}}} \\ \frac{\partial \mathbf{P}_{\mathbf{STAT}}}{\partial \boldsymbol{\theta}} & \frac{\partial \mathbf{P}_{\mathbf{STAT}}}{\partial \mathbf{V}} & \frac{\partial \mathbf{P}_{\mathbf{STAT}}}{\partial \boldsymbol{\theta}_{\mathbf{sh}}} & \frac{\partial \mathbf{P}_{\mathbf{STAT}}}{\partial \mathbf{V}_{\mathbf{sh}}} \\ \frac{\partial \mathbf{V}_{\mathbf{BUS}}}{\partial \boldsymbol{\theta}} & \frac{\partial \mathbf{V}_{\mathbf{BUS}}}{\partial \mathbf{V}} & \frac{\partial \mathbf{V}_{\mathbf{BUS}}}{\partial \boldsymbol{\theta}_{\mathbf{sh}}} & \frac{\partial \mathbf{V}_{\mathbf{BUS}}}{\partial \mathbf{V}_{\mathbf{sh}}} \end{bmatrix} \begin{bmatrix} \Delta\boldsymbol{\theta} \\ \Delta\mathbf{V} \\ \Delta\boldsymbol{\theta}_{\mathbf{sh}} \\ \Delta\mathbf{V}_{\mathbf{sh}} \end{bmatrix} = \begin{bmatrix} \Delta\mathbf{P} \\ \Delta\mathbf{Q} \\ \Delta\mathbf{P}_{\mathbf{STAT}} \\ \Delta\mathbf{V}_{\mathbf{BUS}} \end{bmatrix} \tag{7.23}$$

or

$$\begin{bmatrix} \multicolumn{2}{c}{\mathbf{J}_{\mathbf{old}}} & \begin{matrix}\mathbf{J}_1\\ \mathbf{J}_3\end{matrix} & \begin{matrix}\mathbf{J}_2\\ \mathbf{J}_4\end{matrix} \\ \mathbf{J}_5 & \mathbf{J}_6 & \mathbf{J}_7 & \mathbf{J}_8 \\ 0 & \mathbf{J}_9 & 0 & 0 \end{bmatrix} \begin{bmatrix} \Delta\boldsymbol{\theta} \\ \Delta\mathbf{V} \\ \Delta\boldsymbol{\theta}_{\mathbf{sh}} \\ \Delta\mathbf{V}_{\mathbf{sh}} \end{bmatrix} = \begin{bmatrix} \Delta\mathbf{P} \\ \Delta\mathbf{Q} \\ \Delta\mathbf{P}_{\mathbf{STAT}} \\ \Delta\mathbf{V}_{\mathbf{BUS}} \end{bmatrix} \tag{7.24}$$

In the above equation, $\mathbf{J}_{\mathbf{old}}$ is the conventional power-flow Jacobian subblock corresponding to the angle and voltage magnitude variables of the n-buses. The other Jacobian submatrices can be identified easily from Equation 7.24.

The elements of the subblock **J9** are either unity or zero. However, it is to be noted that when the control objective(s) other than bus voltage control is (are) adopted, the elements of all the null matrices, including **J9** (in Equation 7.24), need to be modified accordingly.

Now, in the proposed model, there are $(n + p)$ buses, including p fictitious power-flow buses. Thus, the quantities to be solved for power-flow are $\boldsymbol{\theta}^{\text{new}}$ and $\mathbf{V}^{\text{new}}$, where

$$\boldsymbol{\theta}^{\text{new}} = \left[\theta_2 \ldots \theta_{n+p}\right]^{\mathrm{T}} \text{ and } \mathbf{V}^{\text{new}} = \left[V_{m+1} \ldots V_{n+p}\right]^{\mathrm{T}} \tag{7.25}$$

Thus, Equation 7.24 is transformed in the proposed model as

$$\begin{bmatrix} \mathbf{JX1} \\ \hline \mathbf{JX2} \\ \hline \mathbf{0} \mid \mathbf{JX3} \end{bmatrix} \begin{bmatrix} \Delta\boldsymbol{\theta}^{\text{new}} \\ \hline \Delta\mathbf{V}^{\text{new}} \end{bmatrix} \begin{bmatrix} \Delta\mathbf{PQ} \\ \hline \Delta\mathbf{P}_{\mathbf{STAT}} \\ \hline \Delta\mathbf{V}_{\mathbf{BUS}} \end{bmatrix} \tag{7.26}$$

where:

$$\Delta\mathbf{PQ} = \left[\Delta\mathbf{P}^{\mathrm{T}} \;\; \Delta\mathbf{Q}^{\mathrm{T}}\right]^{\mathrm{T}} \tag{7.27}$$

Also, **JX1**, **JX2**, and **JX3** are identified easily from Equation 7.26. The elements of the matrix **JX3** are either unity or zero, depending on whether an element of the vector $\mathbf{V}_{\mathbf{BUS}}$ is also an element of the vector $\mathbf{V}^{\text{new}}$ or not. Thus, **JX3** is a constant matrix known *a priori* and does not need to be computed.

Now, it can be shown that in Equation 7.26, the matrices **JX1** and **JX2** can be computed using existing Jacobian codes. The justification of the above statement is shown as follows.

Let us first define for the proposed $(n + p)$ bus system:

$$\mathbf{P}^{\text{new}} = \left[P_2 \ldots P_{n+p}\right]^{T} \tag{7.28}$$

Subsequently, a new Jacobian matrix is computed as

$$\mathbf{J}^{\text{new}} = \begin{bmatrix} \dfrac{\partial \mathbf{P}^{\text{new}}}{\partial \boldsymbol{\theta}^{\text{new}}} & \dfrac{\partial \mathbf{P}^{\text{new}}}{\partial \mathbf{V}^{\text{new}}} \\ \hline \dfrac{\partial \mathbf{Q}}{\partial \boldsymbol{\theta}^{\text{new}}} & \dfrac{\partial \mathbf{Q}}{\partial \mathbf{V}^{\text{new}}} \end{bmatrix} = \begin{bmatrix} \mathbf{J}^{\mathrm{A}} \\ \hline \mathbf{J}^{\mathrm{B}} \end{bmatrix} \tag{7.29}$$

From Equations 7.14 and 7.15 of the proposed model, it can be observed that the expressions for P_i $(2 \le i \le n + p)$ and Q_i $(m + 1 \le i \le n)$ in $\mathbf{P}^{\text{new}}$ and $\mathbf{Q}$, respectively (of Equation 7.29), can be computed using existing power flow codes. Hence, the matrix $\mathbf{J}^{\text{new}}$ can be computed with existing codes for calculating the Jacobian matrix.

From Equations 7.18 and 7.22,

$$\mathbf{P}_{\text{STAT}} = \left[\mathbf{P}_{\text{STAT}1} \ldots \mathbf{P}_{\text{STAT}p}\right]^{\text{T}} = \left[\text{P}_{n+1} \ldots \text{P}_{n+p}\right]^{\text{T}} \tag{7.30}$$

Also, from Equations 7.28 and 7.30,

$$\mathbf{P}^{\text{new}} = \left[P_2 \ldots P_{n+p}\right]^{\text{T}} = \left[\mathbf{P}^{\text{T}} \ \mathbf{P}_{\text{STAT}}^{\text{T}}\right]^{\text{T}} \tag{7.31}$$

$$\frac{\partial \mathbf{P}^{\text{new}}}{\partial \boldsymbol{\theta}^{\text{new}}} = \left[\begin{array}{c} \dfrac{\partial \mathbf{P}}{\partial \boldsymbol{\theta}^{\text{new}}} \\ \hdashline \dfrac{\partial \mathbf{P}_{\text{STAT}}}{\partial \boldsymbol{\theta}^{\text{new}}} \end{array}\right] \tag{7.32}$$

From the above equation, it is observed that the subblock $\partial \mathbf{P}/\partial \boldsymbol{\theta}^{\text{new}}$ is contained within $\partial \mathbf{P}^{\text{new}}/\partial \boldsymbol{\theta}^{\text{new}}$. In a similar way, it can be shown that $\partial \mathbf{P}/\partial \mathbf{V}^{\text{new}}$ is contained within $\partial \mathbf{P}^{\text{new}}/\partial \mathbf{V}^{\text{new}}$. Hence, once the matrix $\mathbf{J}^{\text{new}}$ is computed (using existing codes), the matrix

$$\mathbf{JX1} = \left[\begin{array}{c} \partial \mathbf{P}/\partial \boldsymbol{\theta}^{\text{new}} \ \partial \mathbf{P}/\partial \mathbf{V}^{\text{new}} \\ \hdashline \mathbf{J}^{\text{B}} \end{array}\right]$$

can be very easily extracted from the matrix $\mathbf{J}^{\text{new}}$ using simple matrix extraction codes only. Hence, no fresh codes need to be written for computing **JX1**.

Again, from Equation 7.32, it is observed that the subblock $\partial \mathbf{P}_{\text{STAT}}/\partial \boldsymbol{\theta}^{\text{new}}$ is contained within $\partial \mathbf{P}^{\text{new}}/\partial \boldsymbol{\theta}^{\text{new}}$. In a similar way, it can be shown that $\partial \mathbf{P}_{\text{STAT}}/\partial \mathbf{V}^{\text{new}}$ is contained within $\partial \mathbf{P}^{\text{new}}/\partial \mathbf{V}^{\text{new}}$. Thus, the matrix $\mathbf{JX2} = \left[\partial \mathbf{P}_{\text{STAT}}/\partial \boldsymbol{\theta}^{\text{new}} \ \partial \mathbf{P}_{\text{STAT}}/\partial \mathbf{V}^{\text{new}}\right]$ does not need to be computed—it can be formed from the matrix $\mathbf{J}^{\text{A}}$ of $\mathbf{J}^{\text{new}}$ by matrix extraction codes only, without any need of writing fresh codes. Hence, both **JX1** and **JX2** need not be computed and can be extracted from $\mathbf{J}^{\text{new}}$ (subsequent to its computation using existing Jacobian codes). This reduces the complexity of software codes substantially.

7.4.1 Application of Decoupling

If the power-flow equation for Newton power flow solution could be decoupled, the problem of increased complexity of software codes would be reduced drastically. Let us first explore the possibility of decoupling of Equation 7.26. It is important to note that decoupling, if feasible, could be applied to both Equations 7.24 and 7.26 alike, because the Jacobian of the latter is obtained by realigning the elements of the former. The values of the individual Jacobian elements in Equations 7.24 and 7.26 remain identical—only the codes for their formulation change. This is shown for a generalized nonzero element (in the ath row and the bth column) of one typical Jacobian subblock in Equations 7.24 and 7.26:

$$\begin{aligned} \mathbf{J}_2(a,b) &= \frac{\partial P_{a+1}}{\partial V_{\mathrm{sh}\,b}} = \frac{\partial P_{a+1}}{\partial V_{n+b}} \\ &= \frac{\partial}{\partial V_{n+b}}\left[\sum_{k=1}^{n+p} V_{a+1}V_kY_{(a+1)k}\cos\left(\theta_{a+1}-\theta_k-\varphi_{y(a+1)k}\right)\right] \\ &= V_{a+1}Y_{(a+1)(n+b)}\cos\left(\theta_{a+1}-\theta_{n+b}-\varphi_{y(a+1)(n+b)}\right) \\ &= -V_{a+1}y_{\mathrm{sh}\,b}\cos\left(\theta_{a+1}-\theta_{\mathrm{sh}\,b}-\varphi_{y\mathrm{sh}\,b}\right) \end{aligned} \quad (7.33)$$

because, from Equation 7.10, $\mathbf{Y}_{(a+1)(n+b)} = -\mathbf{y}_{\mathbf{sh}b}$.

Similarly, it can be shown that

$$\frac{\partial \mathbf{P}}{\partial \mathbf{V}^{\mathrm{new}}}(a, n+b-m) = -V_{a+1}y_{\mathrm{sh}\,b}\cos\left(\theta_{a+1}-\theta_{\mathrm{sh}\,b}-\varphi_{y\mathrm{sh}\,b}\right) \quad (7.34)$$

Thus, the element in the ath row and the bth column of the $\mathbf{J}_2$ subblock (Equation 7.24) is transformed to an element in the ath row and the $(n + b - m)$th column of the $\partial\mathbf{P}/\partial\mathbf{V}^{\mathrm{new}}$ subblock of matrix **JX1** (Equation 7.26). This can be shown to be true for elements of all the Jacobian subblocks of Equation 7.24. Therefore, any decoupling philosophy applicable to the Jacobian subblocks of Equation 7.24 would also be applicable to the corresponding Jacobian subblocks of Equation 7.26.

To demonstrate that decoupling is applicable to the relevant Jacobian subblocks of Equation 7.24, and hence also to Equation 7.26, the exact and approximate expressions for the generalized elements of the relevant subblocks of Equation 7.24 are shown in the second and third columns of Table A.1 in the Appendix, respectively. The assumptions made for obtaining the approximate expressions of the Jacobian elements are also highlighted in the third column of Table A.1.

7.4.2 Decoupled Power Flow Equations in the Proposed Model

Equation 7.26 can be rewritten as

$$\begin{bmatrix} \dfrac{\partial P}{\partial \theta^{new}} & \dfrac{\partial P}{\partial V^{new}} \\ \dfrac{\partial Q}{\partial \theta^{new}} & \dfrac{\partial Q}{\partial V^{new}} \\ \hline \dfrac{\partial P_{STAT}}{\partial \theta^{new}} & \dfrac{\partial P_{STAT}}{\partial V^{new}} \\ \hline \theta & JX3 \end{bmatrix} \begin{bmatrix} \Delta\theta^{new} \\ \hline \Delta V^{new} \end{bmatrix} = \begin{bmatrix} \Delta P \\ \Delta Q \\ \hline \Delta P_{STAT} \\ \hline \Delta V_{BUS} \end{bmatrix} \tag{7.35}$$

Applicability of decoupling renders the elimination of the subblocks $\partial P/\partial V^{new}$, $\partial Q/\partial \theta^{new}$, and $\partial P_{STAT}/\partial V^{new}$. Further, using Equation 7.31 in Equation 7.35, the form of the decoupled equations as obtained from Equation 7.35 is

$$\begin{bmatrix} \dfrac{\partial P}{\partial \theta^{new}} \\ \hline \dfrac{\partial P_{STAT}}{\partial \theta^{new}} \end{bmatrix} \begin{bmatrix} \Delta\theta^{new} \end{bmatrix} = \begin{bmatrix} \Delta P \\ \hline \Delta P_{STAT} \end{bmatrix} \text{ or } \begin{bmatrix} \dfrac{\partial P^{new}}{\partial \theta^{new}} \end{bmatrix} \begin{bmatrix} \Delta\theta^{new} \end{bmatrix} = \begin{bmatrix} \Delta P^{new} \end{bmatrix} \tag{7.36}$$

and

$$\begin{bmatrix} \dfrac{\partial Q}{\partial V^{new}} \\ \hline JX3 \end{bmatrix} \begin{bmatrix} \Delta V^{new} \end{bmatrix} = \begin{bmatrix} \Delta Q \\ \hline \Delta V_{BUS} \end{bmatrix} \tag{7.37}$$

In contrast to the existing STATCOM models, the subblocks $\partial P^{new}/\partial \theta^{new}$ and $\partial Q/\partial V^{new}$ (of Equations 7.36 and 7.37, respectively) can be computed using existing Jacobian codes in the proposed model. The exact and approximate expression of a generalized element of the subblock $\partial P^{new}/\partial \theta^{new}$ in

TABLE 7.1 Generalized Expression for Element of Any Arbitrary **P–θ** Jacobian Subblock in Proposed STATCOM Model

Generalized Expression for Element of the Jacobian Subblock $\partial \mathbf{P}^{new}/\partial \boldsymbol{\theta}^{new}$ (a, b ≤ n+p)	
Exact	**Approximation**
$\frac{\partial P_a}{\partial \theta_b} = V_a V_b Y_{ab} \sin(\theta_a - \theta_b - \varphi_{yab})$ if $b \neq a$	$\frac{\partial P_a}{\partial \theta_b} \approx -V_a V_b B_{ab}$ if $b \neq a$
$= -Q_a - V_a^2 Y_{aa} \sin \varphi_{yaa}$ if $b = a$	$(for\ \theta_a - \theta_b \approx 0, \lvert V_a \rvert \approx 1.0, \lvert V_b \rvert \approx 1.0)$
	$\approx -V_a^2 B_{aa}$ if $b = a$

the proposed model is shown in Table 7.1. From the second column of the table, it can be observed that decoupled power flow codes can be reused for solving Equation 7.36. In a similar way, it can be shown that decoupled power flow codes can also be reused for solving Equation 7.37.

Now, the first row corresponding to Equation 7.36 is

$$\frac{\partial P_2}{\partial \theta_2}\Delta\theta_2 + \frac{\partial P_2}{\partial \theta_3}\Delta\theta_3 + \cdots + \frac{\partial P_2}{\partial \theta_{n+p}}\Delta\theta_{n+p} = \Delta P_2 \tag{7.38}$$

After applying the simplifications shown in Table 7.1 to Equations 7.38, the equation becomes

$$-V_2^2 B_{22}\Delta\theta_2 - V_2 V_3 B_{23}\Delta\theta_3 - \cdots - V_2 V_{n+p} B_{2(n+p)}\Delta\theta_{n+p} = \Delta P_2$$

or

$$-V_2 B_{22}\Delta\theta_2 - V_3 B_{23}\Delta\theta_3 - \cdots - V_{n+p} B_{2(n+p)}\Delta\theta_{n+p} = \frac{\Delta P_2}{V_2}$$

or

$$\left[-B_{22} - B_{23} - \cdots - B_{2(n+p)}\right]\left[\Delta\boldsymbol{\theta}^{new}\right] = \frac{\Delta P_2}{V_2}$$

$$\left(\text{because} \lvert V_k \rvert \approx 1.0,\ \forall k,\ 2 \le k \le n+p\right)$$

Proceeding in a similar way, for all the rows and combining them, we get

$$[\mathbf{B}']\left[\Delta\boldsymbol{\theta}^{new}\right] = \left[\Delta\hat{\mathbf{P}}^{new}\right] \tag{7.39}$$

Similarly, by again applying the simplifications (not shown) corresponding to elements of $\partial \mathbf{Q}/\partial \mathbf{V}^{new}$ to Equation 7.37 and combining them,

$$\left[\frac{B''}{C}\right]\left[\Delta \mathbf{V}^{\text{new}}\right]=\left[\frac{\Delta\hat{\mathbf{Q}}}{\Delta \mathbf{V}_{\text{BUS}}}\right] \tag{7.40}$$

where:

$$\Delta\hat{\mathbf{P}}^{\text{new}}=\left[\frac{\Delta P_2}{V_2}\cdots\frac{\Delta P_{n+p}}{V_{n+p}}\right]^{\text{T}},\quad \Delta\hat{\mathbf{Q}}=\left[\frac{\Delta Q_{m+1}}{V_{m+1}}\cdots\frac{\Delta Q_n}{V_n}\right]^{\text{T}}$$

$$B'=\begin{bmatrix} -B_{22} & -B_{23} & \cdots & -B_{2(n+p)} \\ -B_{32} & -B_{33} & \cdots & -B_{3(n+p)} \\ \vdots & \vdots & \cdots & \vdots \\ -B_{(n+p)2} & -B_{(n+p)3} & \cdots & -B_{(n+p)(n+p)} \end{bmatrix}$$

$$B''=\begin{bmatrix} -B_{(m+1)(m+1)} & -B_{(m+1)(m+2)} & \cdots & -B_{(m+1)(n+p)} \\ -B_{(m+2)(m+1)} & -B_{(m+2)(m+2)} & \cdots & -B_{(m+2)(n+p)} \\ \vdots & \vdots & \cdots & \vdots \\ -B_{n(m+1)} & -B_{n(m+2)} & \cdots & -B_{n(n+p)} \end{bmatrix}$$

The matrix C equals the matrix **JX3** of Equation 7.26, with elements being either unity or zero. It is also to be noted that in forming the matrices B' and B'' for the $(n+p)$-bus system, the existing decoupled power flow codes for the original n-bus system can be reused without any difficulty. It is also important to note that while forming the matrix B', all transmission line series resistances, shunt capacitors, and reactors are ignored, and the taps of all off-nominal transformers are set to unity. Similarly, angle-shifting effects of all phase shifters are made to omit from the matrix B'' [27].

7.5 ACCOMMODATION OF STATCOM DEVICE LIMIT CONSTRAINTS

In this chapter, two major device limit constraints of the STATCOM have been considered, which are as follows [44,45,70,71]:

1. Injected (shunt) converter voltage magnitude $V_{\text{sh}}^{\text{Lim}}$
2. Shunt converter current $I_{\text{sh}}^{\text{Lim}}$

The device limit constraints have been accommodated by the principle that whenever a particular constraint limit is violated, it is kept at its specified limit, whereas a control objective is relaxed. Mathematically, this signifies the replacement of the Jacobian elements pertaining to the control objective by those of the constraint violated. The control strategies to incorporate the above two limits are detailed as follows.

1. In this case,

$$V_{sh} = V_{sh}^{Lim} \tag{7.41}$$

If V_{sh}^{Lim} is violated, V_{n+1} is fixed at the limit and the bus voltage control objective is relaxed for the bus to which the STATCOM is connected. The corresponding relaxed bus voltage mismatch is replaced by $\Delta V_{n+1} = V_{sh}^{Lim} - V_{n+1}$. The Jacobian elements are changed accordingly. For each violation of V_{sh}^{Lim}, the corresponding row of the matrix **JX3** is again rendered constant with all elements known *a priori*—all of them equal zero except the entry pertaining to V_{sh}, which is unity. Other elements of the matrix **JX3** are also zero or unity, as demonstrated in Section 7.4. The constancy of **JX3** is maintained for both the Newton power flow and the decoupled power flow solutions, alike.

2. If I_{sh}^{Lim} is violated, I_{n+1} is fixed at the limit and again the bus voltage control objective is relaxed for the STATCOM connected bus (say, bus j). Thus, the corresponding mismatch is replaced by $\Delta I_{n+1} = I_{sh}^{Lim} - I_{n+1}$, where (from Equations 7.3 and 7.6)

$$I_{n+1} = I_{sh} = y_{sh}\left[V_{sh}^2 + V_j^2 - 2V_{sh}V_j\cos\left(\theta_j - \theta_{sh}\right)\right]^{1/2} \tag{7.42}$$

The Jacobian elements are changed accordingly.

It is important to note that the elements of the matrix **JX3** (in Equation 7.25) corresponding to Newton power flow can be computed (not shown) from the partial derivatives of Equation 7.42 with respect to the relevant variables (θ_j, θ_{sh}, V_j, and V_{sh}). However, for decoupled power flow, Equations 7.39 and 7.40 have to be solved. It is to be noted that for enforcing the shunt converter current limit, I_{sh} is decoupled from θ, as shown in the text that follows.

From Equation 7.42,

$$\frac{\partial I_{sh}}{\partial \theta_{sh}} = -\frac{-y_{sh}^2 V_{sh} V_j \sin\left(\theta_j - \theta_{sh}\right)}{I_{sh}} \approx 0$$

$$\text{for}\,|V_j| \approx 1.0,\ |V_{sh}| \approx 1.0,\ \theta_j - \theta_{sh} \approx 0 \text{ and } I_{sh} \neq 0$$

Similarly,

$$\frac{\partial I_{sh}}{\partial \theta_j} = -\frac{-y_{sh}^2 V_{sh} V_j \sin(\theta_j - \theta_{sh})}{I_{sh}} \approx 0$$

Subsequently, the elements of the matrix **C** (same as **JX3**) in Equation 7.40 are computed as follows:

$$\frac{\partial I_{sh}}{\partial V_{sh}} = \frac{2y_{sh}\left[V_{sh} - V_j \cos(\theta_j - \theta_{sh})\right]}{2\left[V_j^2 + V_{sh}^2 - 2V_{sh} V_j \cos(\theta_j - \theta_{sh})\right]^{1/2}} \approx \frac{y_{sh}\left[V_{sh} - V_j\right]}{\left[V_j - V_{sh}\right]} \approx -y_{sh}$$

Similarly,

$$\frac{\partial I_{sh}}{\partial V_j} \approx \frac{y_{sh}\left[V_j - V_{sh}\right]}{\left[V_j - V_{sh}\right]} \approx y_{sh}$$

7.6 SELECTION OF INITIAL CONDITIONS

In this chapter, the initial condition for the shunt voltage source was chosen as $V_{sh}^0 \angle \theta_{sh}^0 = 1.0\angle 0^0$ p.u. following suggestions of [60]. However, during the case studies, it was observed that this initial condition renders the shunt converter current magnitude I_{sh} zero. Consequently, all the Jacobians of I_{sh} (obtained from the partial derivatives of Equation 7.42), which contain I_{sh} term in the denominator, assume indeterminate values. This is shown in the Appendix. Modifying this initial condition to $V_{sh}^0 \angle \theta_{sh}^0 = 1.0\angle -(\pi/9)$ solves this problem, without any observed detrimental effect on the convergence.

7.7 CASE STUDIES AND RESULTS

To validate its feasibility, the proposed modeling strategy was applied to compute the power flow solution of the IEEE 118- and 300-bus systems using the algorithms of both the Newton and decoupled power flow algorithms. In each test system, multiple STATCOMs with two different control functions were included, and studies have been carried out for (1) STATCOMs without any device limit constraints and (2) practical STATCOMs with device limit constraints. In all the case studies, a convergence tolerance of 10^{-12} p.u. has been chosen. Although a large number of case studies confirmed the validity of the model, only a few sets of representative results are presented in Sections 7.7.1 and 7.7.2. In all the subsequent tables, the symbol NI denotes the number of iterations taken by the algorithm for convergence.

7.7.1 Studies of STATCOMs without Any Device Limit Constraints

7.7.1.1 Case I: Control of Bus Voltage

7.7.1.1.1 IEEE 118-Bus System In this system, three STATCOMs have been considered on the buses 69 (STATCOM-1), 85 (STATCOM-2), and 116 (STATCOM-3). The control references chosen are as follows: V_{28}^{SP} = 1.0 p.u. (STATCOM-1), V_{52}^{SP} = 1.0 p.u. (STATCOM-2), and V_{115}^{SP} = 1.0 p.u. (STATCOM-3). The results are shown in Table 7.2. The converged final values of the control objectives (bus voltages in this case) are shown in bold cases.

The bus voltage profiles without and with STATCOMs are shown in Figure 7.3a and b, respectively. Further, the difference between the bus voltage magnitudes with and without STATCOMs is shown in Figure 7.3c. Because the converged power flow solution is identical in both the Newton–Raphson and decoupled power flow algorithms, in this case study and all subsequent case studies, only a single bus voltage profile is plotted corresponding to each case study. From Figure 7.3, it is also observed that in the presence of STATCOMs, the bus voltage profile does

TABLE 7.2 Study of IEEE 118-Bus System with STATCOM (No Device Limit Constraints)

Solution of Base Case Power Flow (by Newton Power Flow Method without Any STATCOM)			
Parameter	**Bus 28**	**Bus 52**	**Bus 115**
V_{BUS} (p.u.)	0.9616	0.956	0.9605
NI	5		

Unconstrained Solution of STATCOM Quantities						
	Newton Power Flow Method			**Decoupled Power Flow Method**		
Quantity	**STATCOM-1**	**STATCOM-2**	**STATCOM-3**	**STATCOM-1**	**STATCOM-2**	**STATCOM-3**
V_{sh} (p.u.)	1.0748	1.0474	1.108	1.0748	1.0474	1.108
θ_{sh}	−18.96°	−18.1°	−18.1°	−18.96°	−18.1°	−18.1°
I_{sh} (p.u.)	0.7478	0.474	1.0803	0.7478	0.474	1.0803
V_{BUS} (p.u.)	**1.0**	**1.0**	**1.0**	**1.0**	**1.0**	**1.0**
NI	5			9		

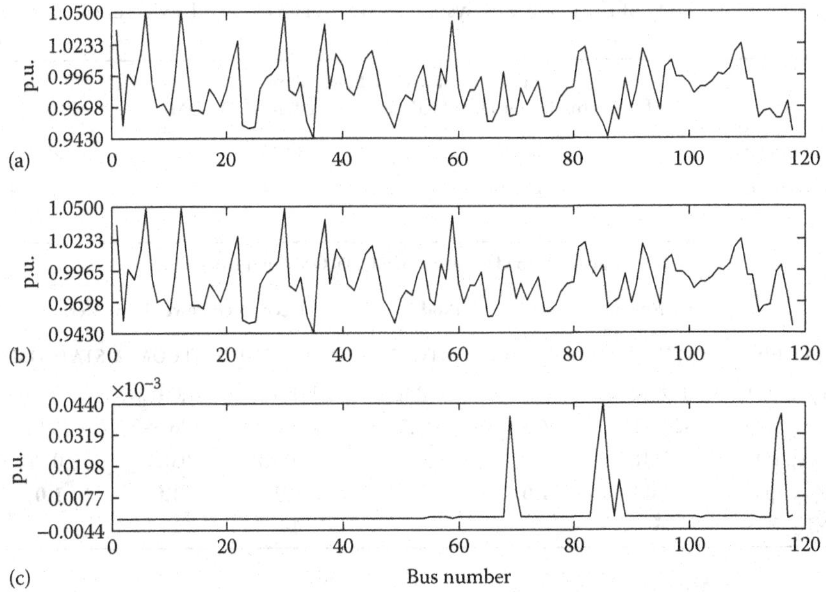

FIGURE 7.3 Bus voltage profile corresponding to the case study of Table 7.2. (a) Bus voltage magnitude without STATCOM; (b) bus voltage magnitude with STATCOM; (c) voltage magnitude difference.

not change much apart from the buses to which the three STATCOMs are connected.

7.7.1.1.2 IEEE 300-Bus System In this system too, three STATCOMs have been considered on the buses 279 (STATCOM-1), 286 (STATCOM-2), and 291 (STATCOM-3). It is important to note that these three buses have been particularly chosen due to the fact that during the base case power flow (power flow without any STATCOM), the magnitude of the voltages at these three buses were observed to be the minimum among all other buses. The control references chosen are as follows: $V_{275}^{SP} = 1.0$ p.u. (STATCOM-1), $V_{282}^{SP} = 1.0$ p.u. (STATCOM-2), and $V_{287}^{SP} = 1.0$ p.u. (STATCOM-3). The results are shown in Table 7.3. The converged final values of the control objectives (bus voltages in this case) are again shown in bold cases.

The bus voltage profile for this case is shown in Figure 7.4. From this figure, it is observed that in the presence of the STATCOMs, the bus

TABLE 7.3 Study of IEEE 300-Bus System with STATCOM (No Device Limit Constraints)

Solution of Base Case Power Flow (by Newton Power Flow Method without Any STATCOM)						
Parameter	**Bus 275**		**Bus 282**			**Bus 287**
V_{BUS} (p.u.)	0.9654		0.9139			0.9243
NI	6					
Unconstrained Solution of STATCOM Quantities						
	Newton Power Flow Method			**Decoupled Power Flow Method**		
Quantity	**STATCOM-1**	**STATCOM-2**	**STATCOM-3**	**STATCOM-1**	**STATCOM-2**	**STATCOM-3**
V_{sh} (p.u.)	1.0012	1.0012	1.0016	1.0012	1.0012	1.0016
θ_{sh}	−23.55°	−26.39°	−25.52°	−23.55°	−26.39°	−25.52°
I_{sh} (p.u.)	0.0118	0.012	0.0164	0.0118	0.012	0.0164
V_{BUS} (p.u.)	**1.0**	**1.0**	**1.0**	**1.0**	**1.0**	**1.0**
NI	6			14		

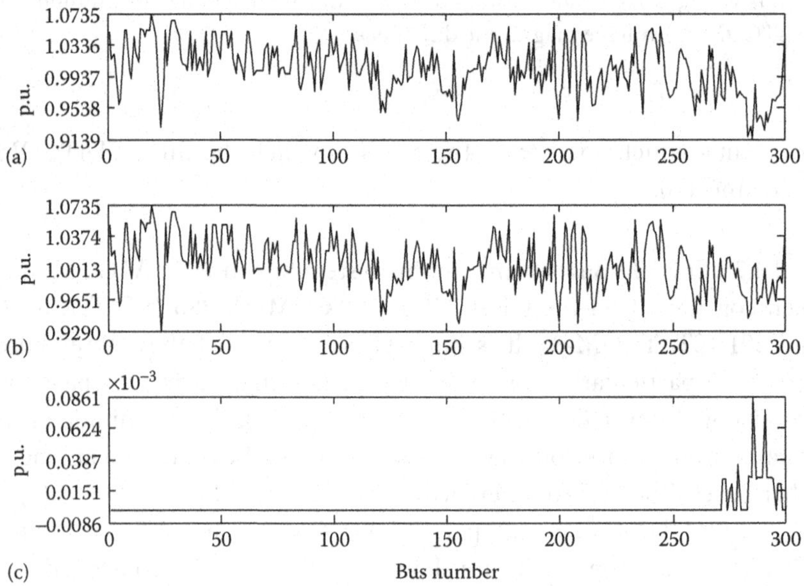

FIGURE 7.4 Bus voltage profile corresponding to the case study of Table 7.3. (a) Bus voltage magnitude without STATCOM; (b) bus voltage magnitude with STATCOM; (c) voltage magnitude difference.

voltage profile does not change much. Peaks of the bus voltage magnitude difference are again observed near the buses to which the three STATCOMs are connected.

7.7.1.2 Case II: Control of Reactive Power Delivered by the STATCOM

7.7.1.2.1 IEEE 118-Bus System For this study also, the same three STATCOMs as described earlier have been considered. The earlier control objective of bus voltage control is now changed to the control of the reactive power delivered by the STATCOMs. The control references chosen are as follows: Q_{STAT} = 0.05 p.u. (STATCOM-1), Q_{STAT2} = 0.04 p.u. (STATCOM-2), and Q_{STAT3} = 0.06 p.u. (STATCOM-3). The results are shown in Table 7.4. The converged final values of the control objectives (the STATCOM reactive power delivered in this case) are shown in bold cases.

The bus voltage profile for this case is shown in Figure 7.5. From this figure, it is again observed that in the presence of the STATCOMs, the bus voltage profile hardly changes except the buses to which the three STATCOMs are connected.

7.7.1.2.2 IEEE 300-Bus System Again, the same three STATCOMs as described earlier have been considered. The control objective is now changed to the control of the reactive power delivered by the STATCOMs. The control references chosen are as follows: Q_{STAT} = 0.01 p.u. (STATCOM-1),

TABLE 7.4 Second Study of IEEE 118-Bus System with STATCOM (No Device Limit Constraints)

	Unconstrained Solution of STATCOM Quantities					
	Newton Power Flow Method			**Decoupled Power Flow Method**		
Quantity	**STATCOM-1**	**STATCOM-2**	**STATCOM-3**	**STATCOM-1**	**STATCOM-2**	**STATCOM-3**
V_{sh} (p.u.)	0.9694	0.964	0.969	0.9694	0.964	0.969
θ_{sh}	–18.27°	–17.38°	–17.39°	–18.27°	–17.38°	–17.39°
I_{sh} (p.u.)	0.0516	0.0415	0.0619	0.0516	0.0415	0.0619
V_{BUS} (p.u.)	0.9642	0.9598	0.9628	0.9642	0.9598	0.9628
Q_{STAT}	**0.05**	**0.04**	**0.06**	**0.05**	**0.04**	**0.06**
NI	6			9		

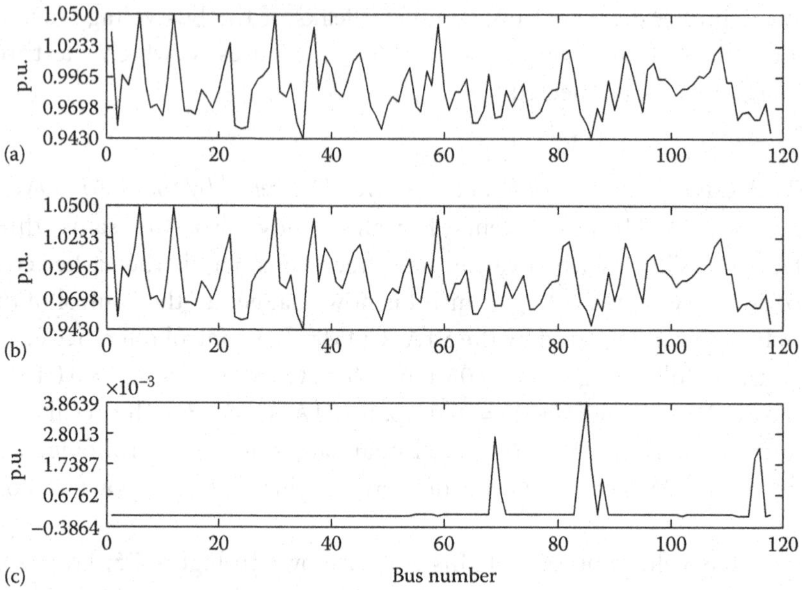

FIGURE 7.5 Bus voltage profile corresponding to the case study of Table 7.4. (a) Bus voltage magnitude without STATCOM; (b) bus voltage magnitude with STATCOM; (c) voltage magnitude difference.

Q_{STAT2} = 0.01 p.u. (STATCOM-2), and Q_{STAT3} = 0.015 p.u. (STATCOM-3). The results are shown in Table 7.5. The converged final values of the control objectives (the STATCOM reactive power delivered in this case) are again shown in bold cases. The bus voltage profile for this case is shown in Figure 7.6. From this figure, it is observed that in the presence of the

TABLE 7.5 Second Study of IEEE 300-Bus System with STATCOM (No Device Limit Constraints)

Unconstrained Solution of STATCOM Quantities						
	Newton Power Flow Method			Decoupled Power Flow Method		
Quantity	**STATCOM-1**	**STATCOM-2**	**STATCOM-3**	**STATCOM-1**	**STATCOM-2**	**STATCOM-3**
V_{sh} (p.u.)	0.9959	0.9885	0.9946	0.9959	0.9885	0.9946
θ_{sh}	−23.51°	−26.4°	−25.53°	−23.51°	−26.4°	−25.53°
I_{sh} (p.u.)	0.01	0.0101	0.0151	0.01	0.0101	0.0151
V_{BUS} (p.u.)	0.9949	0.9875	0.993	0.9949	0.9875	0.993
Q_{STAT} (p.u.)	**0.01**	**0.01**	**0.015**	**0.01**	**0.01**	**0.015**
NI	6			14		

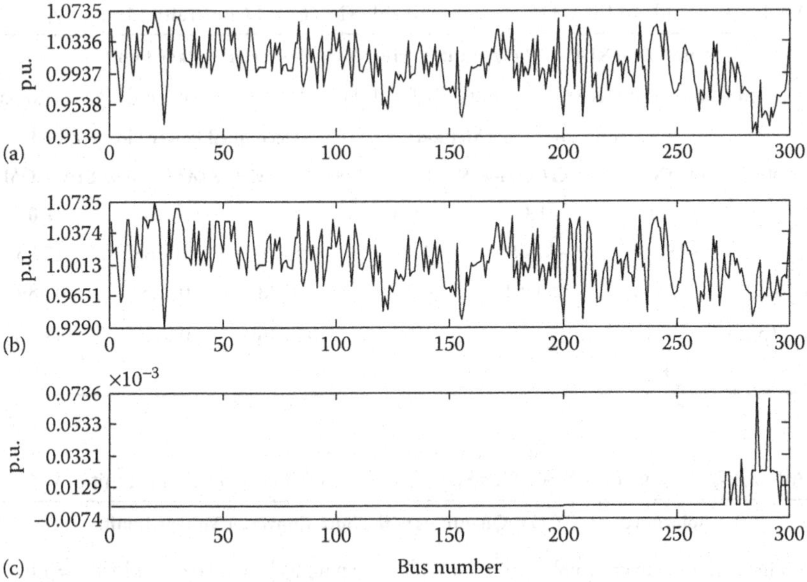

FIGURE 7.6 Bus voltage profile corresponding to the case study of Table 7.5. (a) Bus voltage magnitude without STATCOM; (b) bus voltage magnitude with STATCOM; (c) voltage magnitude difference.

STATCOMs, the bus voltage profile does not change much except the buses to which the three STATCOMs are connected.

7.7.2 Studies of STATCOMs with Device Limit Constraints

In these case studies, various device limit constraints are considered for both the 118- and 300-bus test systems. As already mentioned in Section 7.5, two major device constraint limits have been considered. In all these case studies, the STATCOM control references (prior to enforcement of the specified limit constraints) are kept identical to those in the corresponding case studies pertaining to the bus voltage control (presented in Section 7.7.1). Although a large number of case studies were carried out for implementation of these limit constraints, only a few sets of representative results are presented.

Case 1: In this case, the violations of only the injected voltage of the shunt converter V_{sh}^{Lim} have been studied. To accommodate this violation, the bus voltage control objective has been relaxed. The results for the

TABLE 7.6 Study of IEEE 118-Bus System with STATCOM Constraint Limit of V_{sh}^{Lim}

Solution of STATCOM Quantities with Shunt Converter Voltage Limits

Specified shunt converter voltage limits: $V_{sh1}^{Lim} = 1.0$ p.u., $V_{sh2}^{Lim} = 1.0$ p.u., and $V_{sh3}^{Lim} = 1.0$ p.u.

	Newton Power Flow Method			**Decoupled Power Flow Method**		
Quantity	**STATCOM-1**	**STATCOM-2**	**STATCOM-3**	**STATCOM-1**	**STATCOM-2**	**STATCOM-3**
V_{sh} (p.u.)	**1.0**	**1.0**	**1.0**	**1.0**	**1.0**	**1.0**
θ_{sh}	−18.43°	−17.68°	−17.53°	−18.43°	−17.68°	−17.53°
I_{sh} (p.u.)	0.2538	0.2281	0.289	0.2538	0.2281	0.289
V_{BUS} (p.u.)	0.9746	0.9772	0.9711	0.9746	0.9772	0.9711
NI	8			13		

TABLE 7.7 Study of IEEE 300-Bus System with STATCOM Constraint Limit of V_{sh}^{Lim}

Solution of STATCOM Quantities with Shunt Converter Voltage Limits

Specified shunt converter voltage limits: $V_{sh1}^{Lim} = 1.0$ p.u., $V_{sh2}^{Lim} = 1.0$ p.u., and $V_{sh3}^{Lim} = 1.0$ p.u.

	Newton Power Flow Method			**Decoupled Power Flow Method**		
Quantity	**STATCOM-1**	**STATCOM-2**	**STATCOM-3**	**STATCOM-1**	**STATCOM-2**	**STATCOM-3**
V_{sh} (p.u.)	**1.0**	**1.0**	**1.0**	**1.0**	**1.0**	**1.0**
θ_{sh}	−23.54°	−26.4°	−25.52°	−23.54°	−26.4°	−25.52°
I_{sh} (p.u.)	0.0114	0.0119	0.016	0.0114	0.0119	0.016
V_{BUS} (p.u.)	0.9989	0.9988	0.9984	0.9989	0.9988	0.9984
NI	7			20		

118- and 300 bus systems are given in Tables 7.6 and 7.7, respectively. The converged final values of the shunt converter voltage limit constraints are shown in bold cases.

The bus voltage profiles for these cases are shown in Figures 7.7 and 7.8, respectively. From these figures, it is again observed that in the presence of the STATCOMs, the bus voltage profiles do not change much except at the buses at which the STATCOMs are connected in these two systems.

Case 2: In this case, the violations of only the shunt converter current limits I_{sh}^{Lim} have been considered. To accommodate this, the UPFC SE bus voltage control objective has been relaxed. The results for the 118- and 300-bus systems are given in Tables 7.8 and 7.9, respectively. The converged final values of the shunt converter current limit constraints are shown in bold cases.

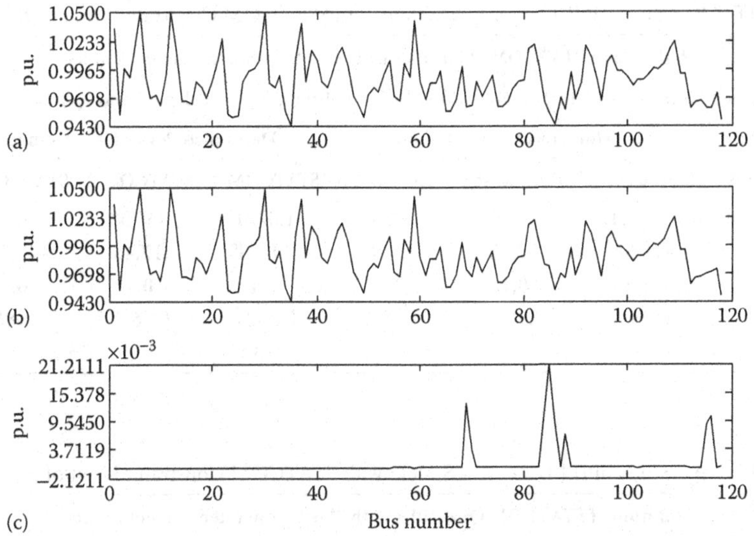

FIGURE 7.7 Bus voltage profile corresponding to the case study of Table 7.6. (a) Bus voltage magnitude without STATCOM; (b) bus voltage magnitude with STATCOM; (c) voltage magnitude difference.

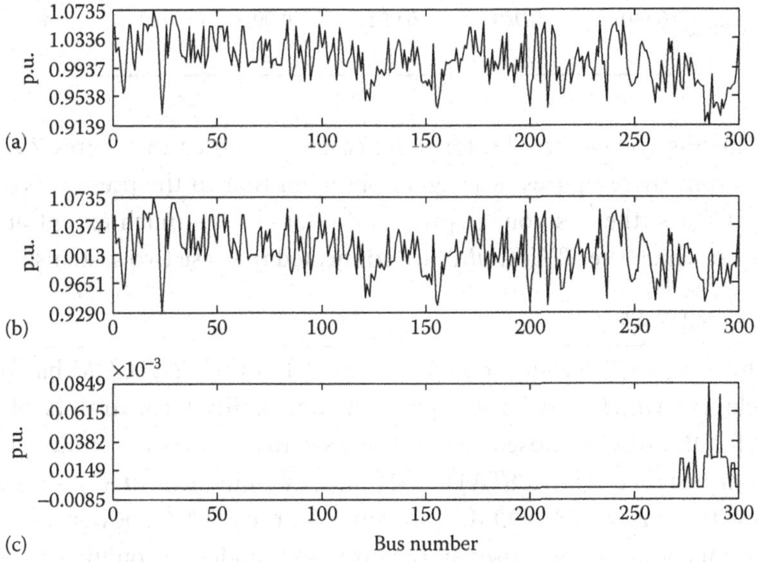

FIGURE 7.8 Bus voltage profile corresponding to the case study of Table 7.7. (a) Bus voltage magnitude without STATCOM; (b) bus voltage magnitude with STATCOM; (c) voltage magnitude difference.

TABLE 7.8 Study of IEEE 118-Bus System with STATCOM Constraint Limit of I_{sh}^{Lim}

Solution of STATCOM Quantities with Shunt Converter Current Limits

Specified shunt converter current limits: $I_{sh1}^{Lim} = 0.4$ p.u., $I_{sh2}^{Lim} = 0.1$ p.u., and $I_{sh3}^{Lim} = 0.7$ p.u.

	Newton Power Flow Method			Decoupled Power Flow Method		
Quantity	STATCOM-1	STATCOM-2	STATCOM-3	STATCOM-1	STATCOM-2	STATCOM-3
V_{sh} (p.u.)	1.0221	0.9753	1.0561	1.0221	0.9753	1.0561
θ_{sh}	−18.58°	−17.48°	−17.78°	−18.58°	−17.48°	−17.78°
I_{sh} (p.u.)	**0.4**	**0.1**	**0.7**	**0.4**	**0.1**	**0.7**
V_{BUS} (p.u.)	0.9821	0.9653	0.9861	0.9821	0.9653	0.9861
NI	7			13		

TABLE 7.9 Study of IEEE 300-Bus System with STATCOM Constraint Limit of I_{sh}^{Lim}

Solution of STATCOM Quantities with Shunt Converter Current Limits

Specified current limits: $I_{sh1}^{Lim} = 0.009$ p.u., $I_{sh2}^{Lim} = 0.005$ p.u., and $I_{sh3}^{Lim} = 0.006$ p.u.

	Newton Power Flow Method			Decoupled Power Flow Method		
Quantity	STATCOM-1	STATCOM-2	STATCOM-3	STATCOM-1	STATCOM-2	STATCOM-3
V_{sh} (p.u.)	0.9928	0.9497	0.9532	0.9928	0.9497	0.9532
θ_{sh}	−23.49°	−26.5°	−25.59°	−23.49°	−26.5°	−25.59°
I_{sh} (p.u.)	**0.009**	**0.005**	**0.006**	**0.009**	**0.005**	**0.006**
V_{BUS} (p.u.)	0.9919	0.9492	0.9526	0.9919	0.9492	0.9526
NI	7			20		

The bus voltage profiles for these cases are shown in Figures 7.9 and 7.10. From these figures, it is again observed that in the presence of the STATCOMs, the bus voltage profiles do not change much except at the buses at which the STATCOMs are connected in these two systems.

7.8 SUMMARY

In this chapter, a Newton power flow model of the STATCOM has been developed, which can handle practical device limit constraints of the STATCOM. The proposed method transforms an existing n-bus power system installed with p STATCOMs into an equivalent $(n + p)$-bus system without any STATCOM. Consequently, the existing power flow and Jacobian codes can be reused in the proposed model, in conjunction with simple codes for matrix extraction. As a result, a substantial reduction in the complexity of the software codes can be achieved. Subsequent application of decoupling techniques renders the Jacobians constant matrices,

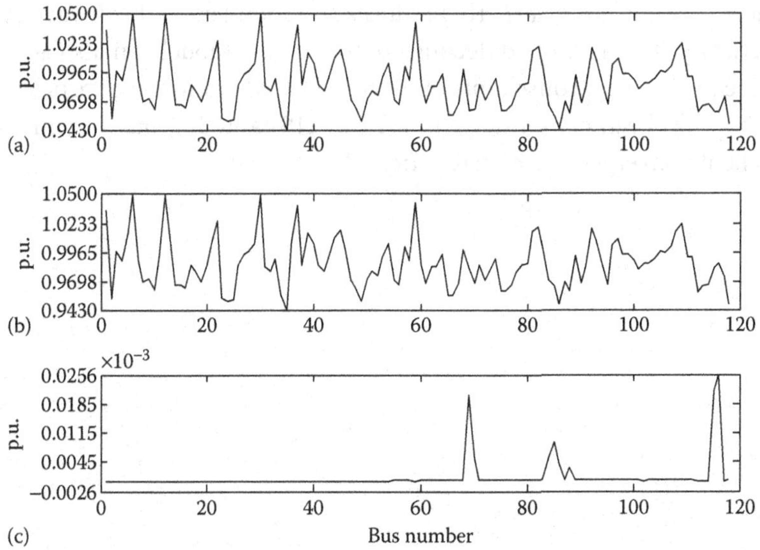

FIGURE 7.9 Bus voltage profile corresponding to the case study of Table 7.8. (a) Bus voltage magnitude without STATCOM; (b) bus voltage magnitude with STATCOM; (c) voltage magnitude difference.

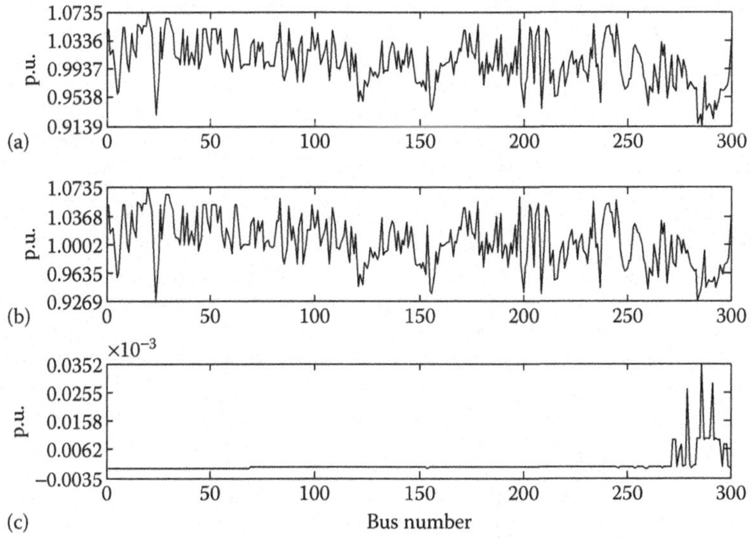

FIGURE 7.10 Bus voltage profile corresponding to the case study of Table 7.9. (a) Bus voltage magnitude without STATCOM; (b) bus voltage magnitude with STATCOM; (c) voltage magnitude difference.

which are known *a priori*. This reduces the complexity of software codes drastically. The developed decoupled power flow model can also account for the device limit constraints of the STATCOM. Validity of the proposed method has been demonstrated on IEEE 118- and 300-bus systems with excellent convergence characteristics.

CHAPTER 8

Newton Power Flow Modeling of Voltage-Sourced Converter Based HVDC Systems

8.1 INTRODUCTION

Over the years, ever-increasing electricity demand has necessitated the requirement of increased transmission capacities by power utilities worldwide. This has been made possible by the development of voltage-sourced converter (VSC)-based HVDC technology with PWM control technique, employing IGBTs and GTOs. VSC-based HVDC transmission systems facilitate the interconnection of asynchronous AC grids along with integration of renewable energy sources such as offshore wind farms. PWM-VSC-based HVDC transmission leads to fast and independent control of both active and reactive powers along with reduction in size and cost of harmonic filters. Moreover, they are immune to the problems of commutation failure [89–90].

Although most of these VSC-HVDC interconnections are two-terminal, their *modus operandi* can also be extended to multiterminal HVDC (MTDC) systems. Unlike a two-terminal HVDC interconnection, an MTDC system is more versatile and better capable to utilize the economic and technical advantages of the VSC-HVDC technology. It is also better suited if futuristic integration of renewable energy sources are planned.

For MTDC operation, one of the terminals is considered as a slack bus at which the DC voltage is specified. Depending on the control specifications adopted, a converter may be termed *master/slave* or *primary/secondary* [90]. Although a master or primary converter operates in the voltage control mode, a slave or secondary converter operates in the PQ or PV control mode.

Further, depending on the locations of the converters, an MTDC system can have two different configurations: back-to-back (BTB) or point-to-point (PTP). In a BTB scheme, all the converters are closely located. However, in a PTP scheme, converters exist at different locations, with the DC terminals interconnected by DC overhead lines or cables. Most of the current HVDC installations are connected in the PTP configuration.

Now, for planning, operation, and control of a power system with multiterminal VSC-HVDC (M-VSC-HVDC) links, power flow solution of the network incorporating them is required. Thus, the development of suitable power flow models of M-VSC-HVDC systems is a fundamental requirement. VSC-HVDC power flow models can follow unified or sequential methods. Unlike unified methods where the AC and DC system equations are simultaneously solved, in sequential methods, they are solved sequentially. Some excellent research works in the area of Newton power flow modeling of VSC-HVDC systems are reported in [89–92]. A comprehensive Newton power flow model of a two-terminal VSC-HVDC system is presented in [89]. An M-VSC-HVDC power flow model applicable for both the BTB and PTP configurations is reported in [90]. A two-terminal VSC-HVDC model suitable for optimal power flow is reported in [91]. A steady-state VSC MTDC model, including DC grids with arbitrary topologies, is reported in [92].

However, most of the above models do not address exclusively the treatment of the converter modulation indices as unknowns. In this chapter, the Newton power flow modeling of an M-VSC-HVDC system is discussed which is based on the unified method [93]. The converter modulation indices appear as unknowns in this model, which is developed from first principles.

8.2 MODELING OF THE PTP VSC-HVDC

Because the PTP configuration is a more generalized one than the BTB, for this analysis, an MTDC system that follows a PTP topology is considered. Figure 8.1 shows an n-bus AC power system network incorporating a VSC-based MTDC system connected in the PTP configuration.

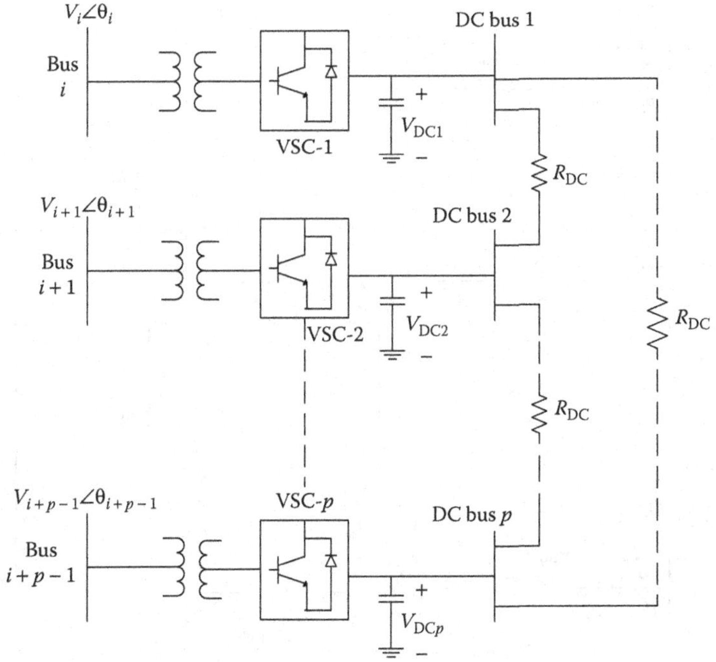

FIGURE 8.1 Schematic diagram of a p terminal PTP M-VSC-HVDC system.

The VSC-MTDC comprises p converters that are connected to p AC buses through their respective coupling transformers. Without loss of generality, it is assumed that the p VSC converters are connected to AC buses i, $(i + 1)$, and so on, up to bus $(i + p - 1)$. The equivalent circuit of Figure 8.1 is shown in Figure 8.2.

In this figure, the VSCs are represented by p fundamental frequency, positive sequence voltage sources. The gth $(1 \leq g \leq p)$ voltage source $\mathbf{V}_{\mathbf{sh}g}$ (not shown) is connected to AC bus $(i + g - 1)$ through the leakage impedance $\mathbf{Z}_{\mathbf{sh}g} = R_{\text{sh}\,g} + jX_{\text{sh}\,g}$ of the gth coupling transformer.

Now, let $\mathbf{y}_{\mathbf{sh}g} = 1/\mathbf{Z}_{\mathbf{sh}g}$. Then, from Figure 8.2, the current through the gth $(1 \leq g \leq p)$ coupling transformer can be written as

$$\mathbf{I}_{\mathbf{sh}g} = \mathbf{y}_{\mathbf{sh}g}(\mathbf{V}_{\mathbf{sh}g} - \mathbf{V}_{i+g-1}) \tag{8.1}$$

In the above equation, $\mathbf{V}_{\mathbf{sh}g}$ is the voltage phasor representing the fundamental frequency, positive sequence output of the gth VSC and is given by $\mathbf{V}_{\mathbf{sh}g} = V_{\text{sh}\,g}\angle\theta_{\text{sh}\,g} = m_g c V_{\text{DC}\,g}\angle\theta_{\text{sh}\,g}$, where m_g is the modulation index of the gth $(1 \leq g \leq p)$ converter and c is a constant that depends on the type of converter. Also, $V_{\text{DC}g}$ is the DC side voltage of the gth converter, which

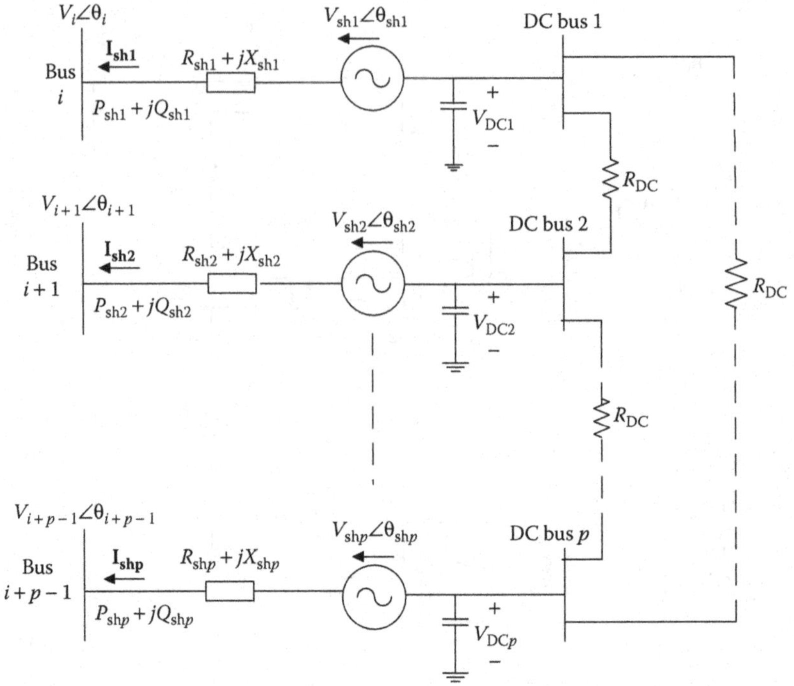

FIGURE 8.2 Equivalent circuit of the p terminal PTP M-VSC-HVDC system.

is connected to the AC terminal bus $(i+g-1)$ whose voltage is represented by the phasor $\mathbf{V}_{i+g-1} = V_{i+g-1}\angle\theta_{i+g-1}$.

Hence, from Figure 8.2, the net current injection at the AC bus $(i+g-1)$ connected to the gth $(1 \le g \le p)$ converter can be written as

$$\mathbf{I}_{i+g-1} = \left(\mathbf{Y}^{\text{old}}_{(i+g-1)(i+g-1)} + \mathbf{y}_{\text{sh}\,g}\right)\mathbf{V}_{i+g-1} + \sum_{k=1,\,k\neq i+g-1}^{n} \mathbf{Y}_{(i+g-1)k}\mathbf{V}_k - \mathbf{y}_{\text{sh}\,g}\mathbf{V}_{\text{sh}\,g} \quad (8.2)$$

where $\mathbf{Y}^{\text{old}}_{(i+g-1)(i+g-1)} = \mathbf{y}_{(i+g-1)0} + \sum_{k=1,\,k\neq i+g-1}^{n} \mathbf{y}_{(i+g-1)k}$ is the self-admittance of bus $(i+g-1)$ for the original n-bus AC system without any converter and $\mathbf{y}_{(i+g-1)0}$ accounts for the shunt capacitances of all the transmission lines connected to bus $(i+g-1)$.

Substituting Equation 8.1 in Equation 8.2 and manipulating, we get

$$\mathbf{I}_{i+g-1} = \sum_{k=1}^{n} \mathbf{Y}_{(i+g-1)k}\mathbf{V}_k - \mathbf{y}_{\text{sh}\,g}\mathbf{V}_{\text{sh}\,g} \quad (8.3)$$

where $\mathbf{Y}_{(i+g-1)(i+g-1)} = \mathbf{Y}^{\text{old}}_{(i+g-1)(i+g-1)} + \mathbf{y}_{\text{sh}\,g}$ is the new value of self-admittance for the bus $(i+g-1)$ with the gth VSC connected.

Thus, with the gth ($1 \leq g \leq p$) VSC connected, the net injected active power at the corresponding AC terminal bus can be written as

$$\begin{aligned} P_{i+g-1} &= \mathrm{Re}\left(\mathbf{V}_{i+g-1}\mathbf{I}^{*}{}_{i+g-1}\right) \\ &= \sum_{k=1}^{n} V_{i+g-1}V_kY_{(i+g-1)k}\cos\left(\theta_{i+g-1}-\theta_k-\varphi_{(i+g-1)k}\right) \\ &\quad - m_g cV_{\mathrm{DC}\,g}V_{i+g-1}y_{\mathrm{sh}\,g}\cos\left(\theta_{i+g-1}-\theta_{\mathrm{sh}\,g}-\varphi_{\mathrm{sh}\,g}\right) \end{aligned} \tag{8.4}$$

because $V_{\mathrm{sh}\,g} = m_g cV_{\mathrm{DC}\,g}$, as already discussed.

In a similar manner, the net injected reactive power at bus $(i + g - 1)$ can be written as

$$\begin{aligned} Q_{i+g-1} &= \sum_{k=1}^{n} V_{i+g-1}V_kY_{(i+g-1)k}\sin\left(\theta_{i+g-1}-\theta_k-\varphi_{(i+g-1)k}\right) \\ &\quad - m_g cV_{\mathrm{DC}\,g}V_{i+g-1}y_{\mathrm{sh}\,g}\sin\left(\theta_{i+g-1}-\theta_{\mathrm{sh}\,g}-\varphi_{\mathrm{sh}\,g}\right) \end{aligned} \tag{8.5}$$

Also from Figure 8.2, the active and reactive power flows at the terminal end of the line connecting the gth VSC to AC bus $(i + g - 1)$ can be written as

$$\begin{aligned} P_{\mathrm{sh}\,g} &= \mathrm{Re}\left(\mathbf{V}_{i+g-1}\mathbf{I}^{*}{}_{\mathrm{sh}\,g}\right) = m_g cV_{\mathrm{DC}\,g}V_{i+g-1}y_{\mathrm{sh}\,g}\cos\left(\theta_{i+g-1}-\theta_{\mathrm{sh}\,g}-\varphi_{\mathrm{sh}\,g}\right) \\ &\quad - V_{i+g-1}^2 y_{\mathrm{sh}\,g}\cos\varphi_{\mathrm{sh}\,g} \end{aligned} \tag{8.6}$$

$$\begin{aligned} Q_{\mathrm{sh}\,g} &= \mathrm{Im}\left(\mathbf{V}_{i+g-1}\mathbf{I}^{*}{}_{\mathrm{sh}\,g}\right) \\ &= m_g cV_{\mathrm{DC}\,g}V_{i+g-1}y_{\mathrm{sh}\,g}\sin\left(\theta_{i+g-1}-\theta_{\mathrm{sh}\,g}-\varphi_{\mathrm{sh}\,g}\right) + V_{i+g-1}^2 y_{\mathrm{sh}\,g}\sin\varphi_{\mathrm{sh}\,g} \end{aligned} \tag{8.7}$$

Again, from Figure 8.2, in the p terminal DC system, the net current injection at the hth ($1 \leq h \leq p$) DC bus, that is, at the DC terminal of the hth VSC, is given as

$$I_{\mathrm{DC}h} = \sum_{j=1}^{p} Y_{\mathrm{DC}\,hj}V_{\mathrm{DC}\,j} \tag{8.8}$$

where $Y_{\mathrm{DC}hj} = 1/R_{\mathrm{DC}hj}$, $R_{\mathrm{DC}hj}$ being the DC link resistance between DC buses h and j. Now, from Figure 8.2, for the gth ($1 \leq g \leq p$) converter, it can be observed that

$$\mathrm{Re}\left(\mathbf{V}_{\mathrm{sh}\,g}\mathbf{I}^{*}_{\mathrm{sh}\,g}\right)=V_{\mathrm{DC}\,g}\left(-I_{\mathrm{DC}\,g}\right)=-\sum_{j=1}^{p}V_{\mathrm{DC}\,g}V_{\mathrm{DC}\,j}Y_{\mathrm{DC}\,gj} \tag{8.9}$$

Substituting Equation 8.1 and manipulating, we get

$$\left(m_g c V_{\mathrm{DC}\,g}\right)^2 y_{\mathrm{sh}\,g}\cos\varphi_{\mathrm{sh}\,g}-m_g c V_{\mathrm{DC}\,g}V_{i+g-1}y_{\mathrm{sh}\,g}\cos\left(\theta_{\mathrm{sh}\,g}-\theta_{i+g-1}-\varphi_{\mathrm{sh}\,g}\right)$$
$$+\sum_{j=1}^{p}V_{\mathrm{DC}\,g}V_{\mathrm{DC}\,j}Y_{\mathrm{DC}\,gj}=0$$

or

$$f_g=0 \tag{8.10}$$

where $1\le g\le p$. Thus, p independent equations are obtained.

Now, for the VSC-HVDC system with p converters, there will be one master or primary converter and $(p-1)$ slave or secondary converters. The master converter is used to control the voltage magnitude of its AC terminal bus, whereas the slave converters operate in the PQ or PV control modes. In the PQ control mode, the slave converters control the active and reactive power flows P_{sh} and Q_{sh} (as given by Equations 8.6 and 8.7, respectively) at the terminal end of the lines connecting the converters to their respective AC terminal buses. Again, without loss of generality, if the qth $(1\le q\le p)$ converter is chosen to be the master converter, the additional equations obtained for the line active and reactive powers of the slave converters can be expressed as

$$P^{\mathrm{SP}}_{\mathrm{sh}\,g}-P^{\mathrm{cal}}_{\mathrm{sh}\,g}=0 \tag{8.11}$$

$$Q^{\mathrm{SP}}_{\mathrm{sh}\,g}-Q^{\mathrm{cal}}_{\mathrm{sh}\,g}=0 \tag{8.12}$$

$$\forall g,\ 1\le g\le p,\ g\ne q$$

In the above equations, $P^{\mathrm{SP}}_{\mathrm{sh}\,g}$ and $Q^{\mathrm{SP}}_{\mathrm{sh}\,g}$ are the specified active and reactive powers, respectively, in the line connecting the gth $(1\le g\le p,\ g\ne q)$ converter to its AC terminal bus $(i+g-1)$, whereas $P^{\mathrm{cal}}_{\mathrm{sh}\,g}$ and $Q^{\mathrm{cal}}_{\mathrm{sh}\,g}$ are their calculated values which can be obtained using Equations 8.6 and 8.7, respectively. Thus, for the $(p-1)$ lines corresponding to the slave converters, we get $(2p-2)$ independent equations.

Now, the master converter controls the voltage magnitude at its AC terminal bus. As discussed earlier in this section, the qth $(1 \leq q \leq p)$ converter is chosen to be the master converter. Then, for the AC terminal bus corresponding to any arbitrary converter, say the gth $(1 \leq g \leq p,\ g = q)$ converter, if V_{i+g-1}^{SP} is the bus voltage control reference and V_{i+g-1}^{cal} is the calculated value of the voltage magnitude at bus $(i+g-1)$. This can be expressed as

$$V_{i+g-1}^{\mathrm{SP}} - V_{i+g-1}^{\mathrm{cal}} = 0 \tag{8.13}$$

$$\forall g,\ 1 \leq g \leq p,\ g = q$$

It may be noted that a slave converter can operate in the PV control mode, in which case it controls the AC bus voltage magnitude, rather than the line reactive power, in which case Equation 8.12 becomes

$$V_{i+g-1}^{\mathrm{SP}} - V_{i+g-1}^{\mathrm{cal}} = 0 \tag{8.14}$$

$$\forall g,\ 1 \leq g \leq p,\ g \neq q$$

Now, similar to AC power flow, a slack bus is chosen for the DC power flow and its voltage is pre specified. It serves the dual role of providing the DC voltage control and balancing the active power exchange among the converters. From Figure 8.2, in the p terminal DC system, the first terminal is chosen as the DC slack bus, by convention. This is represented as

$$V_{\mathrm{DC1}}^{\mathrm{SP}} - V_{\mathrm{DC1}}^{\mathrm{cal}} = 0 \tag{8.15}$$

Instead of specifying $\mathrm{V_{DC1}}$, the modulation index m_1 of the first converter can also be specified as

$$m_1^{\mathrm{SP}} - m_1^{\mathrm{cal}} = 0 \tag{8.16}$$

At this stage, it is worthwhile to take stock of the unknown and the specified quantities. Corresponding to each converter, three new variables come into the picture. These include the converter modulation index m, the converter DC side voltage V_{DC}, and the phase angle θ_{sh} of the converter AC output voltage (phasor). Also, as discussed earlier, the DC side voltage V_{DC1} of the first converter is chosen as a slack bus. Thus, due to incorporation of the VSC-HVDC comprising p converters, $(3p - 1)$ additional variables need to be solved.

Against this, we have p independent equations corresponding to the function f (Equation 8.10) along with $(2p - 2)$ independent equations for the line active and reactive powers (Equations 8.11 and 8.12) corresponding to the $(p - 1)$ slave converters. This gives as $(3p - 2)$ independent equations. Now as the master converter q controls the voltage of its AC terminal bus, the net reactive power injection at that bus is available as a specified quantity. For any arbitrary converter, say the gth $(1 \leq g \leq p)$ converter, this can be expressed as

$$Q_{i+g-1}^{SP} - Q_{i+g-1}^{cal} = 0 \tag{8.17}$$

$$\forall g,\ 1 \leq g \leq p,\ g = q$$

This completes the formulation.

8.3 NEWTON POWER FLOW EQUATIONS OF THE VSC-HVDC SYSTEM

In Figure 8.1, without any loss of generality, if it is assumed that there are r generators connected at the first r buses of the n-bus AC system with bus 1 being the slack bus, then the Newton power flow equation for the AC power system network incorporated with the p terminal HVDC system can be written as follows.

Solve $\boldsymbol{\theta}$, $\mathbf{V}$, and $\mathbf{X}$

Specified $\mathbf{P}$, $\mathbf{Q}$, and $\mathbf{R}$

where:

$$\boldsymbol{\theta} = \left[\theta_2 \ldots \theta_n\right]^T, \mathbf{V} = \left[V_{r+1} \ldots V_n\right]^T$$

$$\boldsymbol{\theta}_{\mathbf{sh}} = \left[\theta_{sh1} \ldots \theta_{sh\,p}\right]^T,\ \mathbf{m} = \left[m_1 \ldots m_p\right]^T, \mathbf{V}_{\mathbf{DC}} = \left[V_{DC2} \ldots V_{DC\,p}\right]^T$$

$$\mathbf{X} = [\boldsymbol{\theta}_{\mathbf{sh}}^T\ \mathbf{m}^T\ \mathbf{V}_{\mathbf{DC}}^T]^T$$

and

$$\mathbf{P} = \left[P_2 \ldots P_n\right]^{\mathrm{T}}, \mathbf{Q} = \left[Q_{r+1} \ldots Q_n\right]^{\mathrm{T}}$$

$$\mathbf{P_{sh}} = \left[P_{\mathrm{sh}2} \ldots P_{\mathrm{sh}p}\right]^{\mathrm{T}}, \mathbf{Q_{sh}} = \left[Q_{\mathrm{sh}2} \ldots Q_{\mathrm{sh}p}\right]^{\mathrm{T}}, \mathbf{f} = \left[f_1 \ldots f_p\right]^{\mathrm{T}}$$

$$\mathbf{R} = \left[\mathbf{P_{sh}^{T}} \mathbf{Q_{sh}^{T}} \mathrm{V}_{\mathrm{i+q-1}}^{\mathrm{T}} \mathbf{f}^{\mathrm{T}}\right]^{\mathrm{T}}$$

Then the basic Newton power flow equation for the AC–DC system is

$$\begin{bmatrix} \multicolumn{2}{c}{\mathbf{J_{old}}} & \frac{\partial \mathbf{P}}{\partial \boldsymbol{\theta}_{\mathbf{sh}}} & \frac{\partial \mathbf{P}}{\partial \mathbf{m}} & \frac{\partial \mathbf{P}}{\partial \mathbf{V_{DC}}} \\ & & \frac{\partial \mathbf{Q}}{\partial \boldsymbol{\theta}_{\mathbf{sh}}} & \frac{\partial \mathbf{Q}}{\partial \mathbf{m}} & \frac{\partial \mathbf{Q}}{\partial \mathbf{V_{DC}}} \\ \frac{\partial \mathbf{R}}{\partial \boldsymbol{\theta}} & \frac{\partial \mathbf{R}}{\partial \mathbf{V}} & \frac{\partial \mathbf{R}}{\partial \boldsymbol{\theta}_{\mathbf{sh}}} & \frac{\partial \mathbf{R}}{\partial \mathbf{m}} & \frac{\partial \mathbf{R}}{\partial \mathbf{V_{DC}}} \end{bmatrix} \begin{bmatrix} \Delta\boldsymbol{\theta} \\ \Delta\mathbf{V} \\ \Delta\boldsymbol{\theta}_{\mathbf{sh}} \\ \Delta\mathbf{m} \\ \Delta\mathbf{V_{DC}} \end{bmatrix} = \begin{bmatrix} \Delta\mathbf{P} \\ \Delta\mathbf{Q} \\ \Delta\mathbf{R} \end{bmatrix} \quad (8.18)$$

where $\mathbf{J_{old}}$ is the conventional power-flow (without incorporating HVDC link) Jacobian subblock given as follows:

$$\mathbf{J_{old}} = \begin{bmatrix} \frac{\partial \mathbf{P}}{\partial \boldsymbol{\theta}} & \frac{\partial \mathbf{P}}{\partial \mathbf{V}} \\ \frac{\partial \mathbf{Q}}{\partial \boldsymbol{\theta}} & \frac{\partial \mathbf{Q}}{\partial \mathbf{V}} \end{bmatrix}$$

In Equation 8.18, $\Delta\mathbf{P}$, $\Delta\mathbf{Q}$, and $\Delta\mathbf{R}$ represent the mismatch vectors. In addition, $\Delta\boldsymbol{\theta}$, $\Delta\mathbf{V}$, $\Delta\boldsymbol{\theta}_{\mathbf{sh}}$, $\Delta\mathbf{m}$, and $\Delta\mathbf{V_{DC}}$ represent correction vectors. It may be noted that m_1 may be removed from $\mathbf{m}$ and $V_{\mathrm{DC}1}$ included in $\mathbf{V_{DC}}$, in which case the elements of the Jacobian matrix in Equation 8.18 would be modified accordingly.

8.4 CASE STUDIES AND RESULTS

The validity of the proposed model was tested on the IEEE 300-bus test system. In this test system, multiple VSC-HVDC networks with different control modes were included and studies were carried out. All the converters were connected to their respective AC terminal buses through converter transformers. The resistances and leakage reactances of all the converter transformers were taken as 0.001 and 0.1 p.u, respectively, for all the case studies. The resistance of each DC link was taken as 0.01 p.u for all the case studies. In addition, the value of *c* for the VSC-based converters was uniformly chosen to be $1/2\sqrt{2}$ [91]. All computations were carried out in MATLAB® on a 2.4 GHz, 4 GB RAM, Intel Core™ Machine with i3–3110 M CPU. A convergence tolerance of 10^{-10} p.u. was uniformly adopted for all the case studies. In all the case studies, NI denotes the number of iterations taken by the algorithm to converge to the specified tolerance. Although a large number of case studies confirmed the validity of the model, only a few sets of representative results are presented in this chapter.

8.4.1 Case Study of IEEE 300-Bus Test System Incorporated with a Three-Terminal VSC-HVDC Network

In this case study, three separate studies are conducted with a three-terminal VSC-HVDC network incorporated in the IEEE-300 bus test system. The three different studies are conducted to demonstrate the versatility of the proposed model. In all the three studies, the VSC-HVDC network is connected among AC buses 266, 270, and 271.

8.4.1.1 Study I: Slave Converters in PQ Control Mode

In this study, the converter connected to AC bus 266 acts as the master converter, whereas those connected to AC buses 270 and 271 act as slave converters. Both the slave converters operate in the PQ control mode. The master converter maintains the voltage magnitude of AC bus 266 at a value of 1.02 p.u. The active powers at the terminal end of the lines connecting the converters to AC buses 270 and 271 are specified as 0.3 and 0.4 p.u, respectively. The line reactive powers for both the slave converters are specified as 0.1 p.u. These specified values are shown in the first row and the third and fourth columns of Table 8.1. The power flow solution is shown in the first row and the fifth to seventh columns of Table 8.1.

8.4.1.2 Study II: Slave Converters in PV Control Mode

This study is conducted on the same AC–DC system but with both the slave converters (connected to AC buses 270 and 271) operated in the PV control mode. Their terminal end line active powers are specified as 0.3 and 0.4 p.u, respectively. The voltages of the AC buses 270 and 271 connected to the slave converters are specified as 1.02 and 1.0 p.u, respectively. The specified quantities for this study are shown in the second row and the

TABLE 8.1 Case Study of IEEE-300 Bus System with a Three-Terminal VSC-HVDC Network

HVDC Link: Connection Details		HVDC Link: Specified Quantities		Power Flow Solution		
				Base case power flow converged in six iterations (NI = 6) $V_i = 1.011\angle{-11.24}$; $V_j = 1.011\angle{-11.32}$; $V_k = 0.998\angle{-17.67}$;		
					HVDC Variables	
Master Converter	Slave Converters	Master Converter	Slave Converters	AC Terminal Buses	Master Converter	Slave Converters
i	j, k	$V_{DCi} = 3$; $V_i = 1.02$;	$P_{sh\,j} = 0.3$; $Q_{sh\,j} = 0.1$; $P_{sh\,k} = 0.4$; $Q_{sh\,k} = 0.1$;	$\theta_i = -11.25$; $\mathbf{V}_j = 1.0202\angle -11.27$; $\mathbf{V}_k = 1.0697\angle -7.81$;	$\theta_{sh\,i} = -15.028$; $m_i = 0.9886$;	$V_{DC\,j} = 2.9989$; $V_{DC\,k} = 2.9988$; $m_j = 0.9722$; $m_k = 0.8805$; $\theta_{sh\,j} = -9.6411$; $\theta_{sh\,k} = -5.8353$;
					NI = 6;	
i	j, k	$V_{DCi} = 3$; $V_i = 1.02$;	$P_{sh\,j} = 0.3$; $V_j = 1.02$; $P_{sh\,k} = 0.4$; $V_k = 1.0$;	$\theta_i = -11.26$; $\theta_j = -11.27$; $\theta_k = -7.3876$;	$\theta_{sh\,i} = -14.96$; $m_i = 1.0099$;	$V_{DC\,j} = 2.9989$; $V_{DC\,k} = 2.9988$; $m_j = 0.9649$; $m_k = 0.94$; $\theta_{sh\,j} = -9.6262$; $\theta_{sh\,k} = -5.0847$;
					NI = 6;	
i	j, k	$m_i = 1.0$; $V_i = 1.02$;	$P_{sh\,j} = 0.4$; $Q_{sh\,j} = 0.1$; $P_{sh\,k} = 0.3$; $Q_{sh\,k} = 0.05$;	$\theta_i = -11.25$; $\mathbf{V}_j = 1.0203\angle -11.25$; $\mathbf{V}_k = 1.046\angle -10.101$;	$\theta_{sh\,i} = -15.014$; $V_{DCi} = 2.978$;	$V_{DC\,j} = 2.9767$; $V_{DC\,k} = 2.9769$; $m_j = 0.9799$; $m_k = 0.9993$; $\theta_{sh\,j} = -9.0775$; $\theta_{sh\,k} = -8.5421$;
					NI = 6;	

Note: For the above case study, $i = 266$, $j = 270$, and $k = 271$; Values of voltage magnitudes and phase angles are in p.u. and degrees, respectively.

third and fourth columns of Table 8.1. The power flow results are shown in the second row and the fifth to seventh columns of Table 8.1.

8.4.1.3 Study III: Modulation Index of Master Converter Specified (Instead of DC Side Voltage)

This study is also conducted on the same AC–DC system as the first two studies. However, in this study, the modulation index m_1 of converter 1 is specified instead of its DC side voltage V_{DC1}. Thus, unlike existing models, the flexibility of selection of either the modulation index or the DC side voltage for the converter exists. The specified quantities for this study are given in the third row and the third and fourth columns of Table 8.1. The power flow results are shown in the third row and the fifth to seventh columns of Table 8.1.

The convergence characteristic plots (variation of mismatch error in p.u. with number of iterations) for the power flows of the base case (without any VSC-HVDC network incorporated) and the three studies of Table 8.1 are shown in Figures 8.3 through 8.6, respectively. From Figures 8.4 through 8.6, it can be observed that the proposed model possesses quadratic convergence characteristics, similar to the base case power flow (Figure 8.3).

The bus voltage profile for the first study (the first row of Table 8.1) is shown in Figure 8.7. From this figure, it is observed that the bus voltage

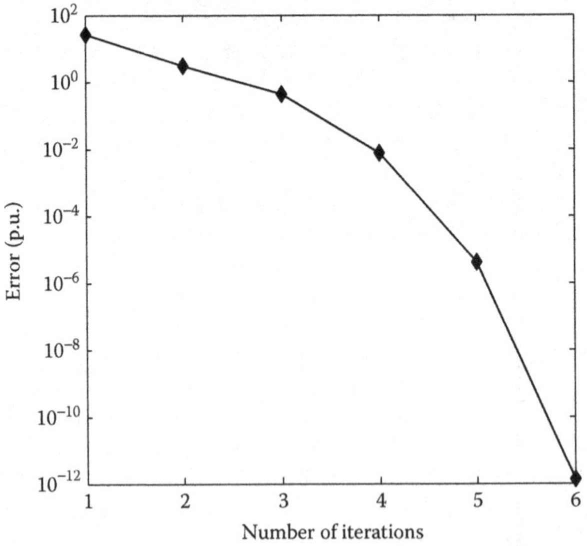

FIGURE 8.3 Convergence characteristics for the base case power flow in IEEE-300 bus system.

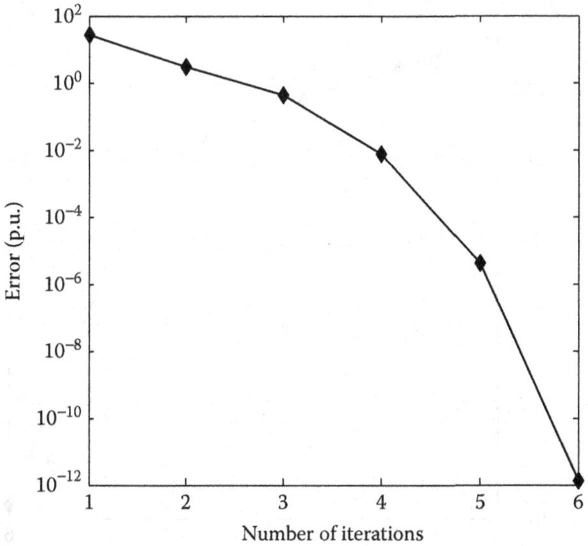

FIGURE 8.4 Convergence characteristics for the first study of Table 8.1.

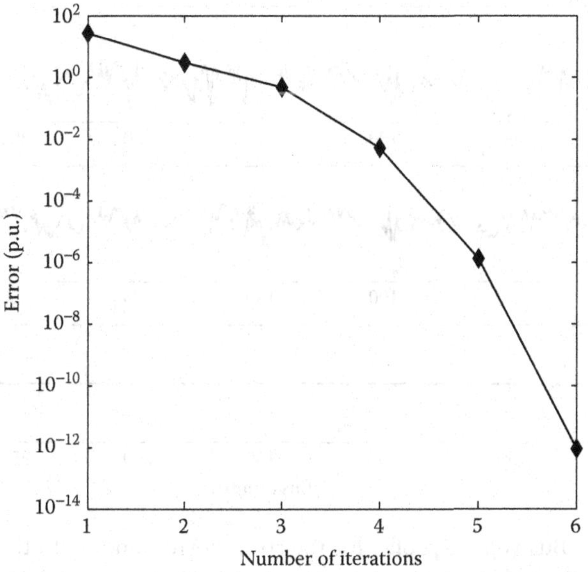

FIGURE 8.5 Convergence characteristics for the second study of Table 8.1.

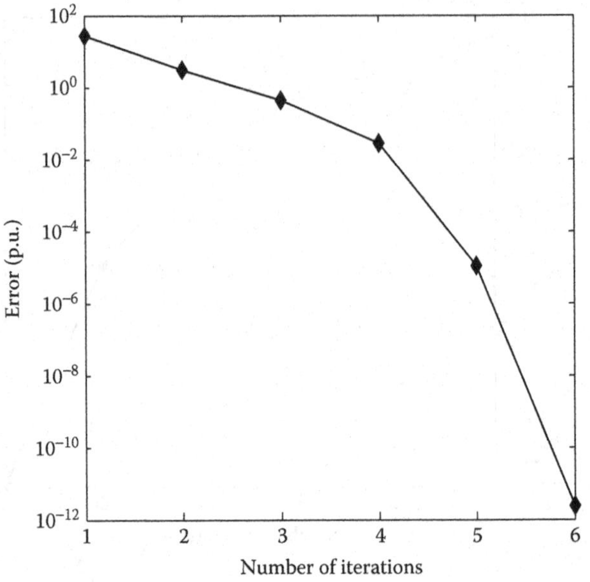

FIGURE 8.6 Convergence characteristics for the third study of Table 8.1.

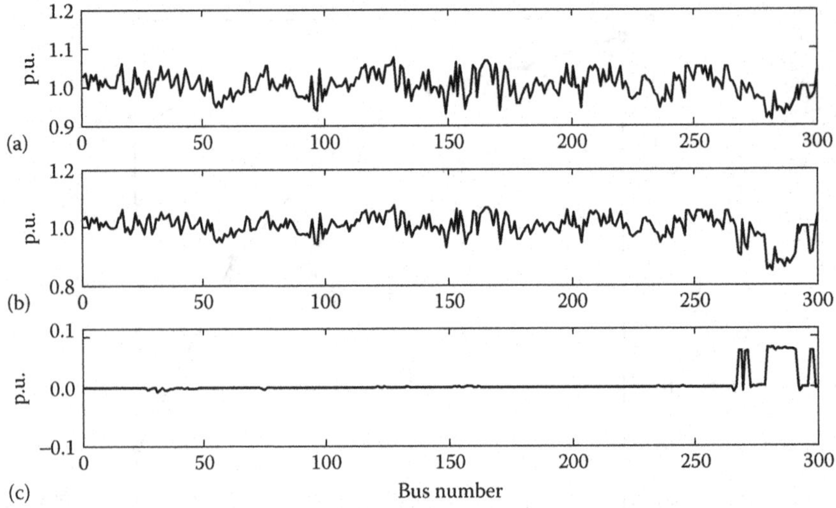

FIGURE 8.7 Bus voltage profile for the study corresponding to the first row of Table 8.1. (a) Bus voltage magnitude of base case; (b) bus voltage magnitude with a three terminal point-to-point VSC-HVDC; (c) bus voltage magnitude difference.

profile hardly changes except for the AC terminal buses at which the VSCs are installed. However, the voltage profiles for the second and third studies of Table 8.1 are not shown here.

8.5 SUMMARY

In this chapter, a unified Newton–Raphson power flow model of an AC power system incorporated with a VSC-based HVDC network has been presented. The modulation indices and the DC side voltages of the converters along with the phase angle of the converter AC side voltage are expressed as state variables in this model. The proposed methodology is implemented by incorporating different topologies of multiterminal DC networks in the IEEE 300-bus test system. To demonstrate the versatility of the proposed model, multiple control modes for the VSC-HVDC are implemented. The power flow solutions and the excellent convergence characteristics validate the feasibility of the model.

Appendix: Derivations of Difficult Formulae

A.1 EXPRESSION FOR THE EFFECTIVE REACTANCE OFFERED BY A STATIC SYNCHRONOUS SERIES COMPENSATOR CONNECTED BETWEEN BUSES *i* AND *j*

From Equation 3.42,

$$Z_{se} = \frac{\mathbf{V}_{se}}{\mathbf{I}_{se}} = \frac{\mathbf{V}_{se}}{Y_{ij}(\mathbf{V}_j - \mathbf{V}_i + \mathbf{V}_{se})} = \frac{V_{se}\angle(\theta_{se} - \varphi_{yij})}{Y_{ij}\left(V_j\angle\theta_j - V_i\angle\theta_i + V_{se}\angle\theta_{se}\right)}$$

or

$$\mathbf{Z}_{se} = \frac{V_{se}\angle(\theta_{se} - \varphi_{yij})}{Y_{ij}\left[V_j(\cos\theta_j + j\sin\theta_j) - V_i(\cos\theta_i + j\sin\theta_i) + V_{se}(\cos\theta_{se} + j\sin\theta_{se})\right]}$$

or

$$\mathbf{Z}_{se} = \frac{V_{se}\angle(\theta_{se} - \varphi_{yij})}{Y_{ij}(a_1 + jb_1)} = \frac{V_{se}\angle\mu}{Y_{ij}(a_1 + jb_1)} \tag{A.1}$$

where:

$$a_1 = \left(V_j\cos\theta_j - V_i\cos\theta_i + V_{se}\cos\theta_{se}\right)$$
$$b_1 = \left(V_j\sin\theta_j - V_i\sin\theta_i + V_{se}\sin\theta_{se}\right) \tag{A.2}$$

and

$$\mu = (\theta_{se} - \varphi_{yij}) \tag{A.3}$$

Thus, from Equation A.1,

$$\mathbf{Z}_{se} = \frac{V_{se}(\cos\mu + j\sin\mu)(a_1 - jb_1)}{Y_{ij}\left(a_1^2 + b_1^2\right)}$$

or

$$\mathbf{Z}_{se} = \frac{V_{se}\left[a_1\cos\mu + b_1\sin\mu + j\left(a_1\sin\mu - b_1\cos\mu\right)\right]}{Y_{ij}\left[a_1^2 + b_1^2\right]} = \frac{\text{Num 1}}{\text{Den 1}} \quad \text{(A.4)}$$

where:

$$\begin{aligned} &a_1\cos\mu + b_1\sin\mu \\ &= V_j\cos\theta_j\cos\mu - V_i\cos\theta_i\cos\mu + V_{se}\cos\theta_{se}\cos\mu \\ &\quad + V_j\sin\theta_j\sin\mu - V_i\sin\theta_i\sin\mu + V_{se}\sin\theta_{se}\sin\mu \\ &= V_j\cos\left(\theta_j - \theta_{se} + \varphi_{yij}\right) - V_i\cos\left(\theta_i - \theta_{se} + \varphi_{yij}\right) \\ &\quad + V_{se}\cos\varphi_{yij} \quad \text{(using Equation A1.3)} \end{aligned} \quad \text{(A.5)}$$

Similarly,

$$\begin{aligned} a_1\sin\mu - b_1\cos\mu = &-V_j\sin\left(\theta_j - \theta_{se} + \varphi_{yij}\right) + V_i\sin\left(\theta_i - \theta_{se} + \varphi_{yij}\right) \\ &- V_{se}\sin\varphi_{yij} \end{aligned} \quad \text{(A.6)}$$

Hence, in Equation A.4,

$$\begin{aligned} \text{Num1} = V_{se}\Big[&\left\{V_j\cos\left(\theta_j - \theta_{se} + \varphi_{yij}\right) - V_i\cos\left(\theta_i - \theta_{se} + \varphi_{yij}\right) + V_{se}\cos\varphi_{yij}\right\} \\ &+ j\left\{V_i\sin\left(\theta_i - \theta_{se} + \varphi_{yij}\right) - V_j\sin\left(\theta_j - \theta_{se} + \varphi_{yij}\right) - V_{se}\sin\varphi_{yij}\right\}\Big] \end{aligned} \quad \text{(A.7)}$$

Thus, using Equation A.2 in the identity $[(x + y + z)^2 = x^2 + y^2 + z^2 + 2xy + 2yz + 2zx]$, we get

$$\begin{aligned} a_1^2 + b_1^2 = &V_i^2 + V_j^2 + V_{se}^2 - 2V_iV_j\cos\left(\theta_i - \theta_j\right) \\ &- 2V_iV_{se}\cos\left(\theta_i - \theta_{se}\right) + 2V_jV_{se}\cos\left(\theta_j - \theta_{se}\right) \end{aligned} \quad \text{(A.8)}$$

Hence, from Equation A.4, the effective reactance offered by the static synchronous series compensator is

$$\begin{aligned} X_{se} &= \text{Im}\left(\mathbf{Z}_{se}\right) = \text{Im}\left(\frac{\text{Num 1}}{\text{Den 1}}\right) \\ &= \frac{V_{se}\left[V_i\sin\left(\theta_i - \theta_{se} + \varphi_{yij}\right) - V_j\sin\left(\theta_j - \theta_{se} + \varphi_{yij}\right) - V_{se}\sin\varphi_{yij}\right]}{Y_{ij}\left[\begin{matrix} V_i^2 + V_j^2 + V_{se}^2 - 2V_iV_j\cos\left(\theta_i - \theta_j\right) - 2V_iV_{se}\cos\left(\theta_i - \theta_{se}\right) \\ +2V_jV_{se}\cos\left(\theta_j - \theta_{se}\right) \end{matrix}\right]} = \frac{c_1}{c_2} \end{aligned}$$

where:

$$c_1 = V_{se}\left[V_i \sin(\theta_i - \theta_{se} + \varphi_{yij}) - V_j \sin(\theta_j - \theta_{se} + \varphi_{yij}) - V_{se} \sin\varphi_{yij}\right] \quad (A.9)$$

$$\begin{aligned} c_2 = Y_{ij}\Big[V_i^2 + V_j^2 + V_{se}^2 - 2V_iV_j \cos(\theta_i - \theta_j) \\ -2V_iV_{se}\cos(\theta_i - \theta_{se}) + 2V_jV_{se}\cos(\theta_j - \theta_{se})\Big] \end{aligned} \quad (A.10)$$

A.2 EXPRESSION FOR THE LINE CURRENT THROUGH THE SERIES CONVERTER OF THE mth UNIFIED POWER FLOW CONTROLLER

As already outlined in Section 4.2, in the proposed model, the series and shunt converters for the mth unified power flow controller (UPFC) ($m \leq p$, where p is the total number of UPFCs) are transformed to fictitious power-flow bus numbers ($n + 2m - 1$) and ($n + 2m$), respectively. Using Equations 4.12 and 4.13, the expression for the net injected current at the ($n + 2m - 1$)th bus (representing the series converter) is

$$\begin{aligned} \mathbf{I}_{n+2m-1} = \sum_{k=1}^{n+2p} \mathbf{Y}_{(n+2m-1)k}\mathbf{V}_k = \mathbf{Y}_{(n+2m-1)i}\mathbf{V}_i + \mathbf{Y}_{(n+2m-1)j}\mathbf{V}_j \\ + \mathbf{Y}_{(n+2m-1)(n+2m-1)}\mathbf{V}_{n+2m-1} \end{aligned} \quad (A.11)$$

Now, from the above equation,

$$\begin{aligned} \mathbf{I}_{n+2m-1} = Y_{(n+2m-1)i}\angle\varphi_{y(n+2m-1)i}V_i\angle\theta_i + Y_{(n+2m-1)j}\angle\varphi_{y(n+2m-1)j}V_j\angle\theta_j \\ + Y_{(n+2m-1)(n+2m-1)}\angle\varphi_{y(n+2m-1)(n+2m-1)}V_{n+2m-1}\angle\theta_{n+2m-1} \end{aligned}$$

or

$$\begin{aligned} \mathbf{I}_{n+2m-1} = Y_{(n+2m-1)i}V_i\angle\{\theta_i + \varphi_{y(n+2m-1)i}\} + Y_{(n+2m-1)j}V_j\angle\{\theta_j + \varphi_{y(n+2m-1)j}\} \\ + Y_{(n+2m-1)(n+2m-1)}V_{n+2m-1}\angle\{\theta_{n+2m-1} + \varphi_{y(n+2m-1)(n+2m-1)}\} \end{aligned}$$

or

$$\begin{aligned} \mathbf{I}_{n+2m-1} = \Big[Y_{(n+2m-1)i}V_i\cos\{\theta_i + \varphi_{y(n+2m-1)i}\} + Y_{(n+2m-1)j}V_j\cos\{\theta_j + \varphi_{y(n+2m-1)j}\} \\ + Y_{(n+2m-1)(n+2m-1)}V_{n+2m-1}\cos\{\theta_{n+2m-1} + \varphi_{y(n+2m-1)(n+2m-1)}\}\Big] \\ + j\Big[Y_{(n+2m-1)i}V_i\sin\{\theta_i + \varphi_{y(n+2m-1)i}\} + Y_{(n+2m-1)j}V_j\sin\{\theta_j + \varphi_{y(n+2m-1)j}\} \\ + Y_{(n+2m-1)(n+2m-1)}V_{n+2m-1}\sin\{\theta_{n+2m-1} + \varphi_{y(n+2m-1)(n+2m-1)}\}\Big] \end{aligned}$$

or

$$\mathbf{I}_{n+2m-1} = p_2 + jq_2 \tag{A.12}$$

where:

$$\begin{aligned} p_2 = \Big[& Y_{(n+2m-1)i} V_i \cos\{\theta_i + \varphi_{y(n+2m-1)i}\} \\ & + Y_{(n+2m-1)j} V_j \cos\{\theta_j + \varphi_{y(n+2m-1)j}\} \\ & + Y_{(n+2m-1)(n+2m-1)} V_{n+2m-1} \cos\{\theta_{n+2m-1} + \varphi_{y(n+2m-1)(n+2m-1)}\}\Big] \end{aligned} \tag{A.13}$$

$$\begin{aligned} q_2 = \Big[& Y_{(n+2m-1)i} V_i \sin\{\theta_i + \varphi_{y(n+2m-1)i}\} \\ & + Y_{(n+2m-1)j} V_j \sin\{\theta_j + \varphi_{y(n+2m-1)j}\} \\ & + Y_{(n+2m-1)(n+2m-1)} V_{n+2m-1} \sin\{\theta_{n+2m-1} + \varphi_{y(n+2m-1)(n+2m-1)}\}\Big] \end{aligned} \tag{A.14}$$

From Equation A.12,

$$\left|\mathbf{I}_{n+2m-1}\right| = I_{n+2m-1} = \left(p_2^2 + q_2^2\right)^{1/2} \tag{A.15}$$

Also, using the identity $\left\{(x+y+z)^2 = x^2 + y^2 + z^2 + 2xy + 2yz + 2zx\right\}$ and from Equation A.13,

$$\begin{aligned} p_2^2 = {} & Y_{(n+2m-1)i}^2 V_i^2 \cos^2\{\theta_i + \varphi_{y(n+2m-1)i}\} + Y_{(n+2m-1)j}^2 V_j^2 \\ & \cos^2\{\theta_j + \varphi_{y(n+2m-1)j}\} \\ & + Y_{(n+2m-1)(n+2m-1)}^2 V_{(n+2m-1)}^2 \cos^2\{\theta_{n+2m-1} + \varphi_{y(n+2m-1)(n+2m-1)}\} \\ & + 2Y_{(n+2m-1)i} Y_{(n+2m-1)j} V_i V_j \cos\{\theta_i + \varphi_{y(n+2m-1)i}\} \\ & \quad \cos\{\theta_j + \varphi_{y(n+2m-1)j}\} \\ & + 2Y_{(n+2m-1)i} Y_{(n+2m-1)(n+2m-1)} V_i V_{n+2m-1} \cos\{\theta_i + \varphi_{y(n+2m-1)i}\} \\ & \quad \cos\{\theta_{n+2m-1} + \varphi_{y(n+2m-1)(n+2m-1)}\} \\ & + 2Y_{(n+2m-1)j} Y_{(n+2m-1)(n+2m-1)} V_j V_{n+2m-1} \cos\{\theta_j + \varphi_{y(n+2m-1)j}\} \\ & \quad \cos\{\theta_{n+2m-1} + \varphi_{y(n+2m-1)(n+2m-1)}\} \end{aligned} \tag{A.16}$$

In a similar way,

$$q_2^2 = Y_{(n+2m-1)i}^2 V_i^2 \sin^2\left\{\theta_i + \varphi_{y(n+2m-1)i}\right\} + Y_{(n+2m-1)j}^2 V_j^2$$

$$\sin^2\left\{\theta_j + \varphi_{y(n+2m-1)j}\right\}$$

$$+ Y_{(n+2m-1)(n+2m-1)}^2 V_{(n+2m-1)}^2 \sin^2\left\{\theta_{n+2m-1} + \varphi_{y(n+2m-1)(n+2m-1)}\right\}$$

$$+ 2Y_{(n+2m-1)i} Y_{(n+2m-1)j} V_i V_j \sin\left\{\theta_i + \varphi_{y(n+2m-1)i}\right\}$$

$$\sin\left\{\theta_j + \varphi_{y(n+2m-1)j}\right\} \tag{A.17}$$

$$+ 2Y_{(n+2m-1)i} Y_{(n+2m-1)(n+2m-1)} V_i V_{n+2m-1} \sin\left\{\theta_i + \varphi_{y(n+2m-1)i}\right\}$$

$$\sin\left\{\theta_{(n+2m-1)} + \varphi_{y(n+2m-1)(n+2m-1)}\right\}$$

$$+ 2Y_{(n+2m-1)j} Y_{(n+2m-1)(n+2m-1)} V_j V_{n+2m-1} \sin\left\{\theta_j + \varphi_{y(n+2m-1)j}\right\}$$

$$\sin\left\{\theta_{n+2m-1} + \varphi_{y(n+2m-1)(n+2m-1)}\right\}$$

Now,

$$\cos A \cos B + \sin A \sin B = \cos\left(A - B\right) \tag{A.18}$$

Using Equations A.16 through A.18 in Equation A.15, we get

$$I_{n+2m-1} = \left(a_1 + a_2 + a_3 + a_4\right)^{1/2} \tag{A.19}$$

where:

$$a_1 = Y_{(n+2m-1)i}^2 V_i^2 + Y_{(n+2m-1)j}^2 V_j^2 + Y_{(n+2m-1)(n+2m-1)}^2 V_{n+2m-1}^2$$

$$a_2 = 2Y_{(n+2m-1)i} Y_{(n+2m-1)j} V_i V_j \cos\left\{\theta_i - \theta_j + \varphi_{y(n+2m-1)i} - \varphi_{y(n+2m-1)j}\right\}$$

$$a_3 = 2Y_{(n+2m-1)i} Y_{(n+2m-1)(n+2m-1)} V_i V_{n+2m-1}$$

$$\cos\left\{\theta_i - \theta_{n+2m-1} + \varphi_{y(n+2m-1)i} - \varphi_{y(n+2m-1)(n+2m-1)}\right\}$$

$$a_4 = 2Y_{(n+2m-1)j} Y_{(n+2m-1)(n+2m-1)} V_j V_{n+2m-1}$$

$$\cos\left\{\theta_j - \theta_{n+2m-1} + \varphi_{y(n+2m-1)j} - \varphi_{y(n+2m-1)(n+2m-1)}\right\}$$

A.2.1 Expression for the Shunt Converter Current of the mth UPFC

Using Equations 4.14 and 4.15, the expression for the net injected current at the $(n+2m)$th bus (representing the shunt converter) is

$$\mathbf{I}_{n+2m} = \sum_{k=1}^{n+2p} \mathbf{Y}_{(n+2m)k}\mathbf{V}_k = \mathbf{Y}_{(n+2m)i}\mathbf{V}_i + \mathbf{Y}_{(n+2m)j}\mathbf{V}_j + \mathbf{Y}_{(n+2m)(n+2m)}\mathbf{V}_{n+2m} \quad \text{(A.20)}$$

where:

$$\mathbf{Y}_{(n+2m)(n+2m)} = \mathbf{y}_{\mathbf{sh}m},\ \mathbf{Y}_{(n+2m)i} = -\mathbf{y}_{\mathbf{sh}m},\ \mathbf{Y}_{(n+2m)j} = 0,\ \mathbf{Y}_{(n+2m)(n+2m-1)} = 0 \quad \text{(A.21)}$$

Now, from Equation A.20,

$$\mathbf{I}_{n+2m} = Y_{(n+2m)i}\angle\varphi_{y(n+2m)i}V_i\angle\theta_i + Y_{(n+2m)(n+2m)}\angle\varphi_{y(n+2m)(n+2m)}V_{n+2m}\angle\theta_{n+2m}$$

or

$$\begin{aligned}\mathbf{I}_{n+2m} = {} & Y_{(n+2m)i}V_i\angle\left\{\theta_i + \varphi_{y(n+2m)i}\right\} + Y_{(n+2m)(n+2m)} \\ & V_{n+2m}\angle\left\{\theta_{n+2m} + \varphi_{y(n+2m)(n+2m)}\right\}\end{aligned} \quad \text{(A.22)}$$

or

$$\mathbf{I}_{n+2m} = \begin{bmatrix} Y_{(n+2m)i}V_i\cos\left\{\theta_i + \varphi_{y(n+2m)i}\right\} + Y_{(n+2m)(n+2m)}V_{n+2m} \\ \cos\left\{\theta_{n+2m} + \varphi_{y(n+2m)(n+2m)}\right\} \end{bmatrix} + j\begin{bmatrix} Y_{(n+2m)i}V_i\sin\left\{\theta_i + \varphi_{y(n+2m)i}\right\} + Y_{(n+2m)(n+2m)}V_{n+2m} \\ \sin\left\{\theta_{n+2m} + \varphi_{y(n+2m)(n+2m)}\right\} \end{bmatrix}$$

or

$$\mathbf{I}_{n+2m} = p_3 + jq_3 \quad \text{(A.23)}$$

where:

$$p_3 = \begin{bmatrix} Y_{(n+2m)i}V_i\cos\left\{\theta_i + \varphi_{y(n+2m)i}\right\} + Y_{(n+2m)(n+2m)}V_{n+2m} \\ \cos\left\{\theta_{n+2m} + \varphi_{y(n+2m)(n+2m)}\right\} \end{bmatrix} \quad \text{(A.24)}$$

$$q_3 = \begin{bmatrix} Y_{(n+2m)i}V_i\sin\left\{\theta_i + \varphi_{y(n+2m)i}\right\} + Y_{(n+2m)(n+2m)}V_{n+2m} \\ \sin\left\{\theta_{n+2m} + \varphi_{y(n+2m)(n+2m)}\right\} \end{bmatrix} \quad \text{(A.25)}$$

From Equation A.23,

$$|\mathbf{I}_{n+2m}| = I_{n+2m} = \left(p_3^2 + q_3^2\right)^{1/2} \tag{A.26}$$

Using Equations A.18, A.24, and A.25 in Equation A.26, we get

$$I_{n+2m} = \left(b_1 + b_2\right)^{1/2} \tag{A.27}$$

where:

$$b_1 = Y_{(n+2m)i}^2 V_i^2 + Y_{(n+2m)(n+2m)}^2 V_{n+2m}^2$$

$$b_2 = 2Y_{(n+2m)i} Y_{(n+2m)(n+2m)} V_i V_{n+2m} \cos\left\{\theta_i - \theta_{n+2m} + \varphi_{y(n+2m)i} - \varphi_{y(n+2m)(n+2m)}\right\}$$

A.2.2 Selection of Initial Conditions for the UPFC Shunt Voltage Source

While enforcing shunt converter current limits, it is observed that the selection of initial conditions for the UPFC shunt voltage source (representing the shunt converter) plays an important role. While enforcing this limit, the sending end bus voltage is relaxed and its Jacobian elements are replaced by those of the shunt converter current magnitude I_{n+2m}.

Now, from Equation A.27, the partial derivatives of I_{n+2m} with respect to the state variable of interest (i.e., $\theta_i, \theta_{n+2m}, V_i$, and V_{n+2m}) can be computed. As a typical example,

$$\frac{\partial I_{n+2m}}{\partial \theta_i} = -\left\{\frac{Y_{(n+2m)i} Y_{(n+2m)(n+2m)} V_i V_{n+2m}}{I_{n+2m}}\right\} \sin\left\{\begin{matrix}\theta_i - \theta_{n+2m} + \varphi_{y(n+2m)i} \\ -\varphi_{y(n+2m)(n+2m)}\end{matrix}\right\} \tag{A.28}$$

It can be shown that all the other partial derivatives contain I_{n+2m} in the denominator.

Now, if the initial conditions for the shunt voltage source were chosen as $V_{\text{sh}m}^0 \angle \theta_{\text{sh}m}^0 = 1.0\angle 0^0$ following [60], it can be observed (using Equation 4.14) that

$$\mathbf{I}_{n+2m} = \mathbf{y}_{\mathbf{sh}m}(\mathbf{V}_{\mathbf{sh}m} - \mathbf{V}_i) = \mathbf{y}_{\mathbf{sh}m}(\mathbf{V}_{n+2m} - \mathbf{V}_i) = \mathbf{0}$$

because, at all load buses, the initial condition for the bus voltage $\mathbf{V}_i$ is usually taken as $1.0\angle 0^0$. This renders all the partial derivatives of I_{n+2m}

indeterminate (from Equation A.28). Modifying the shunt source initial condition to $V_{sh}^0 \angle \theta_{sh}^0 = 1.0\angle -(\pi/9)$ solves this problem, without any observed detrimental effect on the convergence.

A.3 EXPRESSION FOR THE LINE CURRENT THROUGH THE SERIES CONVERTER OF AN INTERLINE POWER FLOW CONTROLLER

The effect of the first series converter of the interline power flow controller (IPFC) is equivalent to an additional (n + 1)th bus without any IPFC. Now, from Equations 5.10 and 5.11, the net injected current at this fictitious (n + 1)th bus is

$$\mathbf{I}_{n+1} = \sum_{q=1}^{n+1} \mathbf{Y}_{(n+1)q}\mathbf{V}_q = \mathbf{Y}_{(n+1)i}\mathbf{V}_i + \mathbf{Y}_{(n+1)j}\mathbf{V}_j + \mathbf{Y}_{(n+1)(n+1)}\mathbf{V}_{n+1} \tag{A.29}$$

or

$$\begin{aligned} \mathbf{I}_{n+1} = {} & Y_{(n+1)i}\angle\varphi_{y(n+1)i}V_i\angle\theta_i + Y_{(n+1)j}\angle\varphi_{y(n+1)j} + V_j\angle\theta_j \\ & + Y_{(n+1)(n+1)}\angle\varphi_{y(n+1)(n+1)}V_{n+1}\theta_{n+1} \end{aligned} \tag{A.30}$$

Equation A.29 is similar to Equation A.11, and further manipulation of terms in Equation A.30 leads to

$$\mathbf{I}_{n+1} = p_4 + jq_4 \tag{A.31}$$

where:

$$\begin{aligned} p_4 = \Big[& Y_{(n+1)i}V_i\cos\{\theta_i + \varphi_{y(n+1)i}\} + Y_{(n+1)j}V_j\cos\{\theta_j + \varphi_{y(n+1)j}\} \\ & + Y_{(n+1)(n+1)}V_{n+1}\cos\{\theta_{n+1} + \varphi_{y(n+1)(n+1)}\}\Big] \end{aligned} \tag{A.32}$$

$$\begin{aligned} q_4 = \Big[& Y_{(n+1)i}V_i\sin\{\theta_i + \varphi_{y(n+1)i}\} + Y_{(n+1)j}V_j\sin\{\theta_j + \varphi_{y(n+1)j}\} \\ & + Y_{(n+1)(n+1)}V_{n+1}\sin\{\theta_{n+1} + \varphi_{y(n+1)(n+1)}\}\Big] \end{aligned} \tag{A.33}$$

From Equation A.31,

$$|\mathbf{I}_{n+1}| = I_{n+1} = \left(p_4^2 + q_4^2\right)^{1/2} \tag{A.34}$$

Squaring Equations A.32 and A.33, then adding and using the trigonometric identity $\cos A\cos B + \sin A\sin B = \cos(A - B)$, we finally get

$$\mathrm{I}_{n+1} = \left|\mathbf{I}_{n+1}\right| = \left(d_1 + d_2 + d_3 + d_4\right)^{1/2} \tag{A.35}$$

where:

$$d_1 = Y_{(n+1)i}^2 V_i^2 + Y_{(n+1)j}^2 V_j^2 + Y_{(n+1)(n+1)}^2 V_{n+1}^2$$
$$d_2 = 2Y_{(n+1)i} Y_{(n+1)j} V_i V_j \cos\left\{\theta_i - \theta_j + \varphi_{y(n+1)i} - \varphi_{y(n+1)j}\right\}$$
$$d_3 = 2Y_{(n+1)i} Y_{(n+1)(n+1)} V_i V_{n+1} \cos\left\{\theta_i - \theta_{n+1} + \varphi_{y(n+1)i} - \varphi_{y(n+1)(n+1)}\right\}$$
$$d_4 = 2Y_{(n+1)j} Y_{(n+1)(n+1)} V_j V_{n+1} \cos\left\{\theta_j - \theta_{n+1} + \varphi_{y(n+1)j} - \varphi_{y(n+1)(n+1)}\right\}$$

A.4 EXPRESSION FOR THE LINE CURRENT THROUGH THE SERIES CONVERTER OF A GENERALIZED UNIFIED POWER FLOW CONTROLLER

Let us consider a most elementary generalized unified power flow controller (GUPFC) with two series and a shunt converter, which is installed in an existing n-bus power system network. It is assumed, without any loss of generality, that the two series converters are connected in series with two transmission lines. The first transmission line is connected in the branch between buses a (sending end [SE]) and b (receiving end [RE]). The second line is connected in the branch between buses c (SE) and d (RE). The shunt converter is connected to bus c.

As already outlined in Section 6.2, in the proposed model, the two series converters are transformed to fictitious power-flow bus numbers $(n + 1)$ and $(n + 2)$. Similarly, the shunt converter is transformed to fictitious power-flow bus number $(n + 3)$.

Using Equations 6.3, 6.4, 6.21, and 6.22, the line current through the first series converter (connected in series with the first transmission line) can be written as

$$\mathbf{I}_{n+1} = -\mathbf{I}_{\mathrm{se}_1} = \alpha_1 \mathbf{V}_{\mathrm{se}_1} - \alpha_1 \mathbf{V}_a + \beta_1 \mathbf{V}_b = \sum_{q=1}^{n+1} \mathbf{Y}_{(n+1)q} \mathbf{V}_q \tag{A.36}$$

where:

$$\mathbf{V}_{n+1} = \mathbf{V}_{\mathrm{se}_1},\ \mathbf{Y}_{(n+1)(n+1)} = \alpha_1,\ \mathbf{Y}_{(n+1)a} = -\alpha_1, \text{and } \mathbf{Y}_{(n+1)b} = \beta_1 \tag{A.37}$$

From Equations A.36 and A.37, we get

$$\begin{aligned}\mathrm{I}_{n+1} = {} & Y_{(n+1)a}\angle\varphi_{y(n+1)a} V_a \angle\theta_a + Y_{(n+1)b}\angle\varphi_{y(n+1)b} V_b \angle\theta_b \\ & + Y_{(n+1)(n+1)}\angle\varphi_{y(n+1)(n+1)} V_{n+1}\theta_{n+1}\end{aligned} \tag{A.38}$$

The above equation is similar to Equation A.30, and further manipulation of terms in the above equation leads to

$$\mathbf{I}_{n+1} = p_5 + jq_5 \tag{A.39}$$

where:

$$\begin{aligned} p_5 = \Big[& Y_{(n+1)a} V_a \cos\{\theta_a + \varphi_{y(n+1)a}\} + Y_{(n+1)b} V_b \cos\{\theta_b + \varphi_{y(n+1)b}\} \\ & + Y_{(n+1)(n+1)} V_{n+1} \cos\{\theta_{n+1} + \varphi_{y(n+1)(n+1)}\}\Big] \end{aligned} \tag{A.40}$$

$$\begin{aligned} q_5 = \Big[& Y_{(n+1)a} V_a \sin\{\theta_a + \varphi_{y(n+1)a}\} + Y_{(n+1)b} V_b \sin\{\theta_b + \varphi_{y(n+1)b}\} \\ & + \mathrm{Y}_{(n+1)(n+1)} V_{n+1} \sin\{\theta_{n+1} + \varphi_{y(n+1)(n+1)}\}\Big] \end{aligned} \tag{A.41}$$

From Equation A.39,

$$|\mathbf{I}_{n+1}| = I_{n+1} = \left(p_5^2 + q_5^2\right)^{1/2} \tag{A.42}$$

Squaring Equations A.40 and A.41, then adding and using the trigonometric identity $\cos A \cos B + \sin A \sin B = \cos(A - B)$, we finally get

$$\mathrm{I}_{n+1} = |\mathbf{I}_{n+1}| = \left(\mathrm{e}_1 + \mathrm{e}_2 + \mathrm{e}_3 + \mathrm{e}_4\right)^{1/2} \tag{A.43}$$

where:

$$\mathrm{e}_1 = Y_{(n+1)a}^2 V_a^2 + Y_{(n+1)b}^2 V_b^2 + Y_{(n+1)(n+1)}^2 V_{n+1}^2$$

$$\mathrm{e}_2 = 2Y_{(n+1)a} Y_{(n+1)b} V_a V_b \cos\{\theta_a - \theta_b + \varphi_{y(n+1)a} - \varphi_{y(n+1)b}\}$$

$$\mathrm{e}_3 = 2Y_{(n+1)a} Y_{(n+1)(n+1)} V_a V_{n+1} \cos\{\theta_a - \theta_{n+1} + \varphi_{y(n+1)a} - \varphi_{y(n+1)(n+1)}\}$$

$$\mathrm{e}_4 = 2Y_{(n+1)b} Y_{(n+1)(n+1)} V_b V_{n+1} \cos\{\theta_b - \theta_{n+1} + \varphi_{y(n+1)b} - \varphi_{y(n+1)(n+1)}\}$$

In a similar manner, the magnitude of the line current through the second series converter can also be written very easily.

A.4.1 Expression for the GUPFC Shunt Converter Current

The shunt converter is connected to SE bus c. In the proposed model, it is transformed to fictitious power-flow bus number $(n + 3)$. The net injected current at this power-flow bus equals the current flowing into the transmission system from this bus and can be written as

$$\mathbf{I}_{n+3} = -\mathbf{I}_{\text{sh}} = \mathbf{y}_{\text{sh}}(\mathbf{V}_c - \mathbf{V}_{\text{sh}}) = \sum_{q=1}^{n+3} \mathbf{Y}_{(n+3)q}\mathbf{V}_q \tag{A.44}$$

where:

$$\mathbf{V}_{n+3} = \mathbf{V}_{\text{sh}},\ \mathbf{Y}_{(n+3)(n+3)} = -\mathbf{y}_{\text{sh}},\ \mathbf{Y}_{(n+3)c} = \mathbf{y}_{\text{sh}}, \text{ and } \mathbf{Y}_{(n+3)d} = \mathbf{0} \tag{A.45}$$

It may be noted that in Equation A.44, $\mathbf{V}_{\text{sh}}$ is the voltage source representing the shunt converter.

Expanding Equation A.44, we get

$$\mathbf{I}_{n+3} = Y_{(n+3)c}\angle\varphi_{y(n+3)c}V_c\angle\theta_c + Y_{(n+3)(n+3)}\angle\varphi_{y(n+3)(n+3)}V_{n+3}\angle\theta_{n+3} \tag{A.46}$$

or

$$\mathbf{I}_{n+3} = Y_{(n+3)c}V_c\angle\left\{\theta_c + \varphi_{y(n+3)c}\right\} + Y_{(n+3)(n+3)}V_{n+3}\angle\left\{\theta_{n+3} + \varphi_{y(n+3)(n+3)}\right\} \tag{A.47}$$

The above equation is similar to Equation A.22, and further manipulation of terms in the above equation leads to

$$\mathbf{I}_{n+3} = p_6 + jq_6 \tag{A.48}$$

where:

$$p_6 = \begin{bmatrix} Y_{(n+3)c}V_c\cos\left\{\theta_c + \varphi_{y(n+3)c}\right\} + Y_{(n+3)(n+3)}V_{n+3} \\ \cos\left\{\theta_{n+3} + \varphi_{y(n+3)(n+3)}\right\} \end{bmatrix} \tag{A.49}$$

$$q_6 = \begin{bmatrix} Y_{(n+3)c}V_c\sin\left\{\theta_c + \varphi_{y(n+3)c}\right\} + Y_{(n+3)(n+3)}V_{n+3} \\ \sin\left\{\theta_{n+3} + \varphi_{y(n+3)(n+3)}\right\} \end{bmatrix} \tag{A.50}$$

From Equation A.48,

$$\left|\mathbf{I}_{n+3}\right| = I_{n+3} = \left(p_6^2 + q_6^2\right)^{1/2} \tag{A.51}$$

Squaring Equations A.49 and A.50, then adding and using the trigonometric identity $\cos A\cos B + \sin A\sin B = \cos(A - B)$, we finally get

$$I_{n+3} = \left(f_1 + f_2\right)^{1/2} \tag{A.52}$$

where:

$$f_1 = Y_{(n+3)c}^2 V_c^2 + Y_{(n+3)(n+3)}^2 V_{n+3}^2$$

$$f_2 = 2Y_{(n+3)c} Y_{(n+3)(n+3)} V_c V_{n+3} \cos\left\{\theta_c - \theta_{n+3} + \varphi_{y(n+3)c} - \varphi_{y(n+3)(n+3)}\right\}$$

A.4.2 Expression for the Bus Voltage on the Line Side of a GUPFC Series Converter

From Figure 6.2, for the first series converter, if $\mathbf{V}_m$ (not shown in figure) is the voltage on the line side, then

$$\mathbf{V}_i - \mathbf{V}_m = \mathbf{V}_{se_1} + \mathbf{I}_{se_1}\mathbf{Z}_{se_1}$$

or

$$\mathbf{I}_{se_1} = \mathbf{y}_{se_1}(\mathbf{V}_i - \mathbf{V}_m - \mathbf{V}_{se_1}) \tag{A.53}$$

Also

$$\mathbf{I}_{se_1} = \mathbf{I}_1 + \mathbf{I}_{10} = (\mathbf{y}_1 + \mathbf{y}_{10})\mathbf{V}_m - \mathbf{y}_1\mathbf{V}_j \tag{A.54}$$

From Equations A.53 and A.54,

$$(\mathbf{y}_1 + \mathbf{y}_{10} + \mathbf{y}_{se_1})\mathbf{V}_m = \mathbf{y}_{se_1}\mathbf{V}_i - \mathbf{y}_{se_1}\mathbf{V}_{se_1} + \mathbf{y}_1\mathbf{V}_j \tag{A.55}$$

or

$$\mathbf{V}_m = \frac{\mathbf{y}_{se_1}\mathbf{V}_i + \mathbf{y}_1\mathbf{V}_j - \mathbf{y}_{se_1}\mathbf{V}_{se_1}}{(\mathbf{y}_1 + \mathbf{y}_{10} + \mathbf{y}_{se_1})} = \frac{\mathbf{y}_{se_1}\mathbf{V}_i + \mathbf{y}_1\mathbf{V}_j - \mathbf{y}_{se_1}\mathbf{V}_{se_1}}{\mathbf{y}_T} \tag{A.56}$$

Let

$$\mathbf{g}_1 = \frac{\mathbf{y}_{se_1}}{\mathbf{y}_T}, \mathbf{g}_2 = \frac{\mathbf{y}_1}{\mathbf{y}_T} \tag{A.57}$$

From Equations A.56 and A.57,

$$\mathbf{V}_m = \mathbf{g}_1\mathbf{V}_i + \mathbf{g}_2\mathbf{V}_j - \mathbf{g}_1\mathbf{V}_{se_1} \tag{A.58}$$

or

$$\mathbf{V}_m = g_1\angle\varphi_{g1}V_i\angle\theta_i + g_2\angle\varphi_{g2}V_j\angle\theta_j - g_1\angle\varphi_{g1}V_{se1}\angle\theta_{se1} \tag{A.59}$$

or

$$\mathbf{V}_m = g_1 V_i \angle(\theta_i + \varphi_{g1}) + g_2 V_j \angle(\theta_j + \varphi_{g2}) - g_1 V_{se1} \angle(\theta_{se1} + \varphi_{g1}) \quad \text{(A.60)}$$

Further manipulation of terms in the above equation leads to

$$\mathbf{V}_m = p_7 + jq_7 \quad \text{(A.61)}$$

where:

$$p_7 = \begin{bmatrix} g_1 V_i \cos\{\theta_i + \varphi_{g1}\} + g_2 V_j \cos\{\theta_j + \varphi_{g2}\} \\ -g_1 V_{se1} \cos\{\theta_{se1} + \varphi_{g1}\} \end{bmatrix} \quad \text{(A.62)}$$

$$q_7 = \begin{bmatrix} g_1 V_i \sin\{\theta_i + \varphi_{g1}\} + g_2 V_j \sin\{\theta_j + \varphi_{g2}\} \\ -g_1 V_{se1} \sin\{\theta_{se1} + \varphi_{g1}\} \end{bmatrix} \quad \text{(A.63)}$$

From Equation A.61,

$$|\mathbf{V}_m| = V_m = \left(p_7^2 + q_7^2\right)^{1/2} \quad \text{(A.64)}$$

Squaring Equations A.62 and A.63, adding and using the trigonometric identity $\cos A \cos B + \sin A \sin B = \cos(A - B)$, we finally get

$$V_{m1} = |\mathbf{V}_{m1}| = \left(h_1 + h_2 + h_3 + h_4\right)^{1/2} \quad \text{(A.65)}$$

where:

$$h_1 = g_1^2 V_i^2 + g_2^2 V_j^2 + g_1^2 V_{se1}^2$$

$$h_2 = 2 g_1 g_2 V_i V_j \cos\{\theta_i - \theta_j + \varphi_{g1} - \varphi_{g2}\}$$

$$h_3 = -2 g_1^2 V_i V_{se1} \cos\{\theta_i - \theta_{se1}\}$$

$$h_4 = -2 g_1 g_2 V_j V_{se1} \cos\{\theta_j - \theta_{se1} + \varphi_{g2} - \varphi_{g1}\}$$

TABLE A.1 Suitability of Decoupling for Elements of Relevant Jacobian Subblocks in Static Compensator

	Typical Elements in Different Jacobian Subblocks	
	Expression for Typical Jacobian Subblock Element	
Jacobian Subblock	Exact	Approximation
$\frac{\partial \mathbf{P}}{\partial \mathbf{V}_{sh}}$	$\frac{\partial P_a}{\partial V_{sh\,b}} = -V_a y_{sh\,b}\cos\left[\theta_a - \theta_{sh\,b} - \varphi_{ysh\,b}\right]$ if STATCOM b is connected to bus a $\frac{\partial P_a}{\partial V_{sh\,b}} = 0$ if STATCOM b is not connected to bus a	$\frac{\partial P_a}{\partial V_{sh\,b}} \approx -g_{sh\,b}$ $\{$for $\theta_a - \theta_{sh\,b} \approx 0$ and $\lvert V_a\rvert \approx 1.0\}$
$\frac{\partial \mathbf{P}_{STAT}}{\partial \mathbf{V}}$	$\frac{\partial P_{STAT\,a}}{\partial V_b} = -V_{sh\,a} y_{sh\,a}\cos\left[\theta_{sh\,a} - \theta_b - \varphi_{ysh\,a}\right]$ if STATCOM a is connected to bus b $\frac{\partial P_{STAT\,a}}{\partial V_b} = 0$ if STATCOM a is not connected to bus b	$\frac{\partial P_{STAT\,a}}{\partial V_b} \approx -g_{sh\,a}$ $\{$for $\theta_{sh\,a} - \theta_b \approx 0$ and $\lvert V_{sh\,a}\rvert \approx 1.0\}$
$\frac{\partial \mathbf{Q}}{\partial \boldsymbol{\theta}_{sh}}$	$\frac{\partial Q_a}{\partial \theta_{sh\,b}} = V_a V_{sh\,b} y_{sh\,b}\cos\left[\theta_a - \theta_{sh\,b} - \varphi_{ysh\,b}\right]$ if STATCOM b is connected to bus a $\frac{\partial Q_a}{\partial \theta_{sh\,b}} = 0$ if STATCOM b is not connected to bus a	$\frac{\partial Q_a}{\partial \theta_{sh\,b}} \approx g_{sh\,b}$ $\{$for $\theta_a - \theta_{sh\,b} \approx 0$, $\lvert V_a\rvert \approx 1.0$, and $\lvert V_{sh\,b}\rvert \approx 1.0\}$
$\frac{\partial \mathbf{P}_{STAT}}{\partial \mathbf{V}_{sh}}$	$\frac{\partial P_{STAT\,a}}{\partial V_{sh\,b}} = 2V_{sh\,a} y_{sh\,a}\cos\varphi_{ysh\,a} - V_c y_{sh\,a}\cos\left[\theta_{sh\,a} - \theta_c - \varphi_{ysh\,a}\right]$ if $b = a$ $\frac{\partial P_{STAT\,a}}{\partial V_{sh\,b}} = 0$ if $b \neq a$	$\frac{\partial P_{STAT\,a}}{\partial V_{sh\,b}} \approx g_{sh\,a}$ $\{$for $\theta_{sh\,a} - \theta_c \approx 0$, $\lvert V_{sh\,a}\rvert \approx 1.0$, and $\lvert V_c\rvert \approx 1.0\}$

Note: STATCOM a is connected to bus c.
STATCOM, static compensator.

References

1. N.G. Hingorani and L. Gyugyi, *Understanding FACTS: Concepts and Technology of Flexible AC Transmission Systems*, IEEE Press, US, 2000.
2. N.G. Hingorani, High power electronics and flexible AC transmission system, *IEEE Power Engineering Review*, 8(7), July 1988, 3–4.
3. N.G. Hingorani, Flexible AC transmission, *IEEE Spectrum*, 30(4), April 1993, 40–45.
4. Y.H. Song and A.T. Johns, *Flexible AC Transmission Systems (FACTS)*, IEEE Press, UK, 1999.
5. R.M. Mathur and R.K. Verma, *Thyristor-Based FACTS Controllers for Electrical Transmission Systems*, IEEE Press, US, 2002.
6. A. Edris, FACTS technology development: An update, *IEEE Power Engineering Review*, 20(3), March 2000, 4–9.
7. E. Larsen, N. Miller, S. Nilsson and S. Lindgreen, Benefits of GTO-based compensation systems for electric utility applications, *IEEE Transactions on Power Delivery*, 7(4), October 1992, 2056–2064.
8. C. Schauder and H. Mehta, Vector analysis and control of advanced static VAR compensators, *IEE Proceedings—Generation, Transmission and Distribution*, 140(4), July 1993, 299–306.
9. L. Gyugyi, Dynamic compensation of AC transmission lines by solid-state synchronous voltage sources, *IEEE Transactions on Power Delivery*, 9(2), April 1994, 904–911.
10. K.K. Sen, STATCOM—STATic synchronous COMpensator: Theory, modeling and applications, *IEEE PES Winter Meeting*, Vol. 2, IEEE, NY, January 31–February 4, 1999, pp. 1177–1183.
11. C.D. Schauder, M. Gernhardt, E. Stacey, T. Lemak, L. Gyugyi, T.W. Cease and A. Edris, Development of a ±100 MVAR static condenser for voltage control of transmission systems, *IEEE Transactions on Power Delivery*, 10(3), July 1995, 1486–1493.
12. C.D. Schauder, M. Gernhardt, E. Stacey, T. Lemak, L. Gyugyi, T.W. Cease and A. Edris, Operation of ±100 MVAR TVA STATCON, *IEEE Transactions on Power Delivery*, 12(4), October 1997, 1805–1811.
13. L. Gyugyi, C.D. Schauder and K.K. Sen, Static synchronous series compensator: A solid state approach to the series compensation of transmission lines, *IEEE Transactions on Power Delivery*, 12(1), January 1997, 406–417.

14. K.K. Sen, SSSC—Static synchronous series compensator: Theory, modeling and application, *IEEE Transactions on Power Delivery*, 13(1), January 1998, 241–246.
15. L. Gyugyi, Unified power-flow control concept for flexible AC transmission systems, *IEE Proceedings—Generation, Transmission and Distribution*, 139(4), July 1992, 323–331.
16. L. Gyugyi, C.D. Schauder, S.L. Williams, T.R. Rietman, D.R. Torgerson and A. Edris, The unified power flow controller: A new approach to power transmission control, *IEEE Transactions on Power Delivery*, 10(2), April 1995, 1085–1097.
17. A.J.F. Keri, A.S. Mehraban, X. Lombard, A. Eiriachi and A.A. Edris, Unified power flow controller: Modeling and analysis, *IEEE Transactions on Power Delivery*, 14(2), April 1999, 648–654.
18. M. Rahman, M. Ahmed, R. Gutman, R.J. O'Keefe, R.J. Nelson and J. Bian, UPFC application on the AEP system: Planning considerations, *IEEE Transactions on Power Systems*, 12(4), November 1997, 1695–1701.
19. C.D. Schauder, E. Stacey, M. Lund, L. Gyugyi, L. Kovalsky, A.J.F. Keri, A.S. Mehraban and A.A. Edris, AEP UPFC project: Installation, commissioning and operation of the ±160 MVA STATCOM (phase 1), *IEEE Transactions on Power Delivery*, 13(4), October 1998, 1530–1535.
20. B.A. Renz, A.J.F. Keri, A.S. Mehraban, C.D. Schauder, E. Stacey, L. Kovalsky, L. Gyugyi and A.A. Edris, AEP unified power flow controller performance, *IEEE Transactions on Power Delivery*, 14(4), October 1999, 1374–1381.
21. A.S. Mehraban, A. Edris, C.D. Schauder and J.H. Provanzana, Installation, commissioning and operation of the world's first UPFC on the AEP system, *Proceedings of the International Conference on Power System Technology*, Vol. 1, IEEE, Beijing, August 18–21, 1998, pp. 323–327.
22. L. Gyugyi, K.K. Sen and C.D. Schauder, The interline power flow controller concept: A new approach to power flow management in transmission systems, *IEEE Transactions on Power Delivery*, 14(3), July 1999, 1115–1123.
23. B. Fardanesh, B. Shperling, E. Uzunovic and S. Zelingher, Multi-converter FACTS devices: The generalized unified power flow controller (GUPFC), *IEEE PES Summer Meeting*, IEEE, Seattle, WA, July 2000, pp. 1020–1025.
24. L. Sun, S. Mei, Q. Lu and J. Ma, Application of GUPFC in China's Sichuan power grid—Modeling, control strategy and case study, *IEEE PES General Meeting*, Vol. 1, IEEE, Toronto, Canada, July 13–17, 2003, pp. 175–181.
25. J.D. Glover, M.S. Sarma and T.J. Overbye, *Power System: Analysis and Design*, Fourth Ed., Thomson Learning, India, 2007.
26. A.R. Bergen and V. Vittal, *Power Systems Analysis*, Pearson Education, 2004.
27. J.J. Grainger and W.D. Stevenson, *Power Systems Analysis*, Tata McGraw-Hill, India, 1994.
28. M.A. Pai, *Computer Techniques in Power Systems Analysis*, Tata McGraw-Hill, India, 2006.
29. W.F. Tinney and C.E. Hart, Power flow solution by Newton's method, *IEEE Transactions on Power Apparatus and Systems*, PAS-86(11), November 1967, 1449–1460.

30. N.M. Peterson and W. Scott-Meyer, Automatic adjustment of transformer and phase-shifter taps in Newton power flow, *IEEE Transactions on Power Apparatus and Systems*, PAS-90(1), January/February 1971, 103–108.
31. W.F. Tinney and J.W. Walker, Direct solution of sparse network equations by optimally ordered triangular factorization, *Proceedings of the IEEE*, 55(11), November 1967, 1801–1809.
32. B. Stott, Decoupled Newton load flow, *IEEE Transactions on Power Apparatus and Systems*, PAS-91(5), September 1972, 1955–1959.
33. B. Stott and O. Alsac, Fast decoupled load flow, *IEEE Transactions on Power Apparatus and Systems*, PAS-93(3), May/June 1974, 859–869.
34. R.A.M. Van Amerongen, A general purpose version of the fast decoupled load flow, *IEEE Transactions on Power Systems*, 4(2), May 1989, 760–770.
35. R.N. Allan and C. Arruda, LTC transformers and MVAR violations in the fast decoupled load flow, *IEEE Transactions on Power Apparatus and Systems*, PAS-101(9), September 1982, 3328–3332.
36. K. Behnam-Guilani, Fast decoupled load flow: The hybrid model, *IEEE Transactions on Power Systems*, 3(2), May 1988, 734–737.
37. J. Arrilaga and N.R. Watson, *Computer Modeling of Electrical Power Systems*, John Wiley & Sons, India, 2003.
38. E. Acha, *Power Electronic Control in Electrical Systems*, Newnes, Elsevier, India, 2006.
39. D.J. Gotham and G.T. Heydt, Power flow control and power flow studies for systems with FACTS devices, *IEEE Transactions on Power Systems*, 13(1), February 1998, 60–65.
40. Y. Xiao, Y.H. Song and Y.Z. Sun, Power flow control approach to power systems with embedded FACTS devices, *IEEE Transactions on Power Systems*, 17(4), November 2002, 943–950.
41. C.A. Canizares, Power flow and transient stability models of FACTS controllers for voltage and angle stability studies, *IEEE PES Winter Meeting*, Vol. 2, IEEE, Singapore, January 23–27, 2000, pp. 1447–1454.
42. C.R. Fuerte-Esquivel and E. Acha, A Newton-type algorithm for the control of power flow in electrical networks, *IEEE Transactions on Power Systems*, 12(4), November 1997, 1474–1480.
43. C.R. Fuerte-Esquivel and E. Acha, Newton-Raphson algorithm for the reliable solution of large power networks with embedded FACTS devices, *IEE Proceedings—Generation, Transmission and Distribution*, 143(5), September 1996, 447–454.
44. X.P. Zhang, E. Handschin and M. Yao, Multi-control functional static synchronous compensator (STATCOM) in power system steady-state operations, *Electric Power Systems Research*, 72(3), December 2004, 269–278.
45. Y. Zhang, Y. Zhang, B. Wu and J. Zhou, Power injection model of STATCOM with control and operating limit for power flow and voltage stability analysis, *Electric Power Systems Research*, 76(12), August 2006, 1003–1010.
46. Z. Yang, C. Shen, M.L. Crow and L. Zhang, An improved STATCOM model for power flow analysis, *IEEE PES Winter Meeting*, Vol. 2, IEEE, Seattle, WA, July 16–20, 2000, pp. 1121–1126.

47. G. Radman and J. Shultz, A new method to account for STATCOM losses in power flow analysis, *Proceedings of the 35th Southeastern Symposium on System Theory*, IEEE, Morgantown, VA, March 16–18, 2003, pp. 477–481.
48. A. Bhargava, V. Pant and B. Das, An improved power flow analysis technique with STATCOM, *International Conference on Power Electronics, Drives and Energy Systems (PEDES-06)*, IEEE, New Delhi, India, December 12–15, 2005, pp. 1–5.
49. X.P. Zhang, Advanced modeling of the multicontrol functional static synchronous series compensator (SSSC) in Newton power flow, *IEEE Transactions on Power Systems*, 18(4), November 2003, 1410–1416.
50. Y. Zhang and Y. Zhang, A novel power injection model of embedded SSSC with multi-control modes for power flow analysis inclusive of practical constraints, *Electric Power Systems Research*, 76(5), March 2006, 374–381.
51. A. Vinkovic and R. Mihalic, A current based model of the static synchronous series compensator (SSSC) for Newton-Raphson power flow, *Electric Power Systems Research*, 78(10), October 2008, 1806–1813.
52. X.P. Zhang, C.F. Xue and K.R. Godfrey, Advanced modeling of the static synchronous series compensator (SSSC) in three-phase Newton power flow, *IEE Proceedings—Generation, Transmission and Distribution*, 151(4), July 2004, 486–494.
53. J. Zhao, J. Guo and X. Zhou, Modeling of static synchronous series compensator in Newton power flow calculation in power system analysis software package (PSASP), *IEEE PES Transmission and Distribution Conference and Exhibition: Asia and Pacific*, 2005, pp. 1–5.
54. K.K. Sen and E. Stacey, UPFC—Unified power flow controller: Theory, modeling and applications, *IEEE Transactions on Power Delivery*, 13(4), October 1998, 1453–1460.
55. S. Arabi and P. Kundur, A versatile FACTS device model for power flow and stability simulations, *IEEE Transactions on Power Systems*, 11(4), November 1996, 1944–1950.
56. A. Nabavi-Niaki and M.R. Iravani, Steady state and dynamic models of unified power flow controller for power system studies, *IEEE PES Winter Meeting*, 96 WM, 257-6 PWRS, Baltimore, MD, January 1996.
57. M. Noroozian, L. Angquist, M. Gandhari and G. Anderson, Use of UPFC for optimal power flow control, *IEEE Transactions on Power Delivery*, 12(4), October 1997, 1629–1633.
58. H. Ambriz-Perez, E. Acha, C.R. Fuerte-Esquivel and A. De La Torre, Incorporation of a UPFC model in an optimal power flow using Newton's method, *IEE Proceedings—Generation, Transmission and Distribution*, 145(3), May 1998, 336–344.
59. T.T. Nguyen and V.L. Nguyen, Representation of line optimisation control in unified power flow controller model for power flow analysis, *IET Proceedings—Generation, Transmission and Distribution*, 1(5), September 2007, 714–723.
60. C.R. Fuerte-Esquivel, E. Acha and A. Perez, A comprehensive Newton-Raphson UPFC model for the quadratic power flow solution of practical power networks, *IEEE Transactions on Power Systems*, 15(1), February 2000, 102–109.

61. C.R. Fuerte-Esquivel and E. Acha, Unified power flow controller: A critical comparison of Newton-Raphson UPFC algorithms in power flow studies, *IEE Proceedings—Generation, Transmission and Distribution*, 144(5), September 1997, 437–444.
62. P. Yan and A. Sekar, Steady-state analysis of power system having multiple FACTS devices using line-flow based equations, *IEE Proceedings—Generation, Transmission and Distribution*, 152(1), January 2005, 31–39.
63. M. Alomoush, Derivation of UPFC DC load flow model with examples for its use in restructured power systems, *IEEE Transactions on Power Systems*, 18(3), August 2003, 1173–1180.
64. X. Wei, J.H. Chow, B. Fardanesh and A.A. Edris, A dispatch strategy for a unified power flow controller to maximize voltage stability related power transfer, *IEEE Transactions on Power Delivery*, 20(3), July 2005, 2022–2029.
65. H. Sun, D.C. Yu and C. Luo, A novel method of power flow analysis with unified power flow controller, *IEEE PES Winter Meeting*, Vol. 4, IEEE, Singapore, January 23–27, 2000, pp. 2800–2805.
66. N.M.R. Santos, V.M.F. Pires and R.M.G. Castro, A new model to incorporate unified power flow controllers in power flow studies, *Transmission and Distribution Conference and Exhibition*, IEEE, Dallas, TX, May 21–24, 2006, pp. 133–140.
67. F. Dazhong, D. Liangying and T.S. Chung, Power flow analysis of power system with UPFC using commercial power flow software, *IEEE PES Winter Meeting*, Vol. 4, IEEE, Singapore, January 23–27, 2000, pp. 2922–2925.
68. X.P. Zhang and K.R. Godfrey, Advanced unified power flow controller model for power system steady-state control, *Proceedings of the IEEE International Conference on Electric Utility Deregulation, Restructuring and Power Technologies*, Vol. 1, IEEE, Hong Kong, April 5–8, 2004, pp. 228–233.
69. A.M.M. Abdel-Rahim and N.P. Padhy, Newton-Raphson UPFC model for power flow solution of practical power networks with sparse techniques, *Proceedings of the IEEE International Conference on Electric Utility Deregulation, Restructuring and Power Technologies*, Vol. 1, IEEE, Hong Kong, April 5–8, 2004, pp. 77–83.
70. J.Y. Liu, Y.H. Song and P.A. Mehta, Strategies for handling UPFC constraints in steady state power flow and voltage control, *IEEE Transactions on Power Systems*, 15(2), May 2000, 566–571.
71. J. Bian, D.G. Ramey, R.J. Nelson and A. Edris, A study of equipment sizes and constraints for a unified power flow controller, *IEEE Transactions on Power Delivery*, 12(3), July 1997, 1385–1391.
72. C.D. Schauder, L. Gyugyi, M.R. Lund, D.M. Hamai, T.R. Rietman, D.R. Torgerson and A. Edris, Operation of the unified power flow controller under practical constraints, *IEEE Transactions on Power Delivery*, 13(2), April 1998, 630–639.
73. J. Yurevich and K.P. Wong, MVA constraint handling method for unified power flow controller in loadflow evaluation, *IEE Proceedings—Generation, Transmission and Distribution*, 147(3), May 2000, 190–194.
74. J.Z. Bebic, P.W. Lehn and M.R. Iravani, P-δ characteristics for the unified power flow controller—Analysis inclusive of equipment ratings and line limits, *IEEE Transactions on Power Delivery*, 18(3), July 2003, 1066–1072.

75. U.P. Mhaskar, A.B. Mote and A.M. Kulkarni, A new formulation for load flow solution of power systems with series FACTS devices, *IEEE Transactions on Power Systems*, 18(4), November 2003, 1307–1315.
76. K.M. Nor, H. Mokhlis and T.A. Gani, Reusability techniques in load flow analysis computer program, *IEEE Transactions on Power Systems*, 19(4), November 2004, 1754–1762.
77. X.P. Zhang, Modeling of the interline power flow controller and the generalized unified power flow controller in Newton power flow, *IEE Proceedings—Generation, Transmission and Distribution*, 150(3), May 2003, 268–274.
78. Y. Zhang, Y. Zhang and C. Chen, A novel power injection model of IPFC for power flow analysis inclusive of practical constraints, *IEEE Transactions on Power Systems*, 21(4), November 2006, 1550–1556.
79. X. Wei, J.H. Chow, B. Fardanesh and A. Edris, A common modeling framework of voltage sourced converters for load flow, sensitivity and dispatch analysis, *IEEE Transactions on Power Systems*, 19(2), May 2004, 934–941.
80. X. Wei, J.H. Chow, B. Fardanesh and A.A. Edris, A dispatch strategy for an interline power flow controller operating at rated capacity, *IEEE PES Power Systems Conference and Exposition*, 2004, IEEE, NY, October 10–13, 2004, 1459–1466.
81. X.P. Zhang, E. Handschin and M. Yao, Modeling of the generalized unified power flow controller (GUPFC) in a nonlinear interior point OPF, *IEEE Transactions on Power Delivery*, 16(3), August 2001, 367–373.
82. B. Fardanesh, Optimal utilisation, sizing and steady-state performance comparison of multiconverter VSC-based FACTS controllers, *IEEE Transactions on Power Delivery*, 19(3), July 2004, 1321–1327.
83. R.L. Vasquez-Arnez and L.C. Zanetta, A novel approach for modeling the steady-state VSC-based multiline FACTS controllers and their operational constraints, *IEEE Transactions on Power Delivery*, 23(1), January 2008, 457–464.
84. S. Bhowmick, Investigations on the development of Newton power-flow modeling of voltage-sourced converter based FACTS controllers, PhD dissertation, Faculty of Technology, University of Delhi, New Delhi, India.
85. S. Bhowmick, B. Das and N. Kumar, An indirect model of SSSC for reducing complexity of coding in Newton power-flow algorithm, *Electric Power System Research*, 77(10), August 2007, 1432–1441.
86. S. Bhowmick, B. Das and N. Kumar, An indirect UPFC model to enhance reusability of Newton power-flow codes, *IEEE Transactions on Power Delivery*, 23(4), October 2008, 2079–2088.
87. S. Bhowmick, B. Das and N. Kumar, An advanced IPFC model to reuse Newton power-flow codes, *IEEE Transactions on Power Systems*, 24(2), May 2009, 525–532.
88. S. Bhowmick, B. Das and N. Kumar, An advanced static synchronous compensator model to reuse Newton and decoupled power-flow codes, *Electric Power Components and Systems*, 39(15), October 2011, 1647–1666.

89. C. Angeles-Camacho, O.L Tortelli, E. Acha, and C.R Fuerte-Esquivel, Inclusion of a high voltage DC voltage source converter model in a Newton Raphson power flow algorithm, *IEE Proceedings—Generation, Transmission and Distribution*, 150(6), November 2003, 691–696.
90. X.P. Zhang, Multiterminal voltage-sourced converter-based HVDC models for power flow analysis, *IEEE Transactions on Power Systems*, 19(4), November 2004, 1877–1884.
91. A. Martinez, C. Esquivel, H. Perez and E. Acha, Modeling of VSC-based HVDC systems for a Newton Raphson OPF algorithm, *IEEE Transactions on Power Systems*, 22(4), November 2007, 1794–1799.
92. J. Beerten, S. Cole and R. Belmans, Generalized steady-state VSC MTDC model for sequential AC/DC power flow algorithms, *IEEE Transactions on Power Systems*, 27(2), May 2012, 821–829.
93. S. Khan and S. Bhowmick, A Novel Power-Flow Model of Multi-terminal VSC-HVDC Systems, *Electric Power Systems Research*, 133, April 2016, 219–227.

Index

Page numbers followed by f and t refer to figures and tables, respectively.

D

E

F

G

H

I

T

Printed in the United States
by Baker & Taylor Publisher Services